**DESIGN OF
FUNCTIONAL
PAVEMENTS**

DESIGN OF
FUNCTIONAL
PAVEMENTS

NAI C. YANG

Chief Civil Engineer
Port of New York Authority

McGRAW-HILL BOOK COMPANY

New York   St. Louis   San Francisco   Düsseldorf   Johannesburg
Kuala Lumpur   London   Mexico   Montreal   New Delhi
Panama   Rio de Janeiro   Singapore   Sydney   Toronto

Library of Congress Cataloging in Publication Data

Yang, Nai C.
  Design of functional pavements.

  1. Pavements—Design and construction. I. Title.
TE251.Y35   625.8   72-5486
ISBN 0-07-072243-9

1234567890 KPKP 765432

The editors for this book were William G. Salo, Jr.,
Jean Ely, and Don A. Douglas, the designer was Naomi
Auerbach, and its production was supervised by George
E. Oechsner. It was set in Baskerville by Scripta
Technica, Inc.
It was printed and bound by The Kingsport Press.

Contents

    6.12. Factors Influencing Settlement 172
    6.13. Methods of Improving Ground................... 175
    6.14. Effect on Design and Performance of Pavement 176

7. MATHEMATICAL MODELS FOR PAVEMENT SYSTEMS 181

A. Equilibrium of Pavement Systems 181
    7.1. General Equilibrium Equations 183
    7.2. Force on Boundary of a Semi-infinite Body 188
    7.3. Extension of Boussinesq Equations 193
    7.4. Displacement along Axis of a Circular Load 194
    7.5. Displacement on Boundary Surface 199
    7.6. Layered Systems 201
    7.7. Systems with Viscoelastic Materials 207
    7.8. Modifications to Layered Systems 210
    7.9. Computer Solutions for Layered Systems 211
    7.10. Finite-element Techniques for Layered Systems 212
B. Stress Analysis of Pavement Composition 219
    7.11. General Plate Theory 219
    7.12. Specific Solutions of General Plate Theory 224
    7.13. Application to Pavement Problems 228
    7.14. Computer Applications to Pavement Problems 231
    7.15. Refinements of Theory 236
    7.16. Inherent Limitations and Conclusion 238
    7.17. Yield-line Theory 241
C. Transient Deflection of Pavement System 244
Appendix 1. Computer Programs 254

8. PROPERTIES OF LANDING GEARS RELATING TO PAVEMENT DESIGN .. 273

    8.1. Structural Characteristics of Landing Gears 279
    8.2. Operational Load of Vehicle 285
    8.3. Distribution and Frequency of Loadings 296
    8.4. Dynamic Impact of Vehicle Load 303
    8.5. Dynamic Resistance of Material 310

9. VEHICLE-PAVEMENT INTERACTION 313

    9.1. Vehicle Vibration 314
    9.2. Transient Vibration 316
    9.3. Steady State of Vibration 318
    9.4. Random Vibration 320
    9.5. Third Level of Forced Vibration 324
    9.6. Damping of Vibration 325
    9.7. Data Acquisition and Reduction 326
    9.8. Experiments on Vehicle-Pavement Interaction 331
    9.9. Transfer Function 338

Preface

With the advent of new giant aircraft, such as the Boeing 747 and the future SST, the time has come for a new look at airport pavement design practice. The busier airport, with its heavier planes, requires better and longer-lived pavement. Furthermore, this problem is not for airports alone. Modern superhighways, with their heavy traffic volume and vehicle load, are also vitally concerned with the problems of design, construction, maintenance, and safe operation.

In present engineering practice, pavement design methods tend to be divided into two groups: (1) empirical or semiempirical methods and (2) theoretical analysis. Both approaches have their merits as well as drawbacks, and many good theories have been developed for pavement design using one or the other approach. However, such practice is inadequate in designing pavements for modern high-speed vehicles and aircraft. First of all, the empirical or theoretical methods have generally been developed independent of each other. There is no correlation between them. Second, these pavement design methods fail to include the parameters of vehicle speed, driver or pilot response, pavement roughness, operational criteria, and maintenance needs. Pavement theories developed in the early automobile age are simply insufficient for today's high-speed transportation.

During the pavement tests for the redevelopment program at Newark Airport, efforts were made to obtain pertinent information which would contribute to the general knowledge of pavement engineering. The conclusions drawn from those tests form the basis for the new pavement design system here presented.

This book deals first with the philosophy of pavement construction, its functional requirements, and the factors governing operational performance. It sets a stage for the future development. After a review of the existing design methods, subgrade condition, quality control, material concept, environmental effects, and mathematical models, a random-vibration theory is introduced to translate pavement surface roughness into vehicle vibration. This translation is governed by the speed, shock absorption, and fundamental frequency of the vehicle.

The vibration problem affects not only the operator's ability to control his vehicle but also the safety and comfort of his passengers. Considering the number of people aboard an airplane and the frequency of multivehicle accidents, it is clearly the public responsibility of an engineer to design a safe and operational pavement.

In addition, the construction of an airport or a superhighway involves a great deal of capital investment. The cost of airport pavement is borne largely by the airport operator and air transportation industry (airlines). The airport users' criteria should have a significant effect in formulating the design concept of modern airport pavements. For highway and road construction, the cost is largely from tax money. There is no reason why a cost-benefit study on such construction should not be conducted.

The materials in this book are largely derived from airport construction, but the philosophy, concept, and methodology contained herein should have a profound influence on the design of modern highway pavements as well.

Nai C. Yang

CHAPTER ONE
Basics of Pavement Design

The scientific approach to the solution of an engineering problem involves first defining its scope and functional purpose and then seeking solutions to the problem within the scope and purpose defined. To illustrate the purpose of pavement construction, consider the difference between a country road which does not have a paved surface and a superhighway paved with a permanent surface. The driver of a vehicle does not care what material the pavement structure consists of, but he is sensitive to the rattling of the vehicle and the safe speed of his travel. On the country road he drives slowly, whereas on the superhighway he can safely go much faster. The driver also notices that on the country road the ride is rather rough, whereas the ride on many superhighway pavements is reasonably smooth. If the driver is in a hurry to reach his destination, the roughness of the pavement may be an important factor in his safety. From this, it can be assumed that pavement design is not necessarily governed by the weight of vehicle; rather, the purpose of a pavement is to provide a functional surface for the safe operation of a vehicle. The dynamic interaction of vehicle vibration and driver response is the primary consideration in judging the functional purpose and the safe standard of a pavement construction.

This analog represents a significant departure from today's pavement design analysis, which emphasizes the weight of vehicle as the primary design consideration. In order to properly define the purpose of pavement construction, the following discussions are introduced.

1.1. GENERAL CONSIDERATIONS

Speed

Prior to the introduction of the automobile, there was no urgent need to design roads similar to modern roadways. Horse-drawn wagons and other slow-moving vehicles did not require a paved road as long as the ground was hard enough for the vehicle to travel on. Although many roads were built in ancient times, such as the Roman Appian Way and the express roads of the early Chinese dynasties, the purpose of the construction was to provide a hard surface for repeated use without interrupting travel speed. With the introduction of the high-speed automobile, unpaved surfaces reduced the efficiency and pleasure of driving. The presence of potholes, roadway erosion, and dust in dry weather added many maintenance and environmental problems.

Initially, concrete pavement was used, and smooth and clean surfaces were achieved. Subsequently, asphalt concrete pavement was introduced. At first, the maximum speed of automobiles was in the range of 30 to 40 mph. Since World War II, the legal speed of vehicles has increased rapidly, from 40 to 60 mph, and even to 80 mph in some states. Many engineers began to realize the inadequacy of pavement design methods because constant complaints were registered by automobile users that their vehicles bounced too much on the pavement at high speeds. This was the first sign that the speed of the vehicle should be an important input in a rational pavement design.

In the last 20 years, aircraft traffic has increased rapidly, culminating in a rush to build airports. As of today, airport engineers still borrow the pavement design concepts developed by highway engineers. No criterion of speed is considered in airport design. Consequently, airport engineers have found that diversified pavement performance has been encountered at airports. In the fast-moving areas, such as runways and taxiways, there have been complaints from pilots about the surface roughness of pavement. In the apron area, the operation of aircraft is less important but the need for maintenance and rehabilitation becomes more critical.

Roughness

In the early years of roadway construction, the surface of concrete pavement was rather smooth but periodic maintenance on expansion joints was required. However, if the pavement surface was not properly constructed or distortion of joints subsequently developed, vehicular vibration was experienced. Such roughness was objectionable because the driver became uncomfortable and, in some cases, lost firm control of the vehicle. Highway engineers have made great efforts to improve pavement construction. Still, in some cases, due to uneven settlement of the subgrade, road surfaces are not smooth and are therefore not desirable.

The undesirability of roughness is a function of vehicle speed and the geometric configuration of the uneven surface. A slow-moving vehicle simply rides up and down on a long undulation of pavement surface, whereas a fast-moving vehicle may register a significant dynamic response. An isolated surface disturbance may not be as critical to vehicle vibration as harmonic undulation would be. Maximum dynamic interaction is encountered at a proper combination of vehicle speed and wavelength of roughness configuration.

Safety

In the early automobile age, traffic accidents were relatively minor—as compared with today's airport situation, in which an accident may cost millions of dollars for the physical damage, not to mention the immeasurable values of service and human lives. If an aircraft is in a state of severe vibration, the pilots cannot read the instruments on the control panel and the aircraft may pitch or overshoot the runway. These are valid complaints, and engineers should design pavement to fulfill the users' requirements. Engineers should not establish the standard of operation and the tolerance on smoothness of a pavement; the users should have the final say on this subject. However, engineers should be guided by a set of operational criteria and standards. They should always keep in mind that there is no substitute for safety and that the pavement should be built, even at a high construction cost, to meet the users' requirements. A safe pavement can be defined as consisting of a smooth, clean surface, providing sufficient traction to brake the vehicle within a defined safe distance. There are other considerations, such as lights, signs, and navigation aids, which are operational requirements; these are peripheral problems relating to pavement design.

Maintenance

When highway traffic is light, it is possible to close a portion of roadway pavement for repair. Today, it is a common experience to find several feet of pavement maintenance causing miles of traffic congestion. There is a tendency to build better pavements in order to reduce the frequency of maintenance. For modern airports, the maintenance of the runways and taxiways is more serious than that of highway pavements. When a runway maintenance program is scheduled, distant airports and the air transportation industry will be informed at least several weeks in advance of the scheduled closedown. For emergency repairs, air traffic can be tied up at distant airports. For instance, the closing down of runway 31L at Kennedy Airport will cause the backup of traffic at London, Paris, Los Angeles, San Francisco, Chicago, and many other airports. The loss of the operational revenue and the inconvenience to the traveling public cannot be accurately measured by money.

Cost

The ultimate goal of all engineering design is the most economical solution to a problem, with maximum utilization of material, equipment, and manpower. Engineers have to utilize their knowledge to develop economical pavement materials and to use these materials where they are most needed. Thus the cost-benefit consideration is the prime objective of a pavement design system. Pavement engineers are concerned not only with the initial capital investment but also with the subsequent maintenance costs. Many property owners in rural areas may, in order to keep the tax rate low, object to having paved roads, and they may be willing to pay the annual cost of filling up the potholes. On the other hand, the maintenance cost for a modern superhighway is entirely different from that for a country road. The detouring and maintaining of traffic flow can be a major difficulty in scheduling pavement repairs. Under such circumstances, the total maintenance costs should include the direct cost of pavement repair, the indirect cost of interruption of operations, and the additional cost (due to the tight schedule) of standby manpower and equipment.

As today's airport construction is paid for primarily by the users, either directly or indirectly, the users should have an opportunity to express their concerns about the functional requirements of the pavement. The pavement engineer is obliged to give the users a price tag for the kind of pavement they desire. Aircraft can take off and land only at certain speeds. There is practically no alternative. Moreover, the future generation of aircraft will have a wider body and heavier gears. Each aircraft will carry up to several hundred passengers. Airport operators have no choice but to build pavements that will meet the speed and safety requirements of airport traffic. In this case, when speed, safety, and interruption of airport operations are taken into consideration, the balance of capital investment and maintenance can be positively defined. In the long run, it may be more economical to have a higher initial construction cost for building a better pavement than to build a weaker pavement and be involved with constant maintenance. This can be seen as a new formula for an economically balanced pavement design. The administrator of the airport may object to paying more in the initial pavement construction. It is the obligation of the pavement engineer to present a valid cost-benefit analysis of the merit of better pavements. By balancing the capital investment and the maintenance costs, an optimal pavement design can be achieved. This may not reflect the best condition of every aspect of pavement performance, but it should be the best possible compromise.

1.2. BASIC ENGINEERING CONSIDERATIONS

Probabilistic Model

All natural events, such as construction, materials, aircraft operations, and human judgment, are subject to random variation. There is no single set of values which can be used to represent a particular event. In modern scientific process, the *degree of reliability* has been introduced to provide a better description of random events. For instance, if we have a supporting system, its probability distribution of events is curve S(p), as shown in Fig. 1.1. The performance of system S(p) will be given by the frequency distribution of the events. The supporting system can be assumed to represent the performance of a pavement system. The forcing function (such as the aircraft load) can be expressed by a similar curve F(p), shown in Fig. 1.1. In all engineering design, a factor of safety greater than 1.0 is always mandatory. This means that the operational requirements of a forcing system will always be less than the supporting system can endure. However, due to the random nature of the probability distribution, a small portion of the supporting system overlaps the functional requirements of the forcing system, such as in the shaded areas *r* and *q* shown in Fig. 1.1. This is the area in which possible failure may occur. As the point of intersection represents the maximum design load and allowable working stress, the reliability *R* of the two interacting systems becomes

$$1 - rq > R > 1 - r - q + rq \qquad (1)$$

The higher the percentage of reliability, the larger the factor of safety of the supporting system. The probability of reliable

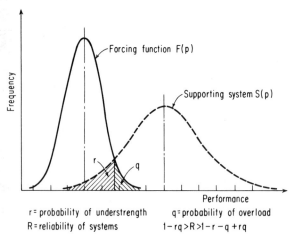

r = probability of understrength   q = probability of overload
R = reliability of systems   1−rq>R>1−r−q+rq

Fig. 1.1   Reliability of systems performance.

interaction is

$$p = P(N_r \geq N_q) \tag{2}$$

where N_r is the reliability of functional performance of the supporting system, N_q is the probability of occurrence of the forcing system at a smaller number than its design service life, and P is a probability function.

Since the performance of pavement is greatly affected by many contributing factors in an unpredictable way, a probabilistic solution may be of some value. The fundamental philosophy of using probabilistic models in engineering analysis is that it is possible to predict the trend of what is likely to happen on the basis of statistical analyses of the past, provided that all contributing factors remain unchanged. In today's rapidly changing technology, airport operations are radically different from those of several years ago. Regression analyses based on previous pavement tests and empirical design charts developed decades ago are of little value to the pavement design for today's aircraft; because the contributing factors have radically changed, a probabilistic solution is not valid.

In the process of an engineering analysis, the physical properties and boundary conditions of a system cannot be clearly defined because of limited data inputs and insufficient observations. It is the power of the probabilistic approach to devise a mathematical model that fits the observed data satisfactorily and that can then be extrapolated for wider interpretation. The major drawback of the probabilistic analysis is that a well-documented statistical model can be used to predict the behavior of a family of events with some degree of confidence within the boundaries of the statistical model but cannot be used to predict the outcome of an individual event in that family.

Deterministic Model

A rational engineering design can be attempted even without the benefit of past experience by determining the physical requirements of the structure based on the anticipated condition of external loads, postulated deformations, stresses in the elements, and the mechanical behavior of materials under various loading and environmental conditions according to the basic laws of mechanics governing motion and force. This fundamental approach has been accepted by the engineering profession for many centuries. Newton's second law of motion and Hooke's law of the linear spring concept of material have become the foundation of modern engineering analyses of vibration and elasticity. The principles are the same, but different details are developed for various boundary conditions. When the deterministic method is used, the internal-external physical relation is established for a particular element. However, the relationship between elements is complicated by the physical and geometrical parameters of the structural system. For that reason, engineers make simplified assumptions so that a theory can fit into the boundary conditions assumed. When the assumptions are valid, deterministic analysis is applicable. However, for a complicated structural system, particularly because of the random nature of its performance, it is not possible to make accurate assumptions to fit the simplified theory. Thus deterministic analysis has its limitations in solving practical problems. The development of the finite-element method in recent years may offer a means of solving practical problems by theory. However, the capacity of the computer remains the major drawback in handling large volumes of iterations, which is a costly and time-consuming process.

Experiment and Simulation

The application of deterministic analysis requires the experimental development of input parameters either in the laboratory or in the field. For a complicated pavement system, tests in the laboratory or in the field do not develop sufficient information for the total system. Consequently, an integrated system of model experiments and computer analysis is used; the process is known as *simulation*. The inputs and outputs are systematically recorded, and the results are objectively analyzed. As all experiments and simulations are limited in scope and observation, it is necessary to use the probabilistic concept to evaluate the reliability of inputs and outputs. This approach involves a three-stage development, from deterministic analysis to experiment and simulation supplemented by probabilistic model, and it has been used extensively for today's complicated engineering systems, such as aircraft, missiles, and response of dynamic forces. The limitation of such a systems approach is primarily determined by the scope of the experiments and the reliability of the instrumentation. Actual data developed from one experiment may not be applicable to another structural system if its characteristics are significantly different from those of the test system. However, the systems approach in combining probabilistic model, deterministic theory, and simulation or experimental tests can result in the development of a much more reliable engineering system.

Observation

Even with the use of the three-stage engineering approach, there is still a possibility that the system may fail. For improving the ultimate reliability of a system, it is necessary to continue field observations during the construction of the pavement system. Any departure from the early experiment and design analysis may require adjustment of the system. Failure to maintain constant observation during the construction and subsequent service life will result in a pavement structure that does not attain its ultimate perfection for the anticipated functional performance. Further, this constant research and feedback will convert past mistakes into valuable experience for future pavement design and construction.

1.3. PURPOSES OF PAVEMENT CONSTRUCTION

The primary purpose of pavement construction is to provide a surface to accommodate the operation of a specified type of vehicle. There are three important functional requirements to be specified by the pavement users: (1) the vehicle will be operating within a defined speed range, (2) the roughness of the pavement surface will not generate a vehicle vibration above a tolerance level, and (3) the safe operation of the vehicle will not be compromised. In addition to these functional requirements, total cost and service life are also governing factors. An optimum design is one that balances the total cost, including capital investment and maintenance, against the performance of pavement construction.

Functional Surface

Translating the users' requirements into reality is actually the job of pavement engineers. Before the translation, however, it is important to study the users' response to and judgment of the performance of pavement. When the pavement construction is completed, the first feedback will come from the pilot or the driver who uses the pavement. His primary concern will be with the safety, comfort, and well-being of his passengers while landing, taking off, or taxiing on the airport pavements. He will be concerned with a clean surface, having no possibility of dust or stones being ingested into his engine. He will also be concerned with visibility, braking traction, hydroplaning, navigation lights, direction signals, and stopping distance, as well as with the geometrical design of the runway-taxiway complex. Among these concerns the most difficult one for an engineer to contend with is the smoothness of a pavement. The engineer must translate the pilot's requirements into a description of physical parameters. The most common physical parameter which can be used to describe the pilot's reaction to aircraft vibration is the vertical acceleration of the aircraft. In dynamic analysis, this is called the *dynamic increment*, $\overline{DI}$, of the aircraft and is closely related to the fundamental frequency f of its spring system and the damping coefficient β of its shock-absorbing mechanism. For a moving aircraft, the dynamic response is also affected by the surface characteristics of the pavement, which can be expressed by its wave geometry, Δ and L; by the progressive change of the pavement surface due to repeated and environmental loading conditions, N; and by the crossing velocity v of the aircraft. For instance, if the aircraft is riding on a grooved surface (short-wave deviations), there is little dynamic increment in the aircraft but a high-frequency noise is registered. On the other hand, if a long-wave roughness is encountered on the pavement surface, slow-moving vehicles will simply ride on the wave surface without registering a noticeable dynamic impact but the response of a fast-moving vehicle may be very significant. In mathematical form, the functional performance of a pavement surface F and the pilot's requirements P can be expressed as

$$F(\Delta, L, N) \; = \; P(\overline{DI}, v, f, \beta) \tag{3}$$

The left side of the equation represents the geometric configuration of the pavement surface which contributes to the pavement's roughness. The right side represents the response of the pilot to the vibration of the aircraft. For a given aircraft operating at a specified speed, there is a definite relationship between the surface geometry of the pavement and the dynamic response of the aircraft at their interface. The pilot's response depends largely on the structural characteristics of the aircraft as well as the human reaction to the vibration. Insofar as the pavement engineer is concerned, the interface vibration is more relevant to the problem of aircraft-and-pavement interaction. The surface geometry of a pavement referred to herein is the profile configuration in the direction of aircraft movement. If the pavement is to be designed for the operation of an aircraft not exceeding a specified level of dynamic response within a specified speed range, then the surface geometry of the pavement should fulfill the following equation:

$$F(\Delta, L, N) \leq k \tag{4}$$

where k is a function of $(\overline{DI}, v, f, \beta)$. When the surface geometry exceeds the k value, the aircraft will generate a vibration over and above the specified tolerance; the pavement surface is then said to experience a *functional failure*. As all pavement surfaces will ultimately reach this tolerance limit after extended use, functional failure is defined as the termination of service life when the vehicle vibration exceeds the designed tolerance.

Optimum Cost

The airport owner, on the other hand, is deeply concerned with the total cost of the pavement. His total costs include initial construction costs IC, maintenance costs MC, operational costs OC, and the cost of contingencies CC, which includes loss of revenue and unforeseen liabilities. He measures the operational surface by the total cost of pavement service and the length of service life. Thus the parameters in cost-benefit analysis become

$$F(N) = IC(N) + MC(N) + OC(N) + CC(N) \tag{5}$$

The minimum value of $F(N)$ represents the most efficient pavement system.

1.4. SYSTEMS APPROACH IN DESIGN ANALYSIS

In today's pavement design, the mechanical transfer of vehicle loadings to the pavement is the primary consideration in design analysis. The method has worked well in some areas but failed in others. As the functional requirements of modern airports vary significantly, airport engineers have to develop a more realistic design system to translate the users' and the airport operator's requirements into a pavement structure which will result in cheaper construction as well as longer service life. In carrying out this task, the engineer must break the design analysis down into many subsystems.

Longitudinal Roughness

Under the functional concept of pavement surface, it becomes necessary to evaluate the longitudinal surface roughness Δ with respect to the number N of load repetitions. Factors in the progressive deformation of a longitudinal surface may be the following:

1. *Construction Tolerance:* The tolerance limit set in the specifications should reflect the construction practice and economic considerations of the project. The tolerance on surface deviation will result in an initial and permanent roughness of pavement performance.

2. *Nonuniform Behavior of Pavement Materials and Subgrade:* Variation of the physical properties of pavement construction is a major contributing factor in the nonuniform deformation of pavement surface.

3. *Nonuniform Distribution of Traffic:* As each load repetition creates a small amount of permanent deformation in pavement, the nonuniform distribution of traffic load will result in uneven cumulative damage of the pavement surface.

4. *Environment:* Since pavements are exposed to environmental forces, such as temperature, moisture, and ground settlement, nonuniform surface deformation is anticipated.

The longitudinal deviation due to construction variance should be a reflection of local construction practices. A tight construction tolerance will result in a high cost of construction. Where labor costs are high and mechanical equipment is extensively utilized, the effective method for controlling construction tolerance is the integration of design analysis with construction practice toward the full utilization of labor and equipment in achieving the best performance of the job. However, no matter how close the tolerance is, there is always an initial roughness on the pavement surface. In pavement design and analysis, engineers deal with the deformations over and above the construction tolerance. The ultimate goal is that the combined surface roughness will not exceed the functional requirements.

As pavement construction involves massive quantities of materials, there are inherent variations in physical properties, strength characteristics, and time response to the operational loadings. The distribution of these inherent variations is complicated by the fact that the degree of variation in the horizontal direction is different from that in the vertical direction. All pavement materials are also nonlinear and nonelastic in their stress-strain relationship. Consequently, the deformation of a pavement surface consists of recoverable deflection, permanent deformation, and creeping with time. The progressive permanent deformation is a function of the load repetitions as well as the stress level in the pavement materials and subgrade. The recoverable deflection, on the other hand, is a function of the elastic properties of the material. A meaningful evaluation will include the relationship of the permanent deformation with the recoverable deflection as well as the stress level in the pavement structure. When the stress increases from one level to another, the rate of permanent deformation will increase. In order to control the permanent deformation, so that it does not exceed a defined limit Δ, it is necessary to relate the permanent deformation to the recoverable deflection and/or to the stress level.

Transverse Deformation

Along the longitudinal wheel path of a pavement, the magnitude of load is constant, and therefore, it is impossible to establish the deformation contour of the pavement. However, for the pavement sections normal to the traffic path (transverse direction), the load is applied only at the wheel path. The deflection basin of a pavement section, in a textbook style, is governed by the basic equilibrium of external load and internal stress-strain of the pavement system. As the deflection basin due to a moving load is a three-dimensional function, the permanent deformation could be measurable in both the longitudinal and transverse directions. The progressive increase of permanent deformation with increasing number of load repetitions is similar to the fatigue behavior of materials. A relation between load repetitions and transverse cumulative deformation D_N may exist, such as

$$D_N = D_1 + D_0 \log N \qquad (6)$$

where D_1 is the initial transverse permanent deformation and D_0 is the rate of increase of transverse permanent deformation per log cycle of load repetitions N. The parameters D_1 and D_0 represent the physical description of the transverse deformation of a pavement. The relation between longitudinal and transverse deformation is governed by the pavement material, construction practice, and operational condition, as well as by environmental factors. The random condition of pavement performance dictates the need of extensive field measurements and full-scale pavement tests to monitor the change in longitudinal deformation as well as transverse deformation at the same number of load repetitions. In defining the functional geometry of the pavement surface, the wavelength L has been used to establish the interaction of pavement and vehicle response. Therefore, the L value should be used as a scale in adjusting the observed data to establish a more meaningful correlation function. The transfer function between the longitudinal deformation $\Delta(N)$ and transverse deformation D_N can be established by multiregression analysis (see Fig. 1.2).

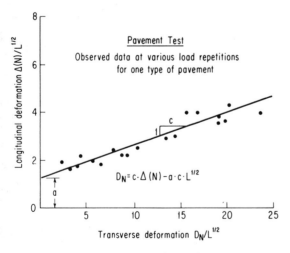

Fig. 1.2 Transfer function of surface deformation.

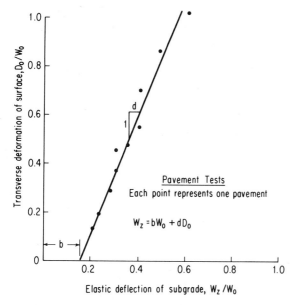

Fig. 1.3   Transfer function of elastic deflection.

Elastic Deformation

In order to utilize deterministic analysis, the permanent transverse deformation should be related to the elastic, or recoverable, deflection of a pavement. In this case, the elastic deflection W_z of a pavement is a function of pavement thickness z, rate of permanent deformation D_0, and the physical properties of the pavement materials and subgrade, which can be represented by a simple parameter W_0. In establishing the relationship between D_0 and W_z, the W_0 value can be used as a scale, and regression analysis involves only dimensionless parameters. As the permanent deformation of a pavement is predominantly governed by the random distribution and characteristics of the pavement materials and subgrade, the correlation between the parameters W_z/W_0 and D_0/W_0 can be obtained for only one type of pavement construction. By varying the thickness z of the pavement, a family of points can be developed and a regression function obtained. Similarly, a family of points can be used to establish the relationship between the stress level σ_z and the deformation parameters D_0/W_0 (see Fig. 1.3).

Elastic Theories

With these three transfer functions—that is, (1) the vehicle response to longitudinal surface geometry, (2) the change of the longitudinal surface in relation to the transverse permanent deformation, and (3) the rate of progressive change of transverse permanent deformation in relation to the elastic deflection or stress level of a pavement—it becomes feasible to use elastic theory to predict the performance of a pavement.

There are three general elastic theories: half-space elastic mass, elastic layers, and viscoelastic analysis. Elastic mass, such as Boussinesq theory in a simple algebraic formulation, requires only the determination of elastic modulus and Poisson's ratio of the subgrade. For a given tire pressure and footprint area, the deflection at depth z can be easily computed. Many studies have indicated that for a pavement thickness z, the surface deformation of that pavement can be reasonably expressed by the W_z value determined by Boussinesq theories. The pavement is assumed to be incompressible and yet as flexible in distributing the load as the subgrade. The layered method can yield more refined data on the stress distribution, the deflection basin, and the boundary conditions of the pavement layers if the inputs are correct. For analyzing an n-layer system, the inputs are n times more than for the Boussinesq method. If the reliability of these inputs is in question (as it often is), the outputs of the layered method may also be in question. The third method involves the determination of the time-dependent properties of the pavement materials. As subgrade always inherits a wide variation, achieving reliability in determining the physical parameters of a time-dependent function is a challenge to the engineering practice.

Both the second and the third methods require long computations and a large amount of testing work. Although several computer programs are available to minimize the engineer's work, the basic drawback still remains in the determination of appropriate input parameters. If the Boussinesq method is used in the computation of subgrade modulus as well as in the evaluation of pavement deflection, the error in the system analysis can be greatly reduced.

Optimization

The system of pavement design does not terminate at the application of elastic theories. At the end of each trial computation, engineers must consider the reliability of functional performance and the total cost of pavement construction as integral parts of pavement analysis. The optimum design is obtained by successive iteration. The functional performance of a pavement $F(p)$ must satisfy the following relationship:

$$F(p) = PS(p) + PSB(p) + PB(p) + PT(p) + PE(p) \qquad (7)$$

where PS = performance of subgrade
   PSB = performance of subbase
   PB = performance of base
   PT = performance of top course
   PE = performance due to environmental conditions

The components of the pavement system must be designed to have an identical reliability, that is, the same degree of performance. This is one of the basic considerations in using the systems approach for pavement design. If any one of the subsystems were to be designed for either a lower or a higher degree of reliability, the system would not be efficient and the cost of construction would not be balanced. Just as no chain is stronger than its weakest link, more reliable physical properties in one subsystem do not increase the performance of the whole system. The same concept should be used in evaluating construction costs. The total construction cost TC is developed as follows:

$$TC(p) = IC(p) + MC(p) + OC(p) + CC(p) \qquad (8)$$

where IC – initial cost
   MC = maintenance costs
   OC = operational costs
   CC = contingency costs

All these cost elements should be allocated on the same probability level. With the functional performance defined and construction costs evaluated, a balanced design is possible.

1.5. ENGINEERING CONSIDERATIONS IN SELECTING MATERIALS

Variations

In analyzing the reliability of a pavement structure, the system can be broken down into component subsystems. Determination of reliability of the whole system depends on the evaluation of each individual subsystem, as shown in Fig. 1.4. If each subsystem contributes equally and harmoniously to the total system, the coefficient of variation of the system can be given by

$$\sigma_s = \sqrt{\frac{\sigma_1^2 + \sigma_2^2 + \cdots}{N-1}} \tag{9}$$

where σ_1, σ_2, etc., represent the coefficients of variation of each subsystem and N is the number of subsystems. However, in pavement construction, each subsystem may have different weights of contribution to the total system. For instance, the contribution of the subgrade is entirely different from the contribution of the top course of the pavement. A more representative coefficient of

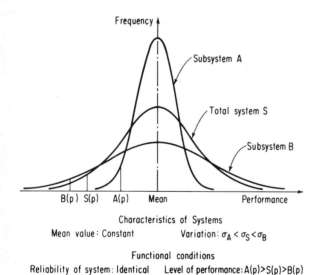

Fig. 1.4   Probability and level of systems performance.

variation for the total system is written as

$$\sigma_s = \sqrt{C_1\sigma_1^2 + C_2\sigma_2^2 + \cdots} \qquad (10)$$

and

$$C_1 + C_2 + \cdots = 1$$

where C_1, C_2, etc., are the weighted factors for the contribution of each subsystem. It should be noted that a subsystem which has a higher contributing factor and/or a higher coefficient of variation will also have a significant effect on the coefficient of variation of the total system. For normal pavement construction, the coefficients of variation are as follows:

Supporting capacity of subgrade	0.20–0.35
Supporting capacity of fill compaction	0.15–0.30
Compressive strength of concrete pavement	0.08–0.15
Stability strength of asphalt-concrete top course	0.10–0.15
Supporting capacity of base courses	0.12–0.25
Tensile strength of reinforcing steel	0.05–0.08

The variation of subgrade will have a significant effect on the performance of a total pavement system.

Effect of Construction

In Fig. 1.5, the solid-line curve represents the probability distribution of an as-built system with a shift of anticipated performance from its designed level. The shift of confidence level of the as-built system is governed by

$$\mu = (1 + k)\sigma_s \qquad (11)$$

where the k value, determined by the normal distribution of random events, indicates the degree of reliability of the as-built system. For instance, if the μ value is equal to 1.0, the reliability of the as-built system is 100 percent as designed. Similarly, if μ is 1.2, 1.5, and 2.0, the corresponding reliability is 93, 87, and 80 percent respectively. If the designed reliability is to be maintained, the service performance of the as-built system should be modified by an identical shift.

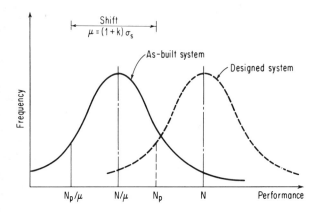

Variation of both systems : Constant
Shift of mean value = shift of confidence level

Fig. 1.5   Confidence level of as-built systems.

Cost-Benefit

The relative reliability of two piles of materials depends largely on the standard deviation of each material if their mean values are identical. For a specified level of confidence, the allowable working stress is equal to the mean value minus a multiple of standard deviations. For a material having a wider distribution, the working stress can be greatly reduced if the reliability is maintained. This demonstrates that a material of greater variation will be less efficient in its utilization. However, material of a large variation usually costs less in pavement construction. The balance of efficient utilization of material and economy in construction cost is a basic engineering consideration in deciding the use of pavement material. For normal pavement construction, the unit costs of pavement materials are likely to be in the following ranges:

Excavation and compaction of subgrade	$0.20–1.50/cu yd
Fill and fill compaction	$0.50–2.00/cu yd
Plain concrete pavement, finished	$0.60–1.00/(in.)(sq yd)
Asphalt-concrete top course, finished	$0.50–0.80/(in.)(sq yd)
Aggregate-base course	$0.20–0.40/(in.)(sq yd)
Stabilized base (local material)	$0.15–0.35/(in.)(sq yd)
Reinforcing steel for concrete pavement	$0.15–0.25/lb

It can be seen that the compaction of subgrade and fill is the most economical element in the cost-performance analysis of a pavement system. The selection of stabilized base is also an important consideration in achieving a cheaper and better pavement.

Structural Performance

The nonlinear load-deformation characteristics of the pavement materials and subgrade are closely related to the nonuniform physical distribution of these materials. The pavement materials usually deform much less than the subgrade at an identical stress level but recover a large portion of its deformation when the load is removed. Because of the repetition of vehicle loads, the accumulated deformation in the subgrade becomes a significant factor in shaping the deformation of pavement surface. The physical characteristics of pavement materials should, therefore, be designed to be compatible with the physical performance of the subgrade. For instance, a base material should be developed to be placed between the concrete slab and the supporting subgrade in order to integrate the performance of two radically different materials. In today's pavement design practice, the selection of pavement materials is limited to concrete, asphalt, and aggregate. Pavement engineers have no opportunity to evaluate the interaction of the pavement structure and its subgrade. The ultimate performance of the pavement structure is, therefore, undetermined.

Stress Analysis

The first step in designing the pavement materials is to determine the bending stress in the pavement layers and then to relate it to such basic characteristics as stress-strain, endurance, durability, ductility, and volumetric change of the material. In performing the stress analysis, the first step is to determine the bending moment in the pavement layers, which is the second derivative of the deflection configuration. The bending stress is then equal to the bending moment divided by the section modulus of the pavement layer. The classic method of solving the bending stress in an elastic plate supported by an elastic foundation was given by Westergaard in the late 1920s. The subgrade is assumed to be a heavy liquid; i.e., subgrade reaction is a linear function of the deformation. For a given subgrade reaction, the relationship between the thickness of a layer and its bending stress can be established. This method has been used for many years in determining the thickness of concrete pavement of a given flexural strength on an assumed subgrade reaction. The fallacy of this approach lies in the thickness determination, which does not reflect the equilibrium condition of the subgrade support. If the thickness of pavement is derived from the elastic theories mentioned in Sec. 1.4, the plate theory can be reasonably used in estimating the physical strength required for each pavement component. By considering the interaction of pavement and subgrade, a new material concept would be evolved for future pavement construction.

Environmental Factors

All airport pavements are exposed to the natural environment, which can destroy pavements even without the application of loads. The strength of material, as determined by stress analysis, is only one of many governing factors in selecting the pavement material. In good engineering practice, an ideal pavement material is (1) able to resist moisture-temperature changes, (2) less sensitive to volumetric change caused by load applications and temperature variation, and (3) ductile, to endure a large deformation. No pavement materials in use today will meet all these requirements. Portland-cement concrete is rigid, having a high modulus of elasticity, and, therefore, possesses undesirable features such as brittleness and high-temperature stress. On the other hand, the lower modulus of elasticity of asphaltic concrete contributes a reduced stability and progressive creeping at elevated temperatures. The limitations imposed by material properties should be reflected in the pavement design but should not govern the design.

For the environmental moisture condition, the moisture may be surface runoff, groundwater, or both. It is important to design a pavement system with an adequate surface drainage as well as proper subsurface drainage. This implies the selection of a well-drained subgrade and the installation of subdrains, thus minimizing capillary water movement and lowering the water table. The material used in the top course of the pavement system should, therefore, be tight enough to prevent the free flow of water but should permit the movement of water vapor. The material for the base courses should be less sensitive to capillary water and should possess a high degree of freeze-thaw resistance.

All pavement materials expand and shrink with fluctuations of temperature. Consequently, all pavements develop cracks in the winter and expand during the summer months. In selecting the pavement material, it is necessary to consider its coefficient of thermal expansion and heat conductivity and their effect on the volumetric change as well as the stress level in the pavement. In designing the pavement system, control joints should be provided to supplement the deficiency in material properties.

In recent years, many modern airports and highways have been located on soft ground, and the general subsidence of the areas has become a pronounced feature of the projects. Pavement systems are subject to long-term deformations; thus the rigidity of a pavement material may become a liability in evaluating the performance of the pavement. The ductility of a material must be considered in formulating the ultimate bending resistance of the pavement. A less rigid material may be more effective in withstanding deformation and also more economical in the construction of the base courses of the pavement system.

There are several other environmental factors governing the selection of pavement materials, such as resistance against oil and chemical corrosions, frictional resistance at wheel braking and turning, and light reflection in aiding visual landing and night driving. An optimum pavement material will not be perfect in all aspects discussed above but will be the best compromise among all requirements.

SUMMARY

The purpose of pavement construction is to provide a functional surface for the safe operation of a type of vehicle at a defined range of speed during the anticipated service life of the pavement. The functional requirements are defined by the users, who—directly or indirectly—assume the full responsibility of paying for the construction. Without an appropriate cost-benefit study, no functional criteria will be reliable.

The role of pavement engineers is primarily that of translating the users' requirements into reality. The first major transfer relates to the interaction of vehicle response and pavement surface. The vehicle response represents the user's specification, and the pavement surface reflects the performance of an end product. The second series of transfer functions is entirely within the domain of the engineering process. The functional surface is characterized by a series of random-wave representations of longitudinal roughness, which is governed by construction tolerance, nonuniform behavior of subgrade, pavement material, and traffic distribution, as well as by environmental factors. Based on the known performance record and test results, the longitudinal roughness can be translated into transverse deformation and, then, the elastic deflection of pavement structure. It becomes feasible to use elastic theories to predict the performance of a pavement.

In the selection of optimum pavement materials, the variation of pavement components, the effect of construction, and environmental factors inherent in the total system must be thoroughly studied. Among all components, the subgrade has the most decisive effect on the performance and the construction cost of a total pavement system. The selection of a stabilized base may offer a possibility for achieving a cheaper and better pavement.

CHAPTER TWO
Development of Design Methods

An achievement surpassing all that has hitherto been accomplished by philosophers, is the creation of a school of men with scientific training and philosophical interests, unhampered by the traditions of the past and not misled by the literary methods of those who copy the ancients in all except their merits.

BERTRAND RUSSELL

The scientific approach to an engineering problem involves, first, reviewing past works relating to the whole or part of the problem. By copying or modifying the appropriate portions of these past works, a new program may be efficiently developed and the repetition of past mistakes prevented. In pavement design, there are probably more than 20 methods which have been developed by various professional groups based on their years of experience in design and construction. The methods vary from chart readings to sophisticated computer analysis. Because of this great diversity, it may be said that pavement design is more like an art than a science.

In reviewing past work, the existing design methods have been divided into three groups, according to the utilization of paving materials. It is an odd way of classifying design approaches, but that is what is happening to the paving industry.

2.1. ASPHALT-CONCRETE PAVEMENT

The early use of coal tar and asphalt in pavement construction was primarily confined to dust control and the surfacing of secondary roads. Through the years, many engineers worked diligently in research and development for the improved use of asphalt material. Their contributions have resulted in the acceptance of asphalt concrete as an important element in pavement structure, and the utilization of deep-strength black base is the ultimate goal of the asphalt paving industry.

The development of the asphalt-pavement design method has been closely related to the growth of the asphalt paving industry. In the early years, the design methods were based on the distribution of wheel load on aggregate base. The Boussinesq type of stress distribution was employed. Subsequently, the presence of cracks on the asphalt surface led to a study of the tensile strength of the surface layer and to the development of layered theories. In the last two decades, many scholars and researchers have devoted their efforts to the refinement of advanced theories and testing procedures. Some of their works are truly out of step with their counterparts in the promotion of empirical methods. In recent years, there have been many construction projects involving asphalt overlays on existing cement-concrete pavement or stabilized base. A new development in pavement design could be anticipated.

Although the design methods may vary in theory or by experience, one thing common to all methods is that the equilibrium of subgrade is always considered. The major problem encountered in asphalt pavement is the rutting of the pavement surface under channelized traffic. All thickness design methods are oriented toward the limited stress in the subgrade. A properly designed asphalt pavement can provide a smooth surface for an extended service life. However, a common drawback of many design methods lies in their inability to analyze the physical strength required for each component of the pavement structure. A brief review of the major design methods is given in the following.

California Method

The California State Highway Department method for the design of asphaltic-concrete pavements has been in use, principally in California and in a number of other Western states, since the late 1940s. It is largely an empirical method based on a procedure developed by Hveem [19,20] and later modified by the results of extensive road tests carried out by the California Division of Highways and the American Association of State Highway Officials (AASHO). The California method relies to a great extent on the stiffness of the foundation materials, which, in turn, determines the ability of the pavement to withstand the applied loads. Results of the AASHO road tests provided several factors which were used to modify the original design equation.

CBR Method

Of all the empirical and semiempirical methods of pavement design, the California Bearing Ratio (CBR) method is not only the most widely known but also the most extensively used in many parts of the world. As the first correlation-based empirical method of flexible pavement design, the CBR method has, since its inception, undergone continuing modifications and extensions, to the point at which it may be considered to have become a truly rational method.

In referring to the degree of acceptance of the CBR method, DeBarros [9] noted: "A review of the technical literature of several countries and of several organizations in the U.S. indicates that the use of the CBR methods and its modifications far exceed the use of all other non-CBR methods combined." This observation still possesses some degree of merit today, although the large-scale use of electronic computers has placed the theoretical and quasi-theoretical methods somewhat on the ascendancy. Nevertheless, the engineering profession still employs the CBR method or modifications of it.

The CBR method was originally developed (and subsequently abandoned) by the California Division of Highways to be used in measuring the strength characteristics of the soil as determined from a simple shear test. The original CBR method was intended primarily for use in highway design but oddly enough, at the time when the CBR method was discarded in the 1940s by the California Division of Highways, the U.S. Army Corps of Engineers adapted and modified it for airfield pavement design.

The improved CBR test procedure includes such factors as soil gradation and Atterberg limits. It thus introduces modern soil mechanics into pavement design. The most significant contribution by the Corps of Engineers was the extension of wheel loads from the original CBR design of 12,000 lb to the 1940s aircraft load of 200,000 lb and the introduction of the concept of multiple wheel load.

Navy Method

In designing the Navy's airfield pavements, Palmer introduced the theory of layered systems supplemented with the method of field evaluation. There are two important assumptions relating to pavement evaluation: (1) the selection of deflection tolerance and (2) the effect of plate rigidity as compared with the rubber-tired wheels. In an early work by Palmer [32], the deflection tolerance is assumed to be 0.2 in.; this was subsequently reduced to 0.1 in. In the Navy's design manual [29], it is assumed to be 0.15 in. Some work should be done to determine the selection of deflection tolerance.

Shell Method

The Shell method is based on Peattie's adaptation of Jones' numerical solution to the three-layer problem for use in the design of flexible pavements. The design curves obtained are based on limiting the vertical compressive strain in the subgrade and the tensile stress in the asphalt pavement layer to such levels as would prevent cracking and excessive deformations in the subgrade and asphalt layers, respectively. The design curves were subsequently modified to include such factors as variations in wheel load, volume of traffic, load repetitions, and fatigue.

Kansas Method

In the original development, Palmer and Barber [31] presented a method for computing the vertical displacement due to a distributed, static load applied over a circular area at the surface of a two-layer system idealized as a homogeneous, isotropic, yielding half-space, with no change in volume during deformation. The settlement is obtained by integrating the vertical strain based on the classic Boussinesq solution for the semi-infinite uniform medium. Due to the nonlinear behavior of the soil medium, the modulus of deformation is used instead of the usual modulus of elasticity. The modulus of deformation is determined by a laboratory test using a triaxial compression device [17].

The Kansas Highway Commission thereafter applied the results of extensive triaxial compression tests, together with some coefficients representing environmental factors, to the theoretical computations and thereby produced a semiempirical design formula for determining pavement thickness.

Early English Method

For pavement design in general, ultimate-strength methods incorporate the failure characteristics of slabs. Nominally, this is effected by using an adequate safety factor which insures the pavement system against failure. The Early English method is one such approach. Introduced in 1944 by Glossop and Golder [14,15], it involves the determination of the thickness of a slab on grade such that when the slab is loaded by a static load distributed over a circular or rectangular area, the pressure on the subgrade multiplied by an adequate factor of safety does not exceed the bearing capacity of the subgrade soil. Originally, the method was developed for flexible airfield pavement construction on soft-to-firm saturated clays, but it was later extended to include rigid pavement construction and other types of subgrade soils.

A vertical stress distribution of the Boussinesq type is assumed both in the pavement and in the subgrade, and the shear strength of the pavement components is neglected. The shear strength of the subgrade is determined by a modified triaxial compression test in which the test specimens are allowed to soften under different overburden pressures before the test.

Canadian Methods

Over the years, developments in pavement design have generally followed two discernible patterns, namely, the theoretical and the empirical design procedures. The theoretical methods, on the one hand, have sought to validate the resultant design equations through experimental work. On the other hand, the empirical methods have sought rationality through statistical methods designed to fit the measured data to a curve whose equation can be used as a basis for predicting behavior under different loading and foundation conditions.

The Canadian Department of Transport (CDT) method [26,39] for the design of flexible pavements is basically an empirical method developed during the analysis of data from extensive plate-bearing tests at Canadian airfields during World War II. The method expresses the required pavement thickness as a function of the applied load and the subgrade strength.

By the use of a step-by-step approach, the CDT method can be used to convert a multilayered system into an equivalent two-layer pavement system. This procedure in effect assumes the *equivalency concept*, in which a certain thickness of one material may be converted into an equivalent thickness of another material.

The Canadian Good Roads Association method for the structural design of flexible pavements was developed based on the experience acquired in one area and extended for the design of another pavement under similar physical and environmental conditions. The elastic-rebound Benkelman-beam deflection value is used not only as a measure of pavement strength but also as a significant parameter in predicting the performance of flexible pavements.

Michigan Method

The Michigan method relies largely on the information obtained in a soil-profile classification and other field studies as a basis for the establishment of design criteria. These studies, when combined with experience in the evaluation of soil, geology, and certain environmental factors, dictate the procedures to be followed in design.

AASHO Method

The American Association of State Highway Officials (AASHO) method of pavement design was a direct outgrowth of the AASHO road-test program. Analysis of the resulting data led to the introduction of a design procedure for flexible and rigid pavements which takes into account such factors as type and volume of traffic, pavement performance and serviceability, fatigue, and the environment. Pavement performance and serviceability are expressed by the *serviceability index*, which is defined as the ability of the pavement to provide a smooth and satisfactory ride at a given time. The scale uses numbers from 1 to 5. Change in the serviceability index is a function of the aforementioned factors, including the pavement thickness and subgrade conditions. The *traffic index* denotes the effect of load repetitions, and the layer-equivalency concept provides a means of expressing the thickness of a given layer in terms of another layer when the traffic index is known.

Asphalt Institute Method

The Asphalt Institute method for the design of asphalt pavements was developed from statistical analyses and theoretical correlations of data from the AASHO road test, with supplementary information from the WASHO road test and the British test roads. It is a quasi-rational method in which the pavement thickness is a function of the volume and type of traffic, the equivalent wheel load, and the strength of the subgrade as expressed by the CBR, Stabilometer R value or the plate-bearing test value. In its present form, the Asphalt Institute method rates the performance of a pavement by the so-called *present serviceability index* (PSI), first introduced by Carey and Irick [5] and defined as the ability of the pavement to provide a smooth and satisfactory ride. Environmental effects have been included, which take into account such factors as drainage, frost, and other cyclic weather conditions. Extensive use of linear regression-analysis techniques, together with the equivalent-layer concept, produced the mathematical formulation, which is presented as design charts to be used in the determination of pavement thickness. Witczak [47] adopted the concept that asphalt pavements behave as multilayered elastic systems, and he calculated stresses and strains against various load and strength criteria. The basic subgrade characteristics were expressed in terms of CBR.

2.2. CEMENT-CONCRETE PAVEMENT

The introduction of portland-cement concrete pavement represented the beginning of an industrial revolution in pavement construction. The smooth paved surface added much joy and safety to the travel of rubber-tired vehicles. Among many contributing factors, the most outstanding one was the development of analytic methods by

Westergaard and the research on physical properties of concrete. The Ohio River Division Laboratory of the U.S. Army Corps of Engineers subsequently extended the design method to airfield pavement and made many modifications and refinements in actual construction. This probably represented the high-water mark of rigid pavement design.

In the AASHO road test, concrete test pavements were generally too strong to yield meaningful information on the failure mode of rigid concrete pavements. No new knowledge was gained from the test. However, there have been many indications that distresses in subgrade supports have resulted in crackings and settlement of concrete pavements. The current practice is to treat concrete pavement as a layered system and to introduce a cement-stabilized base as a transition between the concrete slab and the subgrade support. It took almost twenty years to develop such an approach.

Cement concrete, by its own merit, definitely has an important place in pavement construction. The brittleness of the concrete material and its sensitiveness to temperature stress have caused more pavement distresses than the cracking of pavement due to wheel load. No rigid pavement design is complete without the elaborate design of construction, contraction, and expansion joints. The use of reinforcing steel to control the volumetric change due to temperature and moisture fluctuation remains a controversial and unsettled design subject. No attempt has been made to review the so-called continuously reinforced, or prestressed concrete pavement.

PCA Method

The Portland Cement Association (PCA) method for the design of concrete pavements is based on Pickett and Ray's [35] extension of the Westergaard solution for a distributed static load applied over a circular area at the surface of a homogeneous, isotropic, elastic slab sitting on an elastic subgrade. Three different assumptions for the subgrade material have been investigated by various authors: (1) that it is a dense liquid which behaves like a Winkler foundation, (2) that it is a semi-infinite elastic solid, and (3) that it is an elastic solid layer of a finite thickness.

The theoretical equations for each of these subgrade conditions proved to be too complicated for everyday use; thus Pickett and Ray [35] developed influence charts for deflection and moments for the various subgrade conditions and the loading configurations on the pavement slab. Since the introduction of electronic computers, the Pickett-Ray charts have been replaced by digital computation. The output is much more refined, and optimization of design analysis has become a reality.

Corps of Engineers Method

At the end of World War II, the U.S. Army Corps of Engineers conducted a series of large-scale field tests at the Ohio River Division Laboratories, Mariemont, Ohio. The primary purpose was to determine the validity of Westergaard theories on the design of airfield pavement. Extensive subgrade analysis and instrumentations were conducted during the accelerated traffic tests. The observed pavement behavior led to a modification of the Westergaard theoretical analysis, and a set of design charts was established for the pavement design of military airfields. Subsequently, the design method was extended to the pavements of civil airports. For almost a quarter of a century, the Corps of Engineers manual on rigid pavement [7] has been making a valuable contribution to airfield pavement design.

AASHO Method

During the AASHO road test, the program was practically divided into two camps: one favoring the use of asphalt pavement and the other favoring cement concrete. Similar serviceability and performance equations apply to rigid pavements and flexible pavements. In the case of rigid pavement design, the subgrade strength is the only fundamental parameter in the design equation. The other principal parameters are the load factor, load repetitions, tire pressure, axle type, concrete properties such as strength, modulus of elasticity, and Poisson's ratio, and environmental factors. The approach to rigid pavement design is one of combining theoretical and empirical relations and is practically parallel to the PCA design method.

Subsequently, a critical review was conducted by Vesic and Saxena [45] on the existing theories of structural behavior of rigid pavements and, particularly, the subgrade supports. A performance curve was presented.

Yield-line Method

Swedish investigators, notably Losberg, originated the yield-line method in the early 1960s for use in the design of plain and reinforced-concrete pavements for highway and airport use. The yield-line concept involves the determination of the thickness of a slab sitting on an elastic foundation such that when the slab is loaded by a static load distributed over a small circular area, the maximum radial bending moment along a critical failure line does not exceed the bending moment, which would produce a continuous circumferential crack at a given radius around the loaded area. This bending moment is known as the *ultimate negative bending moment* of the slab section [24,25].

2.3. RULES, REGULATIONS, AND GUIDELINES

The rapid growth of airport operation during the postwar years resulted in the issuance of standards and specifications by the regulatory office in order to achieve a uniform control of quality and design requirements of pavement construction. These technical publications have served as rules, regulations, or code requirements that design engineers must follow. As these technical publications are necessarily broad in scope for use throughout the country, they may not be completely satisfactory for covering a particular situation or a condition peculiar to a certain locality.

Among many governing organizations, the British Department of the Environment, formerly Ministry of Public Building and Works, and the U.S. Federal Aviation Administration (FAA) are the most important regulatory offices. Their technical publications and advisory circulars have practically been the bible in pavement design and construction for the last two decades.

LCN Method

The Load Classification Number (LCN) system was originally developed by the British Ministry of Public Building and Works and has been widely used in the United Kingdom and continental Europe. During the post-World War II years, it became apparent that there was a need for some simple system which would enable the loading characteristics of an aircraft to be readily compared with the load-bearing capacity of a pavement. An extensive series of loading tests was, therefore, conducted on existing pavements of varying compositions and subgrade supports. The purpose of these tests was to determine the load-carrying capacity of a pavement with respect to the load intensity, contact area, and mode of failure.

In order to devise a system in which the capacity of a pavement to carry an aircraft could be expressed as a single number, the concept of a standard load-classification curve was introduced. It is at best an approximation since pavement strengths will vary from point to point and the loading test may show a large variation. The selection of the LCN to classify aircraft as well as pavement is therefore a matter of statistical analysis and engineering judgment. It is not practicable to use the system to accuracies greater than, say, 10 percent.

FAA Method

The original advisory circular on airport paving was issued by the Federal Aviation Agency, which subsequently became an Administration, in the early 1960s, when air transportation was in the midst of rapid expansion. As federal funds participate in the construction of public aviation facilities, the advisory circular was intended to establish general guidelines for upgrading airport facilities. However, the ambiguities in the advisory circular plus the laxity of airport pavement engineers have resulted in a number of unpleasant experiences. In the last few years, a series of changes have been issued by the FAA relating to load repetitions, pavement structure, and material specification as well as to construction-testing procedures. However, there is no significant change in the stated functions and purposes of pavement construction. The current (1970) statement is:

> Airport pavements are constructed to provide adequate support for the loads imposed by aircraft using the airport and to produce a firm, stable, smooth, all-year, all-weather surface, free from dust or other particles that may be blown or picked up by propeller wash or jet blast. In order to satisfactorily fulfill these requirements, the pavement must be of such quality and thickness that it will not fail under the load imposed. In addition, it must possess sufficient inherent stability to withstand, without damage, the abrasive action of traffic, adverse weather conditions, and other deteriorating influences. To produce such pavements requires a coordination of many factors of design, construction, and inspection to assure the best possible combination of available materials and a high standard of workmenship.

Thus, the primary purpose of airport pavement construction is to provide adequate support for the loads imposed by aircraft.

2.4. INTRODUCTION OF FUNCTIONAL CONCEPT

A sophisticated refinement of engineering concept has been introduced for the pavement construction at the New York airports. The new pavement was designed to satisfy the requirements for safe operation of the Boeing 747 aircraft with a growth possibility for future aircraft weighing more than 1 million lb. The base of the Port Authority's new pavement is pozzolanic. Called LCF for the lime, cement, and flyash that combine with sand (and stone where needed) to form a semirigid base, the material spreads like dirt and hardens like concrete but has many of the advantages of asphalt.

A 30-in.-thick LCF base has a load performance equivalent to 18 in. of portland-cement concrete or 50 in. or more of aggregate-based asphaltic concrete. The new pavement is being placed for as little as $7.60 per square yard, compared with about $12 for an equivalent conventional asphaltic concrete and more than $20 for a concrete pavement capable of sustaining the same loads (1968 price).

The important thing is that the final pavement will be far superior to one made of conventional materials. It will remain safer and smoother and require less maintenance. This is because:

• The material is not as rigid as concrete and so will conform more readily to long-term settlement of the subgrade with a minimum of stress induced in the pavement structure.

• At the same time, it is able to withstand more loadings without cracking or rutting than a conventional asphaltic base.

• Its strength increases significantly for at least 5 years, automatically compensating for the almost certain increase in airplane size and weight.

• The design is based on the interaction of aircraft and pavement, and the longer service life is accomplished by limiting deflection and stress in the subgrade. The less deflection, the more satisfactory and durable the pavement.

Basic Research

The new pavement design evolved from a $500,000 research program that required 3 years of intensive effort in the field, in the laboratory, and in basic engineering research.

When a pavement construction is completed, the first feedback will come from the pilot or driver who uses the pavement. His primary concern is with the safety, comfort, and well-being of his passengers while landing, taking off, taxiing, or traveling on the pavements. He is concerned with a clean surface, having no possibility of dust or stones being ingested into his engine. He is also concerned with visibility, braking traction, hydroplaning, traffic lights, direction signals, and stopping distance, as well as the geometrical design of the pavement complex. Among these concerns, the most difficult one for an engineer to contend with is the smoothness of a pavement. The engineer must translate the user's requirements into a description of physical parameters. The most common physical parameter which can be used to describe the user's reaction to vehicle vibration is the vertical acceleration of the vehicle. It is closely related to the fundamental frequency and the damping characteristics of the vehicle. The surface characteristics of a pavement expressed by its wave geometry is acting as a forcing function in causing the vehicle vibration.

The owner, on the other hand, is deeply concerned about the total cost of the pavement, which includes initial construction, maintenance, and operational costs and the cost of contingencies. He measures the value of a pavement by its total cost and benefit of service. A satisfactory operation surface will offer a reasonable length of service life without affecting the total cost.

A key design criterion of New York's new airport pavements is that pavement deterioration must not exceed a level at which the dynamic response of the aircraft to rough pavement will exceed $0.12g$ in the concentrated traffic area and $0.3g$ in the infrequent traffic areas. The pavement is designed to have a service life of 20 years free from major maintenance.

Development Tests

The sites of the New York airports—Kennedy, Newark, and LaGuardia—were originally marshy swamps. To begin the development of the airports, the areas were reclaimed with hydraulic sand fill brought from the sea bottom. This sand represents a paid-for, on-site material. If the sand could be used, significant economies might result. Laboratory tests were begun in 1965 toward this end.

It was soon discovered that the fines in the sand fill had been washed away in the process of pumping the sand from barge to reclaimed site. The pumped sand resembled granulated sugar in that the grains were largely of one size (30 to 50 mesh). The sand was deficient in the fine particles which are vital in interlocking the granulated soil structure, to provide stability under load. In order to introduce, at low cost, the needed fines so the sand could function as a good base course under the pavement, flyash and hydrated lime were added. The combination showed remarkable promise. Flyash, of course, is a waste product of power plants. It is very fine (largely passing a 200-mesh screen) residue from burning pulverized coal and consists mostly of silicon and aluminum compounds. By its reaction with lime, flyash can become an effective cementing agent. The use of lime in engineering construction is almost as old as recorded civilization. The Chinese used slack lime and clay to build the Great Wall about 2,000 years ago. The Romans used slack lime and volcanic ash (the natural counterpart of modern flyash) in such enduring structures as the Appian Way, the aqueducts, and the Coliseum. In fact, the lime-flyash concept repeats nature's process in forming sandstone and conglomerate, as in the Grand Canyon.

After accurate determination of the properties of all ingredients, the laboratory work led to the decision to use, as pavement base, a mixture of hydrated lime, portland cement, flyash, crushed stone (in some of the LCF materials), and sand. When water is added, the mixture hardens as does concrete (though much more slowly). The portland cement acts as an additive to accelerate development of the chemical bond. The development of compressive strength of the LCF mix is directly related to the chemical exchange of lime and flyash as affected by the temperature of curing. Because of slow initial setting, there is no need to finish paving work the same day that the LCF material is mixed. Man and machine time can be conserved.

The next step in the research program was the development of a realistic concept in the design of airport pavements. Theories alone were not sufficient to tackle the problem. A test strip was constructed at Newark Airport. Gages and performance measurements were recorded during the load-simulation test (see Chap. 10). Upon the completion of the field experiment, the findings and recorded data were introduced in the design analysis.

System of Pavement Design

The design's unconventional approach and the basic complexity of the pavement problems required extensive use of systems engineering analysis.

The first subsystem related the dynamic response of the aircraft to pavement roughness. The characteristics of aircraft, mode of operation, density of traffic, and patterns of movement were derived from current airport operation and projected for future traffic growth. The acceptable levels of aircraft vibration at landings, taxiings, and takeoffs were based on comments made in the preceding several years by airport managers and pilots. A random-vibration theory was developed to translate aircraft vibration into pavement surface roughness. The translation is governed by the speed, shock absorption, and fundamental frequency of the aircraft. During the Newark test, the progressive change in longitudinal and transverse roughness was simultaneously recorded at a given number of load repetitions of the test traffic. For each type of pavement, the magnitude of elastic deflection and subgrade stress was directly related to the rate of change in longitudinal roughness of the pavement surface. Thus, the allowable elastic deflection or allowable subgrade stress was established for a given amount of traffic service.

The second subsystem involved the equilibrium of pavement structure on the subgrade. In pavement design, the subgrade immediately below the pavement is the most important governing factor, but its properties are often defined in ambiguous ways. The new design system attempted to describe the stress-strain relation of the subgrade by methods used for classifying common construction materials. There are two elastic theories, namely, Burmister's layered method and Boussinesq's elastic mass. The stress and deflection analysis by Boussinesq is usually on the high side, but the solution of its algebraic equation is rather simple for practical engineering application. Pavement construction always involves the use of a massive volume of road materials, and economy suggests acceptance of a reasonable variation in quality control. Because of this variation in materials, the simpler Boussinesq theory should be accurate enough for practical pavement design computations. In the Newark test, attempts were made to correlate the test results with the Boussinesq theory. It is believed that by using parameters derived directly from tests, the design theories are more meaningful in their practical application.

The third subsystem involved the stress-strain in pavement components. This required the analysis of pavement elements or layers. The progressive reduction of material strength due to load repetitions and the concept of safe working stress were determined by the endurance limit of material, the number of load repetitions, and the quality variation. The general theory that was used for the analysis deals with an idealized condition in which wheel loads are carried by an elastic plate supported by an elastic mass. Several related problems, such as bending stress due to thermal gradient, differential settlement, and horizontal traction, were also analyzed by flexural theories. By combining the pavement stress under all these loading conditions, a proper pavement composition could be developed.

The fourth subsystem related to the basic considerations in optimizing the material concept, construction tolerance, maintenance operation, cost-benefit, and fiscal policy. Here the entire design came together. The results derived from elastic theories were combined with the findings from the yield theory, which defines the lower limit of the plastic state. To achieve a smooth pavement, it is essential to maintain stresses within the elastic state of equilibrium. The degree of quality control and construction tolerance were developed by balancing construction cost against structural requirements. For the base courses in pavement construction in the New York area, materials could have a coefficient of variation in quality control ranging from 0.15 to 0.25. High-quality material was limited to pavement surfacing, where high strength is needed. The use of relatively low-strength base material ensured not only a resilient pavement but economical construction as well. Consideration of maintenance cost at modern airports should include not only the direct cost of repairing the pavement but also the indirect costs due to interruption of service to airport operators, carriers, and the traveling public. Sound fiscal policy is to reduce maintenance cost by increasing the strength of pavement, although this may raise initial capital investment. Another consideration is general local practice in pavement construction. If a pavement design can be in harmony with the best local practice, pavement cost will be substantially lower.

The new design concept has been written in a computer program, which will save considerable computation time. However, the most important step in the whole design system is the engineering judgment of the designer. There is no substitute for the sound judgment of a well-informed engineer. The construction practice and material concept used at New York airports may not be directly applicable to other pavement construction. However, the introduction of a new design concept will lead to a better understanding of the performance of pavements. A detailed general discussion will be given in the following chapters.

REFERENCES

1. Proposed Recommended Practice for Design of Concrete Pavements, *ACI J.,* vol. 28, no. 8, pp. 717-750, 1957.
2. Thickness Design—Asphalt Pavement Structures for Highways and Streets, *Asphalt Inst. Manual Ser.,* no. 1, 7th ed., The Asphalt Institute, College Park, Md., 1963.
3. F. S. Barber, Application of Triaxial Test Results to the Calculations of Flexible Pavement Thickness, *Proc. HRB,* vol. 26, p. 26, 1946.
4. Canadian Good Roads Association Committee on Pavement Design and Evaluation, Pavement Evaluation Studies in Canada, *Proc. Int. Conf. Struct. Design Asphalt Pavements, Ann Arbor, Mich.,* pp. 274-350, 1962.
5. W. N. Carey, Jr., and P. E. Irick, The Pavement Serviceability—Performance Concept, *HRB Bull.* 250, pp. 40-58, 1960.
6. "Benkelman Beam Procedure," Canadian Good Roads Association Technical Publication 12, Ottawa, 1959.
7. "Engineering and Design, Rigid Airfield Pavements," U. S. Army Corps of Engineers Manual EM 1110-45-303.
8. "Revised Method of Thickness Design for Flexible Highway Pavements at Military Installations," U. S. Army Corps of Engineers Technical Report 3-582, 1961.
9. S. T. DeBarros, A Critical Review of Present Knowledge of the Problem of Rational Thickness Design of Flexible Pavement, *Highway Res. Record,* no. 71, pp. 105-128, 1963.
10. G. M. Dormon and C. T. Metcalf, Design Curves for Flexible Pavements Based on Layered System Theory, *Highway Res. Record,* no. 71, 1965.
11. "Airport Paving," Federal Aviation Administration Advisory Circular 150/5320-6A, 1970.
12. C. R. Foster and R. G. Ahlvin, Development of Multiple-wheel CBR Design Criteria, *Proc. ASCE,* vol. 84, May, 1958.
13. C. R. Foster and R. G. Ahlvin, Notes on the Corps of Engineers CBR Design Procedures, *HRB Bull.* 210, pp. 1-12, 1959.
14. R. Glossop and H. Q. Golder, "The Construction of Pavements on a Clay Foundation Soil," Institution of Civil Engineers Road Paper 15, London, 1944.
15. R. Glossop and H. Q. Golder, The Shear Strength Method of the Determination of Pavement Thickness, *Proc. Int. Conf. Soil Mech. Found. Eng., 2nd, Rotterdam, 1948,* vol. 4, pp. 164-167.
16. *Highway Res. Circ.* 112, October, 1970.
17. C. A. Hogentogler, Discussion of Flexible Surfaces, *Proc. HRB,* vol. 20, p. 329, 1940.
18. W. R. Hudson and B. F. McCullough, An Extension of Rigid Pavement Design Methods, *Highway Res. Record,* no. 60, 1963.

19. F. N. Hveem and G. B. Sherman, The Factors Underlying the Rational Design of Pavements, *Proc. HRB,* vol. 28, pp. 101–136, 1948.
20. F. N. Hveem and G. B. Sherman, Thickness of Flexible Pavements by the California Formula Compared to AASHO Road Test Data, *Highway Res. Record,* pp. 142–166, 1963.
21. A. Jones, Tables of Stresses in Three-layer Elastic Systems, *HRB Bull.* 342, pp. 176–214, 1962.
22. Kansas State Highway Commission, Design of Flexible Pavements Using the Triaxial Compression Test, *HRB Bull.* 8, 1947.
23. W. J. Liddle, Application of AASHO Road Test Results to the Design of Flexible Pavement Structures, *Proc. Int. Conf. Struct. Design Asphalt Pavements, Ann Arbor, Mich.,* 1962.
24. A. Losberg, On Load-carrying Capacity of Concrete Pavements, by G. G. Meyerhof, *Proc. ASCE, J. Soil Mech. Found. Div.,* vol. 89, 1963.
25. G. Meyerhof, Load Carrying Capacity of Concrete Pavements, *Proc. ASCE, J. Soil Mech. Found. Div.,* vol. 88, pp. 89–116, 1962.
26. N. W. McLeod, Some Notes on Pavement Structural Design, Parts 1 and 2, *Highway Res. Record,* no. 13, pp. 66–141, 1963, and no. 71 pp. 85–104, 1965.
27. W. W. McLoughlin and O. L. Stokstad, Design of Flexible Surfaces in Michigan, *Proc. HRB,* vol. 26, 1946.
28. "Airfield Design and Evaluation," Ministry of Public Building and Works Technical Publication 109/59, London, 1964.
29. "Airfield Pavements," Department of the Navy NAVFAC DM-21, November, 1967.
30. J. P. Nielsen, Implication of Using Layered Theory in Pavement Design, *ASCE J.,* vol. 96, no. TE4, November, 1970.
31. L. A. Palmer and E. S. Barber, Soil Displacement under Circular Loaded Area, *Proc. HRB,* vol. 20, p. 279, 1940.
32. L. A. Palmer, Special Procedures for Pavement Design, *ASCE Trans.,* vol. 80, paper 2684, 1954.
33. "Design of Concrete Airport Pavement," Portland Cement Association.
34. K. R. Peattie, Stress and Strain Factors for Three-layer Elastic Systems, *HRB Bull.* 342, pp. 215–253, 1962.
35. G. Pickett and G. K. Ray, Influence Charts for Concrete Pavements, *ASCE Trans.,* vol. 116, paper 2425, pp. 49–73, 1951.
36. G. Pickett, M. E. Raville, W. C. Jones, and F. J. McCormick, Deflections, Moments and Reactive Pressures for Concrete Pavements, *Kansas State Coll. Bull.* 65, 1951.
37. O. J. Porter, Foundations for Flexible Pavements, *Proc. HRB,* vol. 22, 1942.
38. O. J. Porter, Development of the Original CBR Method for Highway Design, *Trans. ASCE,* vol. 115, pp. 461–468, 1950.
39. G. Y. Sabastyan, Flexible Airport Pavement Design and Performance, *Proc. Int. Conf. Struct. Design Asphalt Pavements, 2nd, Ann Arbor, Mich.,* 1967.
40. "Shell 1963 Design Charts for Flexible Pavements," Shell Oil Company, 1963.
41. J. F. Shook and F. N. Finn, Thickness Design Relationships for Asphalt Pavements, *Proc. Int. Conf. Struct. Design Asphalt Pavements, Ann Arbor, Mich.,* 1962.
42. J. F. Shook, Development of Asphalt Institute Thickness Design Relationships, *Proc. Ass. Asphalt Paving Technol., Ann Arbor, Mich.,* vol. 33, pp. 187–220, 1964.
43. S. P. Timoshenko and S. Woinowsky-Kreiger, "Theory of Plates and Shells," 2d ed., pp. 52–53, McGraw-Hill Book Company, New York, 1959.
44. W. J. Turnbull and R. G. Ahlvin, Mathematical Expression of the CBR Relations, *Proc. Int. Conf. Soil Mech. Found. Eng. 4th, London,* August, 1957.
45. A. S. Vesic and S. K. Saxena, Analysis of Structural Behavior of Road Test Rigid Pavement, *Highway Res. Record,* no. 291, 1969.

46. W. Van Breemen, Current Design of Concrete Pavement in New Jersey, *Proc. HRB,* vol. 28, pp. 77–101, 1948.
47. M. W. Witczak, "Design Analysis—Full-depth Asphalt Pavement for Dallas-Fort Worth Regional Airport," The Asphalt Institute Research Report 70-3, November, 1970.

CHAPTER THREE
Pavement Support Condition

A. Characterizing Its Physical Properties

From the AASHO road tests, the Corps of Engineers airfield pavement tests, the 1952 and 1967 Newark pavement tests, and many other pavement studies, valuable information has been collected for the design and construction of modern airport pavements.

The most common problem encountered at airports is the surface irregularity of pavement caused by subgrade deformation and the resultant rutting under channelized traffic. When stress in the subgrade exceeds a certain level, the deformation of the subgrade exceeds the ability of the subgrade to rebound. The deflection basin of a pavement will follow the movement of its subgrade. In the Newark tests, it was noted that the permanent deformation of a pavement increases with increasing elastic deformation when the load passes over the pavement. The primary function of the pavement is to distribute the external wheel load through a pavement structure of adequate thickness to the subgrade. If the deflection of a pavement is reduced, the interaction of pavement structure and subgrade can be improved. Therefore, the deformation of the subgrade is of prime importance in designing the pavement.

In the last several decades, many theoretical and empirical methods have been proposed for pavement design. They can be classified into three groups, as follows: (1) those based on consideration of the working capacity of pavement components, (2) those based on consideration of the stress and deflection in the pavement structure and its supporting subgrade, and (3) those based on empirical observations.

The first group is concerned with failure of the pavement components. A factor of safety is selected to compute the allowable working capacity of the components. This approach to pavement design possesses all the advantages common to structural engineering, namely, stress-strain relations, basic equilibrium theory, etc. However, it does not furnish information concerning deformation of the subgrade, which is assumed to perform either as an elastic mass or as having a uniform elastic property to support the pavement elements. Experience indicates that pavement deformation, such as rutting under channelized traffic, is actually the most important factor in pavement performance. The subgrade soil supporting the pavement is not uniform and is not an idealized elastic mass. The working-stress method in its present form is not suitable for pavement design.

The second group of methods deals with the elastic behavior of the pavement under static load conditions. Burmister's layered theory and Palmer's Boussinesq deflection solution have been widely used. Their basic design requires the evaluation of the stress and strain in the pavement material as well as in the subgrade. The variation of the modulus of deformation E and Poisson's ratio μ of the pavement material and subgrade led to the assumption that the pavement system consists of individual layers, each transmitting wheel loads. The outstanding feature of this design method is the basic criterion limiting the pavement deflection. The maximum theoretical deflection of the pavement surface is assumed to be 0.1 in. by Palmer and 0.15 in. by the U.S. Navy method. Due to the mature of the problem, arbitrary assumptions are used in the evaluation of elastic constants and deflection tolerance. This is the basic shortcoming of the second method. In subsequent years, in situ tests performed by the Corps of Engineers with actual airplane loads on an airfield pavement led to the development of a semiempirical method known as the *CBR design curves*. The ambiguity in determining the deformation modulus of pavement components becomes the principal weakness of the second group of design theories and methods.

The third group of methods involves the engineer's introduction of his own experience into pavement design. As a result, the intuition of each individual engineer is the principal approach to the pavement design. The diversified experience of each individual engineer led to the national program of pavement tests conducted by the AASHO. This test cost about 3 years time and more than $20 million. At the conclusion, a design method was deduced from the actual test results by the process of statistical regression and the use of design parameters such as structural number (SN).

3.1. PRESENT METHOD OF SUBGRADE EVALUATION

With these pavement design methods as background, we shall review the existing methods of evaluating subgrade properties. The first method, determining the working capacity of pavement elements, is typically represented by the famous Westergaard method. Professor Westergaard, a theoretical mechanics-oriented researcher, developed this method in the early 1920s, when there was little knowledge about soil mechanics. Based on his structural background, he established the differential equation for an elastic plate, and for convenience in solving that equation, he introduced a spring constant, k, to represent the subgrade soil in determining the stress and moment in the elastic plate. Because of his contribution to solving that differential equation, engineers have been compelled to determine the k value for the subgrade, whether right or wrong. The field-testing method involves the use of a certain size steel plate and the measurement of the deflection under certain load increments. It was arbitrarily assumed that with a 30-in. plate and a deflection of 0.1 in., the deformation and the load represented very closely the actual performance of a pavement. The pavement at that time was primarily designed for highway vehicles and consisted of portland-cement concrete of a thickness ranging from 6 to 9 in.

In view of the advancement of today's soil-mechanics theory, the validity of these tests should be reconsidered. First, there is no such k value in the soil. The k value is a tool for solving the differential equation of an elastic plate and is assumed to represent the supporting soil as a heavy liquid. The 30-in. plate and settlement of 0.1 in. represent the stress-strain property in the subgrade at one particular level below the plate. If the size of the plate or the load intensity is changed, the k value can vary radically, even in the same test at the same location. Because of such an arbitrary testing procedure, the validity of using these data in pavement design should be questioned. Service experience has indicated that many concrete pavements do not break because of deficiency in ultimate strength but, rather, because of the deformation of the subgrade and the resulting loss of full support under the pavement, such as pumping. Many practicing engineers and scholars have tried to modify the testing procedure and establish the relationship between the k value and the size of the plate. The recent study by Vesic and Saxena suggested that the k value be modified by the ratio of E values of the subgrade and of the concrete pavement. However, all these attempts are aimed at refining the k value and making it more accurately represent the theoretical solution of a concrete pavement. Of importance in Vesic and Saxena's study is that the k value is related to Poisson's ratio and Young's modulus of the subgrade. This is a more scientific method of characterizing the physical properties of the materials.

The use of Poisson's ratio and Young's modulus is not new in pavement design. Burmister and the Navy introduced them in designing pavement layers. However, there is no definite, clear-cut method for determining the E modulus or value of the subgrade in the field. As a result of this inability to present a definite method of characterizing the subgrade, the Corps of Engineers advanced the use of the CBR method. The CBR testing procedure was initially introduced by the California Highway Department to determine the bearing value of highway subgrades in comparison with the bearing value of a standard crushed-stone base. This is where the term *California bearing ratio* comes from. Subsequently, the California Highway Department introduced a different but better concept of evaluating the subgrade reaction. They have practically abandoned the original CBR concept. There are many merits to the refinement of the CBR concept. Quantitative data have indicated that subgrade reaction may vary from location to location. Although there is no definite recommendation to describe the variation of the subgrade, the CBR method does offer a procedure which describes the characteristics of the subsoil as a group. Modern soil classification and testing equipment have been introduced in pavement engineering for the first time. Based on the results of extensive tests conducted in the laboratory as well as in the field, a set of pavement design curves was developed by the Corps of Engineers. This has been one of the most significant contributions in the last 20 years of pavement design. However, the lack of a definite method for describing the variation of CBR values has a detrimental effect on the determination of appropriate CBR values to be fitted into the design curve developed by the Corps of Engineers. For instance, the CBR tests at one job have values ranging from 7 to 25 and averaging 15. If the average value, 15, were used in the pavement design, the result would be that 50 percent of the pavement would perform satisfactorily and the other 50 percent would be overstressed under the design load. For modern airports with heavy traffic, such low reliability poses serious problems for pavement engineers.

Another group of engineers trying to improve the test method for determining the E and μ values is the Shell Research Laboratory in Holland. The research of this group is based on nondestructive tests using vibratory forces exerted on the pavement and measuring the acceleration of the pavement, from which the displacement of the pavement is determined. They introduced the term *stiffness*, which is expressed by the load per unit of deformation. They establish a statistical correlation between the stiffness and the E value, which is then introduced into Burmister's layer theory for the pavement computations. It is a refined approach, but the statistical correlation nullifies the theoretical concept of E value.

Because of these diversified approaches to characterizing subgrade properties, the validity of pavement design methods developed through the use of any of these characterizations can be questioned. As a result, many design methods completely ignore the subgrade properties. The structural-number method advanced at the AASHO road test is a complete departure from pavement design based on physical properties of the subgrade and pavement elements. Each component is given a certain structural number for its thickness and types of material. Each pavement requires a certain total structural number to meet the service and loading conditions of the pavement. This number is derived empirically. The AASHO design method represents a complete departure from the basic laws governing the equilibrium of force and reaction. Another empirical method, sponsored by the FAA, also ignores the physical properties of the material. The classification of the subgrade support is dependent primarily on the gradation of the soil particles and the drainage conditions of the area. No engineering knowledge is required in the design of modern aircraft pavements by this method.

The original Corps of Engineers tests which formed the background of the CBR design curves were conducted some 20 years ago. At that time, the maximum weight of the airplane was 250,000 lb. Through the years, the CBR curves have been extrapolated and expanded to an aircraft weight of 350,000 lb. However, the present generation of airplanes already weighs about 700,000 lb. Future generations of the airplane will exceed 1 million lb. With these large aircraft, the validity of such an empirical method should be questioned. In the past, a basic sin of the engineering profession has been the failure to determine the subgrade properties. The simple fact is that the engineer cannot even determine the units of the parameter used to describe the subgrade. We use pounds per cubic inch for the k value, pounds per square inch in the E value, pounds per inch in the Shell stiffness value, a percentage for the CBR value, and no units at all in the FAA method. This is definitely not a scientific approach to developing a rational design method for modern airport pavements. In reviewing the historical background for such confusion today, we note that in the early stages of development of pavement design methods, most work was contributed by engineers with strong structural backgrounds. As a result, the engineers developed methods for determining the stress and strain in the pavement element and paid very little attention to the supporting condition of the subgrade. In the last quarter century, significant contributions have been made by engineers with strong soil-mechanics backgrounds. In many cases, the role of soil mechanics has been overemphasized, and the interaction of soil structures has been belatedly recognized. The combination of structural principles and soil mechanics will be needed for engineers to establish the basic equilibrium of pavement and subgrade under the influence of external loads. With a knowledge of geology and statistical concepts, the natural variation of in situ soils and the random distribution of pavement construction can be more effectively related to the variation of pavement performance. Consequently, the basic stress-strain concept and the variation of natural deposits should be introduced in characterizing the physical properties of the subgrade.

3.2. STATIC LOAD AND DEFORMATION OF SUBGRADE

The modulus of deformation can be determined by a static-load test, either in the laboratory or in the field. For the laboratory test, determination of the modulus of deformation is governed by Hooke's law:

$$E = \frac{\sigma_1 - 2\mu\sigma_3}{\epsilon_1} \tag{1}$$

where E = modulus of deformation of soil sample, psi
σ_1 = vertical principal stress, psi
σ_3 = horizontal confining stress, psi
μ = Poisson's ratio, ratio of volumetric change of soil sample
ϵ_1 = strain in the direction of σ_1, in./in.
$\sigma_1 - \sigma_3$ = a deviator stress in the confined-compression test

For an incompressible material, such as water, $\mu = 0.5$ and the effective stress $\sigma_1 - 2\mu\sigma_3 = \sigma_1 - \sigma_3$. When the subgrade is predominantly of granular soils, Poisson's ratio is not equal to 0.5 and the deviator stress $\sigma_1 - \sigma_3$ will not reflect the actual compressive stress. For an ideal material, having no horizontal displacement due to vertical stress, $\mu = 0$ and the effective vertical stress will be σ_1. Thus, the computed E value can be $(\sigma_1 - \sigma_3)/\epsilon$ for $\mu = 0.5$ and σ_1/ϵ_1 for $\mu = 0$. Without the determination of μ, the test result will be less reliable.

For stiff, clayey soils, the test can be performed in an unconfined condition $\sigma_3 = 0$, and the modulus of deformation $E = \sigma_1/\epsilon_1$. For a granular subgrade, it is impossible to perform any test without a significant confining pressure, which can be far greater than the confining stress under actual pavement construction. As the determination of Poisson's ratio is much more difficult than the determination of stress and strain of a soil sample, the reliability of the E value determined by the static-load test in the laboratory is not very satisfactory.

For static-load tests in the field, the most common method is the plate-bearing test. Detailed testing procedures have been standardized by the ASTM and AASHO. The modulus of deformation of the subgrade can be computed either by Boussinesq's elastic theory:

$$E = \frac{2pa}{W_0} C(1 - \mu^2) \tag{2}$$

or by Burmister's layered theory:

$$E = \frac{2pa}{W_0} CF_w \tag{3}$$

where C = coefficient of plate rigidity ($\pi/4$ for rigid and 1. for flexible)

p = unit pressure on bearing plate

a = radius of bearing plate

W_0 = vertical surface deformation of subgrade

F_w = coefficient of deformation, which is a function of $(h/a, E_1/E, \mu_1, \mu)$ —the thickness, modulus of deformation, and Poisson's ratio of the pavement structure, respectively

As three more parameters are introduced in Eq. (3), it becomes more difficult to determine the E value of the subgrade. Boussinesq's approach, Eq. (2), is applicable only when the bearing plate is seated on the surface of an elastic mass in half-space. Any surcharge around the bearing plate will require the application of a correction factor, which will be discussed in Sec. 3.3, in the form

$$E = \frac{2pa}{W_0} C(1 - \mu^2) \frac{K_{zm}}{K_{0m}} \tag{4}$$

where K_{zm} and K_{0m} are governed by the ratio of depth of confinement z and radius of test load a.

The reliability of computing the E values by various equations is only as good as the reliability (adequacy) of the assumptions behind the parameters involved. The ranges of variation are summarized in Table 3.1. It can be seen that Eqs. (2) and (4) would result in an E value of a narrower range of variation.

TABLE 3-1   Range of Variables in Computing E Value

Equation used	Confinement	Poisson's ratio	Plate rigidity	Range of computed E value		
				Low	High	High/Low
(1)	$(0.0–1.0)\sigma_3/\sigma_1$	0.0–0.5	- - - - -	0.00	1.00	
(2)	No confinement	0.0–0.5	0.78–1.00	0.59	1.00	1.7
(3)	No confinement	0.0–0.5	0.78–1.00	0.04	0.40	10.0
(4)	$(0.0–1.0)\, z/a$	0.0–0.5	0.78–1.00	0.50	1.00	2.0
(4)	$(1.0–5.0)\, z/a$	0.0–0.5	0.78–1.00	0.44	0.71	1.6

The subgrade is not an elastic mass. However, within a limited range of stress-strain characteristics, Hooke's law is approximately valid. In order to utilize Eqs. (1), (2), and (4) for determining the E value, the definition of modulus of deformation should be properly discussed. Figure 3.1 shows a typical load-deformation curve obtained from a conventional plate-load test on the subgrade. The modulus of deformation can be interpreted in several different ways (see Fig. 3.2):

1. *Initial Tangent Modulus:* This is the tangent of the stress-strain curve at zero stress. It represents the stress-strain condition at the beginning of loading. In actual testings, the accuracy of the initial tangent modulus is offset by the seating condition of the bearing plate and the initial friction in the loading system. The test value would have no direct relation to the actual physical properties of the subgrade.

2. *Tangent Modulus:* This represents the slope of the stress-strain curve at the intensity of stress under consideration. It is a precise value when the significant range of stress is known in the pavement designs.

3. *Secant Modulus:* This represents the average stress-strain condition between the initial stress limit and a defined stress limit

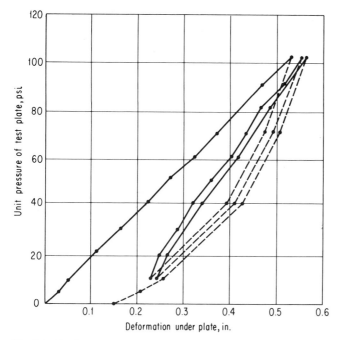

Fig. 3.1 Typical loading and reloading record of plate-bearing test.

encountered in the test. Such an average value of the stress-strain ratio does not yield a precise description of the subgrade.

4. *Rebound Modulus:* This is actually a secant modulus on the unloading cycle of the test. Because there is no initial load adjustment between the bearing plate and the subgrade (i.e., the seating and the initial loading condition), the rebound modulus is rather uniform as compared with the secant modulus. The rebound modulus is also closely related to the permanent and recoverable deformation under repeated loading conditions. It is a very useful parameter in studying the progressive deformation of pavement.

5. *Resilient Modulus of Deformation:* This was advanced by Hveem and refined by Seed and others. It is defined as the ratio of the deviator stress and the recoverable strain, such as the $(\sigma_1 - \sigma_3)/\epsilon_r$ shown in Fig. 3.2. The question of whether or not the deviator stress represents the effective stress (as discussed previously) will have a significant effect on the validity of the resilient modulus.

For ordinary pavement construction, the stress level in the subgrade ranges from 10 to 60 psi. The stress-strain relationship under the pavement is more appropriately represented by the tangent

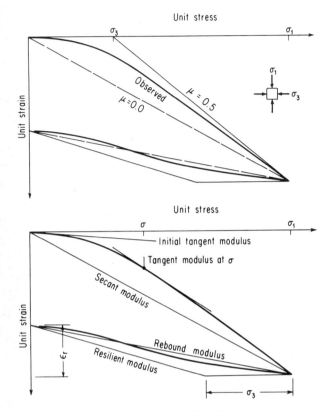

Fig. 3.2 Definition of modulus of deformation.

modulus at higher stress levels. For this study, the term *modulus of deformation* will mean the tangent modulus at a stress level ranging from 40 to 60 psi. Just as the pavement is subject to load and rebounds when vehicles are completely removed, so the range of stress varies from a peak stress to zero. In this case, the term *rebound modulus* will indicate the secant modulus on unloading cycles of the load-deformation curve.

3.3. EFFECT OF CONFINING PRESSURE— THEORETICAL ANALYSIS

Since the Boussinesq pattern of stress is applicable only for the load on the surface of an elastic mass in half-space, the effect of a uniform surcharge load should be investigated independently. In his analysis of a point load inside an elastic mass, Mindlin [1] has advanced the deformation inside the mass (see Fig. 3.3) by

$$w(r, z) = \frac{p_z(1 + \mu)}{8\pi E(1 - \mu)} \left[\frac{3 - 4\mu}{(r^2 + z^2)^{1/2}} + \frac{8(1 - \mu)^2 - (3 - 4\mu)}{(r^2 + 4z^2)^{1/2}} \right. $$

$$\left. + \frac{z^2}{(r^2 + z^2)^{3/2}} + \frac{(3 - 4\mu) z^2}{(r^2 + 4z^2)^{3/2}} \right] \quad (5)$$

and the deformation on the surface of elastic mass by

$$w(r, 0) = \frac{p_z(1 + \mu)}{8\pi E(1 - \mu)} \left[\frac{8(1 - \mu)^2}{(r^2 + z^2)^{1/2}} + \frac{4(1 - \mu)z^2}{(r^2 + z^2)^{3/2}} \right] \quad (6)$$

For a flexible circular test plate, the deformation along the centerline of the plate is

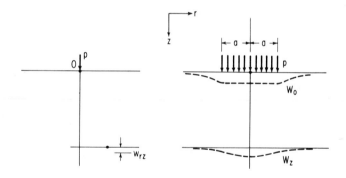

(a) Boussinesq pattern of stress

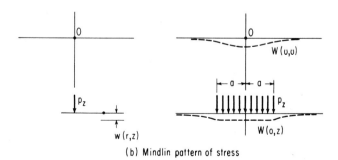

(b) Mindlin pattern of stress

Fig. 3.3 Pattern of stress-strain in elastic mass.

$$W(0, z) = \int_0^a 2\pi r w(r, z) \, dr \qquad (7)$$

which results in

$$W(0, z) = \frac{2p_z a}{E} (1 - \mu^2) K_{zm} \qquad (8)$$

and

$$W(0, 0) = \frac{2p_z a}{E} (1 - \mu^2) K_{0m} \qquad (9)$$

where

$$K_{zm} = \frac{(1 - 2\mu)n^2}{4(1 - \mu)^2} \left[\frac{3 - 4\mu + (2 - 4\mu)n}{(1 + n^2)^{1/2}} \right.$$

$$\left. + \frac{(5 - 12\mu + 8\mu^2) + (17 - 44\mu + 32\mu^2)n^2}{(1 + 4n^2)^{1/2}} - \left(\frac{21}{2} - 26\mu + 16\mu^2 \right)n \right]$$

and

$$K_{0m} = \frac{1}{2(1 - \mu)} \left[\frac{2(1 - \mu) + (1 - 2\mu)n^2}{(1 + n^2)^{1/2}} - (1 - 2\mu)n \right]$$

The n value represents the ratio of depth of embedment z and the radius of test plate a. Thus $n = z/a$.

Under the Boussinesq pattern of stress, the surface deformation under a flexible load plate is

$$W_0 = \frac{2pa}{E} (1 - \mu^2) \quad \text{for} \quad C = 1 \qquad (10)$$

and the deformation at a depth z below the center of the plate is

$$W_z = \frac{2pa}{E}(1 - \mu^2)K_{zb} \qquad (11)$$

It is noted that K_{zb} and K_{0m} are identical. When $p = p_z$, the deformation at depth z caused by the load on the surface is equal to the deformation at the surface caused by the same load at depth z. The validity of this reciprocal theorem demonstrates the compatibility of the Boussinesq and Mindlin patterns of stress.

For the plate-bearing test under a surcharge depth z, the observed deformation is $W(0, z)$ and the effective test load is p_z, which will be different from the effective load p in the Boussinesq condition.

When the surface deformation under the Boussinesq pattern of stress, W_0, is equal to the surface deformation under the Mindlin pattern of stress, $W(0, 0)$, it demonstrates that the load p_z and p are of equal effectiveness in causing the surface deformation of the elastic mass:

$$\frac{2p_z a}{E}(1 - \mu^2)K_{0m} = \frac{2pa}{E}(1 - \mu^2) \qquad (12)$$

or

$$p_z = \frac{p}{K_{0m}} \qquad (13)$$

By substituting the above equation, Eq. (8) can be rewritten for the correction factor of confining pressure of the plate-bearing test:

$$E = \frac{2pa}{W(0, z)}(1 - \mu^2)\frac{K_{zm}}{K_{0m}} \qquad (4)$$

The value of K_{zm}/K_{0m} represents the confinement factor for the surcharge depth z and has been computed for various μ values (plotted in Fig. 3.4). The correction factor is not very sensitive when the depth of surcharge is deeper than the radius of the bearing plate, and the error in assuming a μ value is also insignificant.

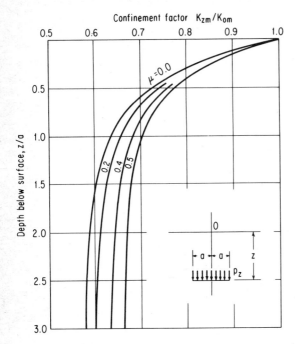

Fig. 3.4   Confinement factor for Mindlin pattern of stress.

3.4. TESTING PROCEDURE AND COMPILATION OF RESULTS

All tests conducted in the field were confined to plate-bearing loads. No triaxial tests were performed in the laboratory. The testing procedure was similar to the ASTM standards for determination of the k value except that 16- and 24-in.-diameter plates were used. The 24-in.-diameter plates were used for testing the open surface of the subgrade, either with or without confinement, while the 16-in. plates were buried in the pavement. An 8-in. hole was subsequently drilled through the pavement to the embedded plate. Test loads were applied on one group of bearing plates prior to the simulated traffic test, and another, identical group of bearing plates was subjected to load tests after the simulated traffic test. It was intended to obtain information about the effect of confining pressure as well as the number of traffic repetitions. A typical test record for 16- and 24-in. plates is shown in Fig. 3.5. In the stress range of 40 to 60 psi, the tangential slope of the load-deformation curve is graphically determined and the parameter $2pa/W_0$ is therefore computed. For the

24-in. plate seated on the surface of the subgrade, a rigid plate correction factor of $\pi/4$ is used. Poisson's ratio μ is assumed to be equal to zero. For the plate test below the surface of the subgrade, another correction factor [refer to Eq. (4) of the Mindlin pattern of stress] is also applied to the parameter $2pa/W_0$ in converting the confined load condition to the Boussinesq pattern of stress.

A rosette of SR-4 strain gages was installed to monitor the bending stress in each of the embedded 16-in. plates during the test. If the plate rigidity is assumed to be $\pi/4$ when the bending stress f in the plate is equal to zero (rigid plate) or 1 when the bending stress equals its yield stress f_y (flexible plate), the coefficient w can be expressed by

$$ C = 1 - \left(\frac{1-\pi}{4}\right)\frac{f}{f_y} \tag{14} $$

The observed C values range from 0.83 to 0.87, which is close to the rigid-plate condition.

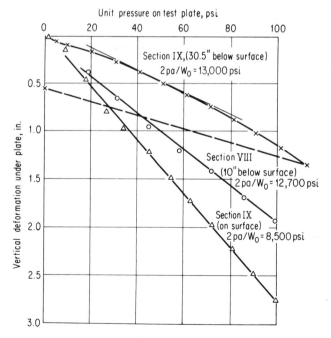

Fig. 3.5 Typical record of plate-bearing test.

TABLE 3.2 Results of Bearing-plate Test on Subgrade

Section	$2pa/W_0$ psi	Depth z, in.	z/a	Mindlin coefficient $K_{zm}/K_{0m}*$	Assumed rigidity of plate	Adjusted E value, psi
IA	9,200	0	0	1.0	0.785	7,200
I	5,400	0	0	1.0	0.785	4,200
II	11,000	0	0	1.0	0.785	8,600
III	7,400	0	0	1.0	0.785	5,800
IV	10,200	0	0	1.0	0.785	8,000
V	6,500	0	0	1.0	0.785	5,100
VI	8,400	0	0	1.0	0.785	6,600
VII	7,000	0	0	1.0	0.785	5,500
VIII	8,900	0	0	1.0	0.785	7,000
IX	8,500	0	0	1.0	0.785	6,700
X	7,500	0	0	1.0	0.785	5,900
IA	13,200	8	0.67	0.69	0.785	7,100
III	13,500	10	0.83	0.66	0.785	7,000
IV	14,900	7.5	0.62	0.70	0.785	8,200
V	10,300	9.5	0.79	0.66	0.785	5,400
VI	11,400	10	0.83	0.66	0.785	5,900
VII	13,800	8	0.67	0.69	0.785	7,500
VIII	12,700	10	0.83	0.66	0.785	6,600
XI	12,100	6	0.50	0.73	0.785	6,900
XII	11,700	6	0.50	0.73	0.785	6,700
XIII	11,300	10	0.83	0.66	0.785	5,800
XIV	14,000	11	0.92	0.65	0.785	7,200
XIV	11,800	10	0.83	0.66	0.785	6,100
XIV	14,600	7.5	0.62	0.70	0.785	8,000
XV	14,900	7	0.58	0.71	0.785	8,300
I	21,300	23.4	2.92	0.58	0.835	10,400
II	18,200	12.6	1.57	0.61	0.848	9,400
V	11,900	19.0	2.37	0.59	0.854	6,000
VII	20,000	20.4	2.55	0.59	0.849	10,100
IX	13,000	30.5	3.81	0.57	0.848	6,300
X	14,000	36.1	4.50	0.56	0.834	6,500
XI	14,500	22.7	2.84	0.58	0.873	7,300
XII	13,900	19.2	2.40	0.59	0.861	7,100

Arithmetic mean: $\bar{E} = 7,000$ psi
Standard deviation: $\sigma = 1,300$ psi
Coefficient of variation: $v = 0.186$

*Assumed Poisson's ratio: $\mu = 0$.

The adjusted E values are tabulated in Tables 3.2 and 3.3 for the subgrade and the mixture of stone screenings, respectively.

3.5. EFFECT OF CONFINING PRESSURE— TEST RESULTS

In Tables 3.2 and 3.3, the values of $2pa/W_0$ have been used in evaluating the modulus of deformation. If the test load is applied on the surface of the subgrade having Poisson's ratio $\mu = 0$ and the plate is sufficiently flexible to follow the curvature of subgrade deformation, the value of $2pa/W_0$ represents the modulus of deformation of the subgrade under Boussinesq stress-strain conditions. In the actual field test, improper seating of the plate, together with a slight disturbance of the subgrade surface, may seriously affect the

TABLE 3.3   Results of Bearing-plate Test on Mixture of Stone Screenings

Section	$2pa/W_0$ psi	Depth z, in.	z/a	Mindlin coefficient $K_{zm}/K_{om}*$	Assumed rigidity of plate	Adjusted E value, psi
IA	9,900	0	0	1.0	0.785	7,800
III	9,400	0	0	1.0	0.785	7,400
IV	13,500	0	0	1.0	0.785	10,600
V	9,600	0	0	1.0	0.785	7,500
VI	8,800	0	0	1.0	0.785	6,900
VII	11,500	0	0	1.0	0.785	9,000
VIII	11,300	0	0	1.0	0.785	8,900
XIII	14,400	0	0	1.0	0.785	11,300
XIV	11,000	0	0	1.0	0.785	8,600
XV	12,800	0	0	1.0	0.785	10,000
XV	16,500	0	0	1.0	0.785	13,000
IA	13,800	2	0.17	0.88	0.785	9,500
I	20,300	6	0.50	0.73	0.785	11,600
VIII	15,100	9	0.75	0.67	0.785	8,000
IV	21,600	11.3	1.41	0.62	0.866	11,600
VI	17,200	13.2	1.65	0.61	0.849	8,900
VIII	19,500	14.2	1.78	0.60	0.834	9,800

Arithmetic mean: $\bar{E}$ = 9,400 psi
Standard deviation: σ = 1,700 psi
Coefficient of Variation: v = 0.181

*Assumed Poisson's ratio: $\mu = 0$.

deformation reading of the plate test. It has been a common practice to set the test plate several inches below the surface. The computed value of $2pa/W_0$ from such a test can be different from the Boussinesq stress-strain condition. As indicated by Eq. (4), the computed $2pa/W_0$ value is much higher than the actual E value. Similar results were experienced in the determination of k values.

In the Newark test program, attempts were made to isolate the effects of various correction factors. Eleven bearing-plate tests were performed on the surface of the subgrade. The parameter and the coefficient of plate rigidity were computed separately. The second group of tests involved 14 bearing plates nested 6 to 11 in. below the surface of the subgrade. A Mindlin confinement factor was computed in adjusting the effect of surcharge. The third group of tests demonstrated the actual loading condition under the pavement. As given in Tables 3.2 and 3.3, the coefficient of plate rigidity varies within a very narrow band and is not significant in affecting the variations of the parameter $2pa/W_0$. In Fig. 3.6, the value of $2pa/W_0$ is plotted against the depth parameter z/a. The theoretical relation, as

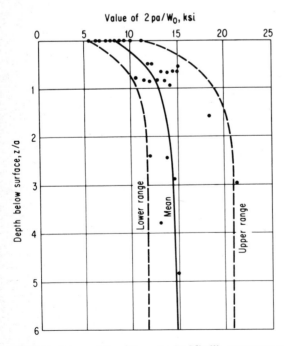

Fig. 3.6 Range of $2pa/W_0$ value by Mindlin stress pattern.

given by the Mindlin pattern of stress, is also plotted for the boundary condition of Poisson's ratio. The random nature of the tests precludes a good fit between the theoretical computation and the actual observation. However, the Mindlin pattern of stress for $\mu = 0$ does represent the best average of all tests. The use of the Mindlin confinement factor should be a valuable refinement in the determination of the subgrade modulus.

3.6. EFFECT OF REPEATED LOADINGS

At the completion of the traffic test, load tests were performed on eleven 16-in.-diameter plates buried underneath the test pavement and located in the middle of the wheel path. The test results are summarized in Table 3.4. For the tests conducted prior to the traffic load, the ratio between the tangent modulus and the rebound modulus of the subgrade was about 0.67, with a coefficient of variation of 12 percent. This indicates that the rebound characteristics of the subgrade were reasonably uniform. A similar ratio and coefficient of variation were also observed in the tests after the traffic loads. It would seem that the number of load coverages did not have a significant effect on the rebound characteristics of the subgrade.

On the other hand, the tangent modulus and rebound modulus increased significantly after 3,000 to 5,000 load coverages. The average rate of increase was 1.62 times, with a coefficient of variation ranging from 25 to 30 percent. This indicates that the change of modulus of deformation was rather erratic.

According to the results of the Newark pavement tests, the cumulative deformation D_N of a pavement was found to be a function of the load repetitions N in the form

$$D_N = D_1 + D_0 \log N \qquad (15)$$

where D_1 and D_0 are the initial, and rate of, pavement deformation. As the permanent deformation can be closely related to the elastic deflection of a pavement, the rate of increase of modulus of deformation of the subgrade can be expressed by the rate of change of pavement deformation, in the form

$$E_N = E_1(1 + k \log N) \qquad (16)$$

TABLE 3.4 Results of Bearing-load Test on Buried Plates

| Section | Load coverages N (1) | Value of $2pa/W_0$ | | | | Rebound ratio | | Effect of traffic | |
| | | No traffic | | After traffic | | | | | |
		Tangent, psi (2)	Rebound, psi (3)	Tangent, psi (4)	Rebound, psi (5)	(2)/(3) (6)	(4)/(5) (7)	(4)/(2) (8)	(5)/(3) (9)
I	5,013	21,300	28,000	24,700	40,000	0.763	0.617	1.16	1.43
II	5,013	18,200	27,900	32,000	40,300	0.654	0.794	1.76	1.45
V	2,981	11,900	22,200	29,800	39,700	0.535	0.752	2.50	1.79
VII	2,981	20,000	27,900	29,800	33,300	0.714	0.813	1.36	1.19
IX	5,019	13,000	22,300	17,000	25,900	0.581	0.654	1.31	1.16
X	5,021	14,000	22,000	18,900	32,000	0.637	0.592	1.35	1.46
XI	5,021	14,500	17,800	26,900	52,500	0.813	0.510	1.85	2.95
XII	4,985	13,900	22,500	30,000	42,900	0.617	0.699	2.16	1.91
IV	2,981	21,600	37,900	26,400	35,500	0.775	0.741	1.22	1.27
VI	2,981	17,200	27,900	27,500	41,600	0.617	0.662	1.60	1.49
VIII	4,985	19,500	30,200	32,000	46,700	0.645	0.685	1.64	1.55
Mean		16,800	26,000	26,800	39,100	0.668	0.683	1.63	1.61
Standard deviation		3,510	5,460	4,960	7,290	0.083	0.086	0.40	0.48
Coefficient of variation		0.209	0.211	0.185	0.187	0.124	0.126	0.245	0.298

During the traffic test, the movement of the vehicle was chan-nelized in a very narrow path. A small number of test runs, less than 1 percent of the total traffic coverages, traveled at a distance of about 1 ft off the wheel path. The effective channelized traffic, according to the normal probability distribution, would be 85 percent of the total load coverages. In Table 3.4, the average value of log $0.85N$ is 3.55 and the average value of E_N/E_1 is 1.62. Therefore, the k value in Eq. (16) is equal to 0.17.

In evaluating the performance of a new pavement, the original modulus of deformation prior to traffic load should be used. For existing pavement, the extended service condition will be governed by the higher modulus of deformation, E_N, when the Nth load coverage has been applied to the pavement. The subsequent pavement deformation will be encountered at a smaller magnitude, and the subsequent service condition will be improved accordingly, if the environmental conditions do not damage the pavement structure. According to one school of thought, thin pavement can be economically upgraded by releveling the pavement surface when the modulus of deformation of the subgrade has been improved by the compaction of actual aircraft loads. However, for many busy airports it is not feasible to take runways and taxiways out of service for an extended maintenance program. As wide variation is anticipated in the higher modulus of deformation, the variation in rutting depth and surface deviation remains a serious problem in achieving a smooth airport pavement.

3.7. OPTIMIZATION OF E VALUE FOR PAVEMENT DESIGN

As shown in Table 3.2, the range of the E value is from 4,200 to 10,400 psi. The range of these test values represents the natural variation of the subgrade and errors in sampling and testing. If the engineer were to utilize these test results for design, it would be very difficult to separate the error due to sampling and testing from that due to variation of the natural subgrade. Determining the optimum E value for the design requires study of the cost-benefit ratio and the reliability of service. For instance, according to Table 3.2, the mean value of this test is 7,000 psi. Based on this value, the corresponding pavement thickness is 29.5 in. for a particular type of aircraft. For the same aircraft, the same operation, and the same functional requirements, the corresponding thickness is 31.5, 34, and 24 in. for E values of 5,700, 4,200 and 10,400 psi, respectively. The variation of pavement thickness is plotted in Fig. 3.7. In the Newark area, the construction cost of the pavement is assumed to be $0.70/(in.)(sq yd) for the 4-in. wearing surface and $0.30/(in.)(sq yd) for the base and subbase materials. Using this unit price, the unit cost of the pavement is plotted in Fig. 3.8, shown as the initial cost. The cost varies from $11.80/sq yd to $8.80/sq yd for E values of 4,200 and 10,400 psi, respectively. According to normal probability distribution, the tail area below that design figure is the possibility of failure. Any test above the value used indicates that the design is on the safe side. Based on the normal distribution of probability, the percentage of reliability is 50 percent for an E value of 7,000 psi and 84.1, 98.4, and 0.5 percent for E values of 5,700, 4,200, and 10,400 psi, respectively. The probabilities of failure for E values of 4,200, 5,700, 7,000, and 10,400 psi are 1.6, 15.9, 50, and 99.5 percent, respectively. During the service life of the pavement, constant repair not only will affect the direct cost for the replacement but also will involve the loss of revenue due to the out-of-service pavement and

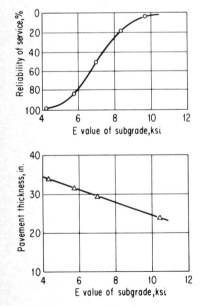

Fig. 3.7   Reliability of thickness design.

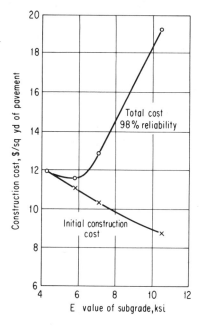

Fig. 3.8 Construction and maintenance cost of pavement.

the additional cost of the construction for the repair work, which is usually higher than the initial construction cost per unit. If all pavements are assumed to be eventually strengthened by overlay to a common standard of 98 percent reliability, the percentage of overlay pavement and cost analysis can be as shown in Table 3.5. The total cost is plotted in Fig. 3.8. Based on this analysis of the cost ratio, it can be seen that the most economical combination for this design is an E value equal to the mean value minus one standard deviation, or 5,700 psi. This cost-ratio analysis is applicable to the subgrade and construction costs only in the New York area. For pavement jobs in other localities, similar studies should be conducted to determine the optimum cost ratio. The optimum E value thus determined for the design will reflect the minimum total cost for initial construction and replacement.

The conventional method for determining the subgrade modulus usually uses the mean value of all the tests. As shown in Fig. 3.7, it can be seen that this is not a very reliable and economical method for pavement design; the cost of initial construction may be 5 or 10 percent lower than with the optimum design. The major drawback of this method is that the interruptions of traffic due to maintenance constitute a very serious problem for today's airport operations.

TABLE 3.5 Cost and Reliability Analysis

	E value assumed			
	4,200 psi	5,700 psi	7,000 psi	10,400 psi
Mean and standard deviation	$\bar{E} - 2.15\sigma$	$\bar{E} - 1.0\sigma$	$\bar{E}$	$\bar{E} + 2.60\sigma$
Reliability of assumed E value to be smaller than actual, %	98.4	84.1	50.0	0.5
Pavement thickness for assumed E value, in.	34.0	31.5	29.5	24.0
Initial construction cost, per sq yd:				
Top course	$ 2.80	$ 2.80	$ 2.80	$ 2.80
Base and subbase	9.00	8.25	7.65	6.00
Subtotal	11.80	11.05	10.45	8.80
Replacement for 98% reliability:				
Area to be overlayed, %	—	13.9	48	97.5
Thickness of overlay, in.	—	2.5	4.5	10.0
Direct replacement cost, per sq yd*	—	$ 0.26	$ 1.62	$ 7.32
Indirect replacement cost, per sq yd†	—	$ 0.13	$ 0.81	$ 3.66
Total cost for 98% reliability, per sq yd	$11.80	$11.44	$12.88	$19.78
Cost ratio	1.03	1.00	1.12	1.73

*(0.98 − 0.841) × (34.0 in. − 31.5 in.) @ $0.75/(sq yd)(in.) = $0.26/sq yd.
†Assumed loss of pavement service is 50 percent of direct replacement.

Indirect costs for replacement might be much higher than the direct replacement costs assumed in this study. The total cost for the pavement construction will be much higher if the E value is selected on the high side.

When cost analysis is introduced in pavement design, it requires a reliable statistical analysis. This means that a sufficient number of tests must be available to determine the mean value and standard deviation. Because of the expense involved in plate-bearing tests, a limited number of tests are often used to develop the subgrade data for the pavement design. Therefore, the statistical reliability is poor and the pavement design is not reliable. On the other hand, if a large number of tests are available, the range of variation is more reliable. The test data can be grouped together according to the use of pavement, and an optimum E value can be developed for each group. The pavement design can therefore be more efficient.

B. Nondestructive Test of Subgrade and Pavement

3.8. PURPOSE OF NONDESTRUCTIVE TEST

A nondestructive test is one from which the necessary information can be obtained to define the physical properties of a sample without destroying it. For subgrade and pavement construction, the non-destructive test requires a large mechanical setup to duplicate the vehicle load without destroying the subgrade and pavement. In the discussion of subgrade reaction and its physical characteristics, it has been pointed out that the natural variation of the subgrade will result in a scattered load-deformation response. Any meaningful inter-pretation of such scattered results requires quantitative data for optimizing the design inputs. The conventional plate-bearing tests in the field or cylinder tests in the laboratory have their inherent limitations. The plate-bearing test requires rather long testing periods in the field. This is reflected not only in the cost of operations but also in the interference with airport operations. Therefore, the second purpose of nondestructive tests is to reduce testing time and minimize interference with airport operations. For laboratory cylinder tests, the problem is one of reproducing field conditions in the laboratory. The third purpose of the nondestructive test, therefore, is to improve the quality of the test so that it can reproduce actual field conditions. For the conventional plate-bearing test, the loads are, at best, only a reproduction of the static load. The effect of a moving load can be quite different from that of a static load. The fourth purpose of the nondestructive test is to simulate the effect of moving loads. With these four purposes in mind, the selection of a nondestructive test can be more effective.

3.9. REVIEW OF PAST EXPERIENCE

The use of the nondestructive test is not new to the engineering profession. Benkelman introduced a steel beam to measure the deflection of pavement under a wheel load. This method, introduced some 30 years ago, is well known to pavement engineers. It was the first use of a nondestructive test to measure the static deflection of a pavement under a wheel load. The deflection measured by the Benkelman beam does not reflect the effect of a moving load.

In the second generation nondestructive tests, a modification of the Benkelman beam was developed by Dyna-Flex. A vibrator is introduced as a forcing function, and a series of geophones is used to measure the ground acceleration at a certain distance from the vibratory force. The signals of the geophones are connected to an analog-type integrator; the double integration of ground acceleration reflects the deflection of the monitoring point. When the distance from the vibrator is known, a deflection profile can be plotted. This machine, from the point of view of instrumentation, is a great improvement over the Benkelman beam. It does not require the actual wheel load to determine the deflection curve. However, there are two limitations to this test. First, the measurements of deflection are determined from the ground acceleration, which is a result of forced vibration at a specific frequency. Consequently, the magnification of the ground acceleration reflects only one phase of the vibration of the response system. The second limitation derives from the fact that the maximum deflection always coincides with the point of load application, where the geophones cannot be placed. An error might be introduced in estimating the maximum deflection under that loading.

In the late 1950s, the Corps of Engineers at the Waterways Experiment Station, Vicksburg, in cooperation with the Shell Research Laboratory in Amsterdam, developed a heavy mechanical vibrator to measure the ground vibration of the subgrade and the pavement. The machine consists of two components (see Fig. 3.9). The input, or forcing function, consists of a pair of eccentrically

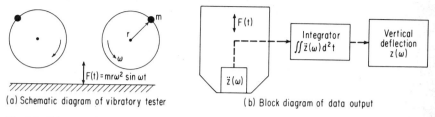

(a) Schematic diagram of vibratory tester   (b) Block diagram of data output

Fig. 3.9 Diagram of nondestructive test.

rotating drums having an eccentricity r and rotating at an angular velocity ω. The horizontal eccentric forces are canceled by each other, resulting in only a vertical vibratory force $F(t)$ on the pavement. The forcing function assumes a sine wave in the form

$$F(t) = mr\omega^2 \sin \omega t \tag{17}$$

The frequency of the force is $\omega/2\pi$, and the amplitude is equal to $mr\omega^2$. The output of that machine is monitored by an accelerometer attached to the bottom of the machine and processed by a double integrator. The response at the point of load application can be determined. When the pavement is vibrated over a selected range of frequencies and forces, the output is the wave propagation in the pavement as well as in the subgrade and the dynamic acceleration $\ddot{z}$ of the pavement surface. The double integration of the ground acceleration is the deflection, in the form

$$z(\omega) = \iint \ddot{z}(\omega)\, dt^2 \tag{18}$$

Foster and Heukelom [2] reported that the correlation between CBR and dynamic E value is considerably scattered, ranging from $E=50(\text{CBR})$ to $E=200(\text{CBR})$, with a mean value of $E=110(\text{CBR})$. The theoretical relation should be $E=10(\text{CBR})$. This indicates that the dynamic E value obtained from this vibratory machine is much larger than that obtained by the static method. Since the Foster and Heukelom report, there has been little additional work done with nondestructive tests of subgrade and pavement.

3.10. THEORETICAL ANALYSIS

In order to make this discussion of nondestructive tests more meaningful, it is necessary to review the theoretical background of the tests. For all plate-bearing tests or laboratory cylinder tests, the forcing function is a static load. In dynamic or vibration analysis, the influence of the static load can be defined as a constant amplitude for an infinite time. By applying the Fourier series of a time function, the static load can be expressed as a scrics of sinusoidal time functions in the form

$$F(t) = \frac{2}{\pi} \sum_{\omega=0}^{\infty} \frac{F}{\omega} \sin \omega t \qquad (19)$$

This means that a quasi-static load $F(t)$ is equivalent to the summation of all periodic functions with their amplitude modified by the frequency. For an elastic system whose load-deformation function obeys Hooke's law, the deformation under all periodic functions having their amplitude modified by the frequency is

$$W_0 = \frac{2}{\pi} \sum_{0}^{\infty} \frac{z(\omega)}{\omega} \qquad (20)$$

where $z(\omega)$ represents the deformation (output of the nondestructive test) under the periodic function $F \sin \omega t$ (the forcing function on the nondestructive test).

In static-load analysis, the surface deformation of a half-space elastic mass, as expressed by the Boussinesq equation, is

$$W_0 = \frac{2pa}{E} = \frac{2F}{\pi a E} \qquad (21)$$

By substituting Eq. (20) in Eq. (21), the E value of the elastic mass can be determined:

$$E = \frac{F}{a \sum_{0}^{\infty} \frac{z(\omega)}{\omega}} \qquad (22)$$

This is the basic relation governing the modulus of deformation by the method of nondestructive testing.

There are practical difficulties in the conduct of low-frequency vibratory tests. Consequently, the range of forcing frequency ω, from zero to infinity, cannot be fulfilled by the nondestructive test. If the test is conducted at a steady state of vibration, the peak response of the ground vibration $z(u)$ represents the dynamic response to the forcing function, such as

$$z(u) = FH(u) \tag{23}$$

The output of the steady state of vibration will not be influenced by the harmonic characteristics, $(\sin ut)$ of the forcing function, but will be closely related to the magnification factor $H(u)$, which depends on the ratio of frequency in the form $u = \omega/p$ and the coefficient of damping β. For evaluating the summation in Eq. (22), the above equation can be rewritten as

$$\int_0^\infty \frac{z(u)}{u}\, du = F \int_0^\infty \frac{H(u)}{u}\, du = \frac{F}{2} \ln \frac{1 - 2\beta^2}{\beta^2} \tag{24}$$

and

$$\int_1^\infty \frac{z(u)}{u}\, du = F \int_1^\infty \frac{H(u)}{u}\, du = \frac{F}{2} \ln \frac{1 + \beta}{\beta} \tag{25}$$

For $\beta < 1.0$, the following approximate relation exists:

$$\sum_0^\infty \frac{z(u)}{u} = 2 \sum_1^\infty \frac{z(u)}{u} \tag{26}$$

By substituting Eq. (26) in Eq. (22), the E value can be given by

$$E = \frac{F}{2a} \frac{1}{\sum_1^\infty [z(u)/u]} \tag{27}$$

This means that the low-frequency range of the nondestructive test can be neglected. The data-collection process could be started at the first-mode resonant vibration. The fundamental frequency of the response function becomes a scalar, and the summation of dynamic response is processed from $u = 1, 2, \ldots$ to ∞. As the dynamic response of the ground structure decreases rapidly with increasing frequency of the forcing function, the omission of some high frequencies may not seriously affect the accuracy in determining the E value of the elastic mass. Equation (27) is an important theoretical relation in transferring the dynamic test results into the static-load deflection.

3.11. FIELD EXPERIMENTS

In conjunction with the pavement tests at Newark Airport, a nondestructive test was conducted. A detail discussion of this test is given in the following.

Requirements of Tester

With the above theoretical analysis, it is necessary to review the basic requirements of the vibrator tester. The frequency of the vibratory machine should have a range of between 4 and 200 cps. Because of mechanical limitations, the tester consists of three mechanical systems: one for low frequencies (between 4 to 20 cps), one for normal frequencies (16 to 80 cps), and one for the high-frequency range (60 to 200 cps). The low-frequency vibrator requires a very heavy mass for generating the required force. If the forcing function can be maintained at a constant level by adjusting the eccentricity r with the frequency ω, then the integration of the test results can more effectively reflect the static load.

The load capacity of the test machine is also an important consideration. Under a steady state of vibration, the magnification of the resonance is equal to $1/2\beta$. For a lightly damped system, $1/2\beta$ is normally greater than 10. This means that the response function can be more than 10 times the forcing function. For today's airport construction, the maximum wheel load of aircraft is less than 50,000 lb per wheel. Considering the influence of other wheels, the single equivalent wheel load is likely to be in the range of 60,000 to 70,000 lb. Therefore, the upper range of the forcing capacity of the vibrator must be about 6,000 lb. The upper range of the capacity should be closely related to the optimum range of the tester. For ordinary vibratory machines, the middle half of the rated capability is the most reliable with respect to linearity and resolution. Therefore, the rated capacity of the machine should be 1.3 times the upper range of the forcing capacity. This means that the rated load capacity of the vibrator should be on the order of 8,000 lb.

In conducting any nondestructive tests, several basic procedures should be followed. First, the tests should be conducted in a steady state of vibration. When this state of vibration is established, then the ground acceleration can be monitored. The ground acceleration, expressed by $\ddot{z}(\omega)$, is picked up by an electric signal, which is introduced into an analog-type computer to perform the double integration for determining the vertical ground displacement:

$$z(\omega) \;=\; \iint \ddot{z}(\omega)\, dt^2 \qquad\qquad (18)$$

The machine is then shifted to another frequency, and a new forcing function of a constant amplitude is generated. The same steady state of vibration and double integration are repeated. By changing the frequency of the forcing function, a plot of ω against $z(\omega)$ can be established. (See Fig. 3.10.)

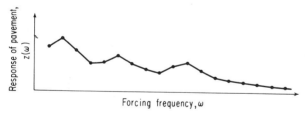

Fig. 3.10 Frequency and response function of vibratory test (constant force).

Error

Although the amplitude of the forcing function is kept within the middle half of the machine capacity, the reliability of the machine output still depends on the inherent error of the tester. All outputs of the tester consist of the true value R of the test plus the mechanical or instrumental error E, expressed as $R + E$. If the output is double-integrated, this result is $(R + E)^2$, which can be expressed as $R(R + 2E)$. This means that after double integration, the error of the test is twice as much as that from the direct machine output. On the other hand, if the ratio of the two outputs is utilized, the error may be reduced from the ratio of the true values. Thus,

$$\frac{R_1 + E}{R_2 + E} = \frac{R_1}{R_2}(1 + E_r) \qquad (28)$$

where R_1/R_2 is the true ratio of the experiment and E_r is the error in the ratio:

$$E_r = \frac{R_2 - R_1}{R_2 R_1} E \qquad (29)$$

expressed as a fraction of E. For R_2 equal to R_1, the error in the observations can be eliminated. If R_2 is equal to $2R_1$, the error of the ratio is reduced to one-half of the error of the direct output. Therefore, more reliable data processing depends on the ratio of the outputs.

Field Tests

In the fall of 1968, the Port Authority and Shell Research Laboratory conducted a joint research program, evaluating the existing pavement at three airports—Kennedy, Newark, and LaGuardia. The Shell machine used at the Port Authority's airports was in the second generation of equipment development. The early version of the Shell machine resembled the tester used at the Waterways Experiment Station in Vicksburg. The present model was a self-powered truck-mounted machine equipped with a complete range of monitoring instruments. The testing machine consisted of one pair of heavy vibrators and one pair of light vibrators. An eccentric weight was attached to the inside of each vibrator drum, and the weights were placed in opposite positions so that the horizontal forces of the rotating drums canceled each other. The resultant vertical harmonic load was applied to the pavement surface through a steel contact plate. An input-load range of from 500 to 4,000 kg was obtained by adjusting the eccentric weight of the vibrator. A slot was built in each drum of the vibrator. The load adjustment could be made while the vibrator motors were in operation (one of the outstanding features of the Shell machine). The machine had an operational frequency range of from 5 to 20 cycles for the low-frequency vibrator. The medium-range vibrator had a frequency range of from 16 to 80 cycles.

A smaller machine of high-frequency vibrators was also included but was separate from the heavy machine. The maximum vibration force was 1,000 kg, and the operational frequency range was from 60 to 200 cps. The vibrating force was transmitted through a 12-in.-diameter plate onto the pavement surface. In the housing of the contact plate were three load cells to monitor the quasi-static load imposed on the pavement. In contact with the loading plate, an accelerometer monitored the acceleration of the ground vibration. The amplitude of the ground vibration was calculated by double integration of the g measurement through an analog computer.

Since the instrumentation and capacity of the Shell machine almost completely fulfilled the requirements of this study, an arrangement was made to have the Shell machine perform dynamic tests on the Newark test pavements. The test pavements also had a complete range of static loading tests for comparison.

Experiment on Subgrade

The experiment was conducted with the Shell tester on subgrade reclaimed from the original marshland. From the surface to a depth of about 10 ft, the subgrade consists of hydraulic sand fill. The gradation of grain size is heavily concentrated between the No. 30 to No. 50 sieve sizes, with less than 3 percent of the particles passing the No. 200 sieve and less than 10 percent retained on the No. 10 sieve. The density of the sand is in the range of 108 to 112 lb/cu ft. Below the sand fill, a meadow mat 3 to 6 ft thick consists of a mixture of silt, sand, and decayed vegetation. Below the meadow mat, the basement material consists of red clayey sand. It is an original deposit, well compacted, and possibly preloaded by the glaciers. The vibration test was conducted on the subgrade, with the machine directly on the sand. At 4 to 20 cps, the heavy vibrator was used, and for 16 to 80 cps, the medium vibrator was used. The amplitude of the forcing function was constant and equal to 500 kg. At each frequency selection, a steady state of vibration was attained. It takes about 2 to 3 sec to achieve the steady state condition, but the vibrator was kept in a steady-state vibration for about 20 to 30 sec. At 20 sec, the ground acceleration was monitored, the integration performed, and the ground displacement recorded. The resolution of the output was in the range of ±1 percent. The frequency interval of the testing runs was from 1 to 2 cycles in the low-frequency test and 3 to 4 cycles in the high-frequency test. The results of the frequency-vs.-displacement chart are shown in Fig. 3.11. The peak response is at 7 cps, which possibly represents the rate of deflection in the meadow mat, which is about 8 to 10 ft below the ground peak. The second peak of resonance is not as clear as the first one and occurs at 17 cps. It could be the sand fill over the meadow mat. The third resonance, encountered at 52 cps, could be the basement material. In evaluating the deflection of the pavement, the deflection at 17 cps may be more significant than that at 7 cps because the stress in the fill sand will be much higher than the stress in the meadow due to aircraft loading. The results of the Shell test on the subgrade are tabulated and processed for $z(u)/u$ in Table 3.6. The computed E value from the dynamic test is given as 15,300 psi. At the same location, a static plate-bearing test was performed. The test results are shown in Fig. 3.12. The computed parameter $2pa/W$ is equal to 14,700 psi. The difference between 15,300 and 14,700 represents a discrepancy of only 2 percent, which is within the normal accuracy of such testing. It can be seen that the dynamic testing method can be used to replace the static plate-load test. The plate-bearing test shown in this case took about 1½ days to complete, at a direct cost of about $800. The nondestructive test with the Shell tester took only about 25 min: 10 min to set up the equipment, 10 min to run the test, and 5 min to remove the

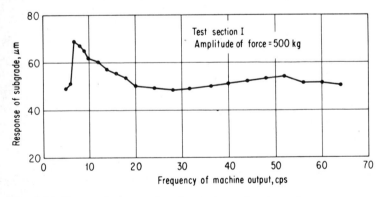

Fig. 3.11   Frequency-response function—test on subgrade sand.

TABLE 3.6 Shell Test on Subgrade Test Section I, Newark Airport

Amplitude of forcing function: $F = 500$ kg
First mode of resonant: $p =$ 7 cps

Frequency ω, cps	$u = \omega/p$	Response $z(u)$, μm	$z(u)/u$, μm
7	1.0	69.0	34.5
14	2.0	57.6	28.8
21	3.0	49.0	18.3
28	4.0	48.4	12.1
35	5.0	50.0	10.0
42	6.0	51.2	8.5
49	7.0	52.6	7.5
56	8.0	50.6	6.3
63	9.0	50.0	5.6

$$\sum \frac{z(u)}{u} = 131.6 \quad \text{say, } 150.0\,\mu\text{m}$$

$$\text{Modulus of deformation } E = \frac{F}{2a}\frac{1}{\sum z(u)/u}$$

$$= \frac{500 \times 2.2}{2 \times 6}\frac{1}{150 \times 10^{-6} \times 39.4} = 15{,}300 \text{ psi}$$

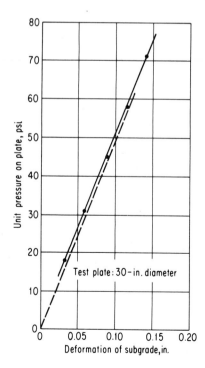

Test plate: 30-in. diameter

Fig. 3.12 Result of plate-bearing test on subgrade.

equipment. The cost of such a test is about $30. In terms of time and money, nondestructive tests can be very appealing to the practical engineer.

Experiment on Test Pavements

The next experiment was conducted on test-pavement section I. The test pavement consists of a 3-in. asphaltic-concrete top course and a 9-in. plant-mix, asphalt-stabilized stone base. The subbase consists of 6 in. of compacted stone screenings. The subgrade is the same as that described in the previous experiment. The testing procedure was the same as outlined previously except that the amplitude of the forcing function was increased. The results are plotted in Fig. 3.13. In interpreting the test results, the deflection and displacement readings in the frequency range of 15 to 25 cps were used and the ratio of the readings was employed to eliminate the error of observation. The deflection of the subgrade for the static load is equal to W_0. From the previous test, the deflection of the subgrade at 18 cycles is equal to 55 μm, or 0.11 μm/kg. The deflection of the pavement using the Boussinesq method is equal to W_z, and in the dynamic test it is equal to 38 μm at 18 cps and a forcing load of 1,000 kg, or 0.038 μm/kg. For a given ratio W_z/W_0, the ratio z/a can be determined by the

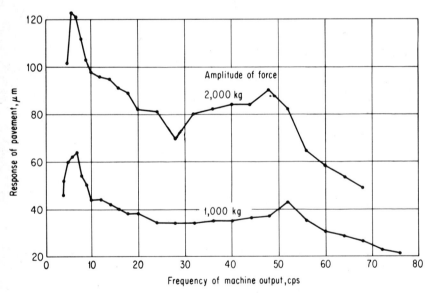

Fig. 3.13 Frequency-response function—test on section I.

Boussinesq method. In this case, for W_z/W_0 equal to 0.345, z/a is equal to 2. As the a of the Shell testing machine is equal to 6 in., the thickness of the consolidated layers is $2a = 12$ in. The actual thickness of the pavement consists of 3-in. asphaltic concrete and 9-in. asphaltic macadam. The test results are in good agreement with actual pavement installations. In this case, it can be seen that the ratio of the Shell test results can be effectively used in determining the thickness of the pavement if the subgrade deflection and the pavement deflection are known.

Experiment on Taxiway Bridge

A third experiment was conducted on the taxiway bridge over the Van Wyck Expressway at Kennedy International Airport. The bridge deck consists of a concrete slab on steel beams of composite-type construction. The width of the bridge is about 82 ft, consisting of 15 steel beams spaced at 6-ft intervals. The beams are 130 ft long, supported on two abutments with a center pier in the middle. Therefore, the effective span is 65 ft on each side of the center pier. The Shell tester was placed along the centerline of the bridge and located in the middle span between one abutment and the center

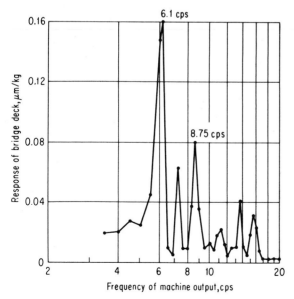

Fig. 3.14   Frequency-response function—test on Van Wyck taxiway bridge.

pier. The second location of the machine was at the quarter-point of span from the center pier. At test location 1, the Shell machine was excited from 3½ to 30 cycles at increments of about ½ cps. At each test frequency, the maximum vibration under the machine was monitored and plotted as in Fig. 3.14. It can be seen that the first mode of vibration of the structure is at 6.1 cps. The second mode of vibration is at 7.4 cps, and the third mode is at about 8.75 cps. The next series of tests was conducted with the machine at location 1 and monitoring the resulting acceleration of the concrete deck at 40 different points along the longitudinal centerline of the bridge. At 6.1 cps, the deflection of one span was equal to the other span. That is, the two spans deflected in harmonic motion. When the machine was put at the quarter-point of span, location 2, using the same 6.1 cycles, the two spans vibrated harmonically, as when the machine

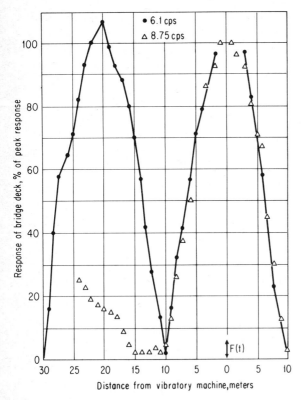

Fig. 3.15 Frequency-response function of a continuous bridge span.

was put at location 1 except that the amplitude was different. Subsequently, the machine was put at location 1 and the exciting frequency was set at 8.75 cps, corresponding to the third mode of vibration. The deflection in the span where the machine was located was the same as that at the vibratory force of 6.1 cps, but the adjacent span did not vibrate in harmony with the test span (see Fig. 3.15). It can be seen that the resonant vibration and steady state of vibration are very important concepts in studying the nondestructive tests.

The results of the Newark experiments demonstrated that the nondestructive test is a useful tool for evaluating the modulus of deformation of the subsoil, the thickness of existing pavement, and the natural frequency of the response function. However, in conducting the test it is necessary to monitor the response of a forcing function of wide frequency range. Using the ratio of the tests can more accurately reflect the actual test values. It is anticipated that a more reliable interpretation of existing pavement conditions and optimization of the pavement design can result from the use of nondestructive tests.

REFERENCES

1. R. D. Mindlin, Forces at a Point in the Interior of a Semi-Infinite Solid, *Physics,* vol. 7, pp. 195–202, 1936.
2. W. Heukelom and C. R. Foster, Dynamic Testing of Pavements, *ASCE Trans.,* vol. 127, pp. 425–457, 1962.

CHAPTER FOUR

Quality Control and Construction Tolerance

Before discussing the subject of quality control, we shall review the flow characteristics of an engineering project. The project is initiated by a planning group, which defines the functional requirements of the project. The engineer uses these functional requirements in establishing the engineering concept intended to translate the functional requirements into reality. In the translation, the engineer determines (1) the engineering feasibility of the project and (2) a realistic estimate of the cost of the project. The project is then referred back to the planning group, either for modification or for acceptance, as conforming to the original functional requirements. Upon acceptance, the program enters the design stage. In the process of design, the engineer constantly refers to the original functional requirements and the related engineering concept established for the project. During the process, modifications are often required to improve the efficiency and practicality of the original concept and to translate the engineering thinking into a viable contract document.

A contract document consists of two parts: the written portion, or specifications, and the attachments, usually contract drawings. Legally, if there is any conflict between the written specifications and the drawings, the contents of the written specifications govern. Too often, engineers spend considerable time in computations, planning, and preparation of contract drawings but not enough time in checking the contents of the specifications, which usually consist of two parts: the legal outline for the execution of the job and the technical part, which specifies details of the construction. When contracts are advertised and the bids are taken, the project is awarded for construction.

In the process of translating the engineering concept into reality, the performance of the contractor is an important factor. To maintain the structural integrity of the project, engineers usually exercise their prerogative of inspecting materials, labor, and performance of the contract. The engineer, based on his inspection results, can reject certain portions of the work. Therefore, a high quality of job performance can be achieved. Such an action, the rejection of construction, is a serious penalty to the contractor, especially if some of the engineering judgment is questionable. There is a clear and legal case for the engineer's formulation of the decision criteria and construction tolerances. This is the subject to be elaborated in detail: the decision criteria, quality of construction, and its related problems of material inspections, in-place performance, validity of sampling, testing, and interpretation of these results. By carefully evaluating each parameter, a sound judgment may be developed for getting the quality of construction the engineer wants and needs.

4.1. REVIEW OF TODAY'S PHILOSOPHY OF CONTRACT DOCUMENTS

The contract documents prepared for the construction program are in actuality a recipe rather than a specification. They spell out the details of the contractor's operations, the equipment to be used for those operations, and the type of material to be used in the project, as well as the end product. For instance, in road construction the specifications require certain paving lengths, widths, and compositions, with exceptions noted for certain areas. The use of certain manufacturers' compactors, spreaders, and mixers of specific size or capacity is often specified. Requirements are given for certain numbers of passes for compaction and certain percentages of density by a specific testing method. Such specifications or contracts are known as *performance contracts.* The contractor gets paid for the materials and labor used in the performance of his job to the requirements specified in the contract. There is no room for variation or flexible tolerance. Everything has to be done according to the specifications. The present experience indicates that such contracts can be successfully managed.

Such success depends especially on the skill of the engineers who design, supervise, and inspect the project. To back up their formulation of acceptance criteria, there is an elaborate program of sampling, testing, and field inspections. During the construction, a field laboratory is set up to test the soil, concrete, and asphalt, with more elaborate testing, such as that for steel and other high-quality materials, performed in a permanent laboratory. Instruments used in the testing process have improved significantly in recent years, and consequently, there is an associated improvement in testing quality. Based on the testing results, the resident engineer in the field and the project engineer in the office are in a better position to make on-the-spot engineering judgments, to decide whether or not a certain portion of the contract meets the specified performance.

The success of a project also largely depends on the cooperation of the contractor. Most contractors are conscientious about their engineering performance. Such cooperation is a keystone in building a successful engineering project. Another feature making a performance contract successful is the use of construction materials of established behavior and performance. For instance, most contractors know how to handle concrete, steel, asphalt, and many functional construction materials. The construction procedure is also reasonably well established for roads, dams, bridges, tunnels, and buildings. The use of established design practices, materials of a known variation, good labor performance, and established construction procedures eliminates many of the uncertainties and helps to ensure the success of a performance contract.

The accelerated changes in today's construction practice and procedures, however, have been a significant factor in the creeping breakdown of conventional performance contracts. The engineer knows little about the performance of new construction materials. The result of an early development test may justify the use of a new material, but on the actual construction job, the performance may be different from that on the tests. In recent years, because of the increase in labor cost, more efficient and powerful equipment has been developed for construction. The construction method has a significant effect on the design concept of many engineering projects. For instance, some 30 years ago, pile-driving equipment was more or less standardized at a 5,000-lb hammer. The pile lengths were limited to 100 ft, with load capacities of less than 100 tons. Today's pile hammer is three or four times heavier, the length of piles has increased to 200 to 300 ft, and the capacity of the pile has become much larger. The increased pile capacity has a definite influence on the design of pile foundations.

Most of the difficulties in maintaining today's performance concept of construction have arisen from errors of sampling, testing, and inspection and from rigidity in interpreting the results. Because of the rapid increase of engineering construction, there is a serious shortage of engineering manpower. As a result, inadequate on-the-spot engineering judgment has too often been experienced in the inspection of contracts. This creates legal problems between the supervising engineer and the contractor who executes the job. The most difficult aspect of the construction process to reconcile between these two groups is the construction tolerance in the performance contract, which seldom reflects the limitation of the construction process with respect to the variations in material properties and random conditions of construction performance.

The contractor's responsibility for the success of the construction becomes a twofold obligation: for job performance on the one hand and for cost control on the other. This conflict becomes a major problem for today's contractor. His primary interest is in making a profit. If there is a disagreement about job performance, the engineer legally must have documented evidence to enforce the contractor's compliance with the specifications. Too often, where there is disagreement, the quality of a job must be compromised to maintain the progress of the construction program.

4.2. DEVELOPMENT OF A QUALITY-ASSURANCE PROGRAM

In the past, quality assurance has always depended on the engineer's judgment in deciding whether an item has passed or failed in accordance with a specified quality of engineering construction. The rigidity introduced into the system has become impractical and/or legally unenforceable. This creates an unusual relationship between the engineer who supervises the construction and the designer who specifies the structure. The quality-assurance program, then, in common terminology, is more or less a grade-score system in lieu of the pass-and-fail system. The quality of a job is graded at a certain level, and the job is accepted on the basis of its grade score. Subsequently, the contractor is paid for the construction according to his grade scores. There are three basic questions to be answered by the quality-assurance program: First, what do we want? Second, how do we order it? And third, how do we determine that we have gotten what we want? The first two questions are related to the design concept and the formation of the contract documents. In the design concept, engineers must realize that construction cannot be classified in one simple grade—right or wrong, pass or fail. There should always be a variation along the optimum grade level. This means that the engineer should introduce the variances associated with materials and construction into the design; tolerances should be clearly spelled out in the contract documents. Room for variation, combined with a quality trend, defines the requirements of the engineer and the manner in which the job is to be performed by the contractor. By introducing variation and tolerance, we can more realistically know what we want and order it.

The third question, that of how to determine what we have gotten, requires a system of testing, inspection, and data analysis to determine the actual trend of the performance and the variation of the construction work. To make this program successful, there are four basic requirements. First, the quality criteria accepted by the engineer in the design and spelled out in the contract documents should be realistic. The contractor should be able to deliver what the engineer asks for. Often, the design engineer uses quality criteria which are not realistic, and it becomes difficult for the contractor to comply with them; the project may then become a controversial engineering design.

Assuming that the quality criteria are realistic, the second requirement to ensure the success of the program involves development of a valid test program which effectively indicates the quality of the job. In order to develop valid quality-control tests, engineers should use their experience to plan an efficient testing procedure.

Not only should all necessary tests be performed, but also all tests performed should be necessary. As test results become available, the data should be processed in a form for meaningful interpretation. The engineer's interpretation of test results determines his decision to accept or reject, or his grading of the quality of the construction.

Third, the engineer's decision-making rules are not normally under question. They are very important, however, with respect to the contractor's compliance with the contract. The contractor's responsibility is legally and contractually defined in terms of "substantial compliance." This means that in some instances, the job may not fully comply with all the specifications but may still be acceptable. It is not until work is rejected that the question of interpretation of the substantiality of the compliance becomes the subject of controversy.

Fourth, to avoid such a controversy, it is advantageous for the engineer to use a more specific description of the required degree of compliance. The engineer's needs can best be detailed by the statistical approach.

All natural (random) events result in a gaussian (normal) distribution curve, such as that shown in Fig. 4.1. In this figure, the abscissa represents the distribution of quality and the ordinate represents the frequency of occurrence of that quality. When a batch of events is said to be of *good uniformity,* the distribution is peaked at its center. If the events are said to be of *poor uniformity,* the distribution is more widely scattered. Mathematically, there are three parameters used to describe the distribution of natural events. They are the arithmetic mean

$$\bar{x} = \frac{1}{n}\sum x \tag{1}$$

the standard deviation

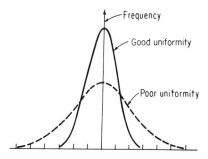

Fig. 4.1 Normal distribution curve.

$$\sigma = \sqrt{\frac{1}{n} \sum (x - \bar{x})^2} \qquad (2)$$

and the coefficient of variation

$$v = \frac{\sigma}{\bar{x}} \qquad (3)$$

It is very common to describe events as having an arithmetic mean of unity. Then the standard deviation and the coefficient of variation become identical. In this context we shall consider σ as the coefficient of variation unless otherwise specified. With the use of statistical analysis, the engineer can more reliably judge the quality of new construction if limited test results are available. Engineers with limited knowledge concerning new jobs can apply their experience from other jobs to extend their judgment on the performance of new jobs.

4.3. VALIDITY OF QUALITY-CONTROL CHARTS

To be assured of the validity of a quality-test program, the error inherent in the test program must be known. This error, for a given confidence level, can be written as

$$\sigma_m = \frac{t\sigma}{\sqrt{n}} \qquad (4)$$

where t, the standard score of the test, relates to the confidence level at which the test was conducted: $t = 3$ at 99.8 percent confidence, $t = 2.33$ for 98 percent, and $t = 2$ for 95.4 percent confidence. At a 50 percent confidence level, $t = 0.674$. The number of sampling operations in the program is represented by n. It can be seen that the error decreases with increased sampling. σ is the coefficient of variation of the test program and consists of the variations due to materials, sampling, testing, compilation, etc., in a sequential order of occurrence but constituting an integral part of the whole program. It is written in the form

$$\sigma = \sqrt{\sigma_1^2 + \sigma_2^2 + \sigma_3^2 + \cdots} \qquad (5)$$

A logical sampling operation must consider three things: First, the reliability of the program must be sufficient to meet the needs of the engineer, but no more. Increased reliability costs money, and the engineer should "purchase" only as much as he needs. Second, the sampling process depends on the homogeneity of the materials or construction being tested. Materials more prone to variation will ordinarily require more extensive sampling. Third, sampling depends on the costs involved in the testing operations and on the possible costs due to rejection of materials or construction.

For a construction job, these sampling operations must be performed on two levels. The materials for a job must be divided into a certain number of lots or batches of some size, each representative of the project. These lots must then themselves be sampled. It is the samples from the lots that go to the laboratory.

The size of the lot depends to a great extent on the value of the material. For low-cost materials, such as sand for a fill operation, the size of the lot can be large—for instance, a barge containing 5,000 cu yd of sand. On the other hand, for high-cost materials, such as the concrete used in a building, the lot size should be smaller—say, one truckload of concrete or perhaps two. In the same manner, for materials with a high cost ratio between testing and the value of the material, the lot sizes are apt to be large to bring the total testing cost more in line with the material value.

The size of a lot can be large for materials which are homogeneous, such as crushed stone. There is assurance that the lot is representative of the whole quantity of material used in the project. On the other hand, a glacial till, or conglomerate, is not homogeneous and therefore the lot size should be smaller.

Similarly, determination of the number of samples to be tested from a batch must consider the reliability and homogeneity of the material, as well as the testing cost. In judging the validity of the tests, the range of the test results is a significant parameter for decision making. In normal building construction practices, the number of samples to be tested should be large enough to reflect a range of test results having a variation from the mean of ± 3 standard deviations. This means that the tail area of the normal distribution curve beyond 3σ is 0.1 percent, or a confidence of test results of 99.8 percent. Such a high level of confidence will be a significant factor in determining the number of tests to be conducted. Keep in mind that as the number of samples tested increases, the range of the results will also increase but the mean of the tests will more accurately describe the actual mean of the material. For pavement construction, the variation of the subgrade is usually high. To achieve a test range of 99.8 percent confidence will require more sample testing than the material warrants by its large σ. For this reason, the confidence level, or reliability, of the test should be augmented with construction practice and economic considerations.

In Fig. 4.2, for a constant mean value $\bar{x}$, the level of confidence for a small standard deviation is much higher than for a large standard deviation. For a curve with a coefficient of variation of 0.1 and a range of test results of ± 2 standard deviations, for $x = (\bar{x} \pm 2\sigma)$, the confidence level is 95.4 percent. For the same material having the same mean but a larger standard deviation, 0.2, the same range of test results becomes the mean ± 1 standard deviation ($\bar{x} \pm 1\sigma$). The confidence level becomes 68.2 percent. To improve the confidence level of the test results, the σ value should be reduced. In order to reduce the coefficient of variation in the lot, the testing methods for the materials, the sampling methods, and the testing operations should be improved; that is, the quality of the testing should be upgraded.

After describing the quality criteria and discussing the validity of the testing process, we must now establish a system of quality control. As discussed previously, in the process of construction and testing, there are variations in the lot or batch which must be examined. The variation in the lot is indicated by the range or the standard deviation of the tests within the lot. So the first system in quality control is to determine the range or the standard deviation of the tests within each lot. The reliability of a construction depends on

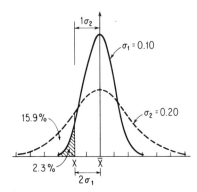

Fig. 4.2 Degree of confidence vs. variation.

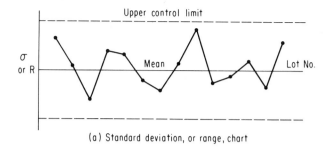

(a) Standard deviation, or range, chart

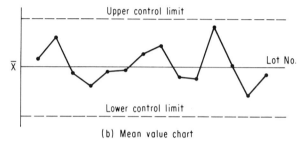

(b) Mean value chart

Fig. 4.3 Construction of quality-control charts.

the confidence level of the quality-control tests. The range or standard-deviation control chart is shown in Fig. 4.3a. In this chart the central line is the mean value of the range or the standard deviation of the test results. In the process of control, a lower control limit and upper control limit should be defined. The variation of the range or the standard deviation denotes the fluctuation of the confidence level in construction. It should be augmented with a mean-value chart which denotes the trend of quality of the lots tested (see Fig. 4.3b). The lower and upper control limits of the control charts should be defined according to the construction tolerance contemplated in the original design. The construction of control charts is based primarily on the analysis of past job performances, i.e., the mean value $\bar{x}'$ of the lots, mean range R' of the lots, and mean standard deviation $\bar{\sigma}'$ of the lots. The charts are then useful for comparing the trend of present and future projects. Without such past records, no meaningful control charts can be established until experience is gained through the trial-and-error process.

The construction of the range chart (R chart) or standard deviation chart (σ chart) is based on the probability of dispersion. For instance, when a die is thrown, the probability of a certain number's occurrence is 1/6. If the die is thrown twice, the probability is 2 times 1/6, or 1/3. The following table gives an example of the probability of dispersion of random samplings. The coefficients are

	No. of tests per lot		
	2	4	9
Centerline of R chart	1.128	2.059	2.970
Upper control limit of R chart ..	3.686	4.698	5.394
Lower control limit of R chart ..	0.000	0.000	0.546
Centerline of σ chart	0.564	0.798	0.914
Upper control limit of σ chart ..	1.843	1.808	1.609
Lower control limit of σ chart ..	0.000	0.000	0.219

multiples of the standard deviation of past performance $\bar{\sigma}'$ for a confidence level of 99.8 percent. The correction factors for 98.0 and 68.2 percent confidence levels are 0.777 and 0.333, respectively. It can be seen that the larger the number of tests in the lot, the larger the tolerance of the range between the upper and lower control limits [1].

On the mean-value control chart, the centerline is the mean value of the job, $\bar{x}'$. The upper limit of the control line is equal to $\bar{x}' + A\bar{\sigma}'$; the lower control limit is equal to $\bar{x}' - A\bar{\sigma}'$. The A value is based on the level of confidence contemplated in the testing program and the number of samples to be tested for each lot. It can be expressed as

$$A = \frac{t}{\sqrt{n}} \qquad (4a)$$

where t and n are the same parameters as given in Eq. (4). For a confidence level of 99.8 percent in the test result, the corresponding t value is 3.0 and the A value is 1.5, if the number of tests is four per lot.

For quality control, two charts should be used simultaneously. The R chart or σ chart controls the variation within the lots, that is, the short-term variation of construction. The $\bar{x}$ chart represents the long-term change of construction. If there is significant variation in the R chart or σ chart, that variation is local, possibly due to fluctuation of material supply, testing procedure, and/or construction method. On the other hand, the variation of the $\bar{x}$ chart is more serious because long-range variations of the job are indicated.

In addition to the regular $\bar{x}$ and σ (or R) control charts, a confidence chart may be introduced. The confidence chart, a plot of $\bar{x} - t\sigma$ or $\bar{x} + t\sigma$, is a combination of the σ chart and the $\bar{x}$ chart. The variation of $\bar{x}$ denotes the fluctuation of the average quality of the lots, and $t\sigma$ represents the construction tolerance. If the original minimum design quality is x_D (see Fig. 4.4) having a confidence level such that 15.9 percent of the job quality will have a value below x_D, the mean value of the job quality should be $\bar{x}_D = x_D + t\sigma_D$, where t is equal to 1.0. During the construction process, the quality control of the lots should be aimed at keeping $\bar{x}_c - t\sigma_c$ better than or equal to x_D, the minimum quality level. The ratio σ_c/σ_D depends on the size of the lots and the number of samples per lot. A detailed discussion of this ratio will be given under subsequent headings. The $\bar{x}_c - t\sigma_c$ values should be noted as being the construction tolerance in the specifications. The confidence chart denotes total quality of the product, and infringement of the minimum quality level indicates an infringement of the specifications. While meant as a supplement to the $\bar{x}$ and σ control charts, the confidence chart can in some cases replace them entirely.

4.4. STATISTICAL APPROACH TO QUALITY CONTROL

In reality, the statistical approach to quality control does not solve the problems involved in achieving a better quality of construction,

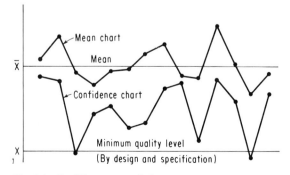

Fig. 4.4 Confidence control chart.

but it is an efficient tool which indicates the trend of construction quality. When the trend is persistently above or below the control values $\bar{x}$, R, and σ, it is an indication that something is wrong in the system. With this early warning, the engineer and the contractor have an opportunity to adjust the system appropriately to put it back into normal operation.

The construction of control charts is based primarily on data from jobs of established standards. The use of such charts ensures only that new construction will comply with old standards with some degree of confidence. Thus, when the construction system or method is new, the quality-control program may not be an efficient tool. However, the power of the statistical method is that (1) the trend of construction can be indicated by continuous sampling data, (2) early signs of deviation can be detected on the control charts, and (3) more efficient utilization of testing facilities and inspection manpower can be realized.

On the other hand, there are many serious drawbacks to the utilization of the statistical method. First, it requires a crew of trained technicians to process a large number of samples for meaningful analysis and interpretation. Without this background, the volume of test results will be useless. Second, the statistical method provides information only on the trend of the quality; particular elements which may be malfunctioning are not detected unless the quality-control program is broken down into quality assurance for each element. Such a control program can become very large and expensive. The scope of the program is, therefore, determined by the tradeoff between the cost of the rejection of certain elements in the system and the cost of quality control for that element.

4.5. QUALITY ASSURANCE AND ITS EFFECT ON ENGINEERING DESIGN

The foregoing discussion refers mainly to quality control during construction. In reality, such programs will not be effective unless the concept of quality assurance is incorporated into the original engineering design concept. In the original design, the engineer should have a definite concept of the mean value and variation of the overall project as well as the variation of structural components. Since the materials and construction are variables, the confidence level of structural performance depends on the reliability in selecting the working capacity of a structural system and the probability of the external loads. If the statistical concept is employed in the original design, the work capacity is normally expressed as the mean value of the system minus a multiple of the standard deviation of that system ($\bar{x} - t\sigma$). The result is considered to be the allowable working conditions. The coefficient of variation represents the scattering of quality in the system. The constant t used in deriving the working capacity is related mainly to the confidence level of safety of the system.

During the construction, the actual performance of the job may not necessarily coincide with the trend anticipated in the original design. For instance, if the average value $\bar{x}$ is persistently on the lower side of the central line but not below the lower control limit, the quality of this job may not be defined as poor but the trend of the construction is on the lower side of the quality-assurance program. Under this condition, the structure may not fail but the confidence level of the design is reduced. In order to maintain the integrity of the structure for the designed level of confidence, it is necessary to shift the allowable working capacity to a lower level. The shift is indicated in Fig. 4.5, in which $\bar{x}_D$ represents the anticipated mean value of the system in the original design and x_D represents the original design capacity for a certain designed level of confidence. Because of persistently low readings, below the center-line of the control chart, the distribution has shifted toward the lower side by a distance μx_D. If the allowable working capacity remains at x_D, the level of confidence of the system will be reduced. If the original confidence level is maintained, the working capacity x_c should be reduced to

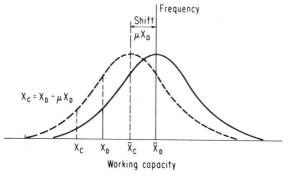

Fig. 4.5 Shift of working capacity due to lower mean value.

$$x_c = x_D - \mu x_D \tag{6}$$

The second term in the right side of the equation is the shift of mean value of the system.

In determining the causes of persistently low or persistently high readings, there are three important possibilities which should be reviewed prior to the adjustment of the allowable working capacity:

1. There are always errors in testing and inspection. The validity of the testing and inspection program should be examined when such variations are experienced.

2. During the construction, there is always the possibility of improper construction practices either in materials handling or in labor performance or in both. In this case, engineers should review the legal requirements of the contractor to perform the job and should reexamine the inspection procedures.

3. If all these factors prove to be reasonably normal, then the cause of the variation may be due to unrealistic loading conditions or impractical construction tolerances. Under these circumstances, it may be assumed that the shift is a natural event, representing the normal variation of material and labor performance under the economic conditions of the construction.

This natural shift should be referred back to the engineer for change of his requirements for design. This can be done by lowering the centerline of control, revising its design requirements changing the job specifications, and/or asking for different construction performance at a change of total cost. In many cases it may be found that lowering the working capacity is most economical and practical for the design.

The engineer must have a definite design concept concerning the factor of safety and construction tolerance for the engineering project. In Fig. 4.6, the solid line depicts the probability distribution of the structural system (response system) responding to an external load (forcing system). In the design, engineers should consider the mean value $\bar{x}$ of the structural system and the working capacity x. The broken line in the figure represents the distribution of the forcing or external load system. At any given degree of confidence from which x is determined, the structural system is assigned a certain degree of risk when the resistance of the structural system is below the defined working capacity or the forcing system is beyond the anticipated distribution. The point of intersection of the probability-distribution curves represents the equilibrium of occurrence of the two systems. The tail area under these two systems, expressed by $r + q - rq$, represents the upper limit of probability of "failure" due to the interaction of the forcing response systems. Here, "failure" means the failure of the response system to meet the designed confidence level of operation. The "failure" does not denote an actual structural failure. If the probability of "failure" under two systems is smaller than the original design probability of failure, the interaction is said to be satisfactory. On the other hand, if the probability of "failure" under the two systems exceeds the probability of failure of the original design, then the response system is said to be not adequate for the designed level of confidence to support the external load. The structural system is said to be understrengthed or poorly constructed.

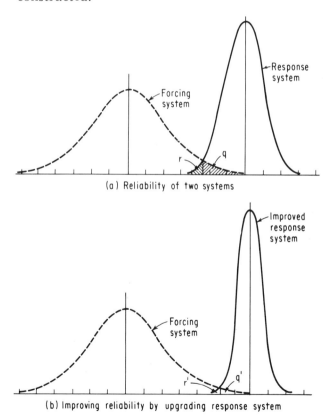

(a) Reliability of two systems

(b) Improving reliability by upgrading response system

Fig. 4.6 Reliability of two interacting systems.

If the forcing system does not change, the method of improving the reliability of the response system is to increase the mean value of the system $\bar{x}$ or to reduce the coefficient of variation σ or both. Increase of the mean values or reduction of the coefficient of variation means an increase in construction costs as well as inspection expenses. This should be considered at the original design stage. For economical designs, the engineer should not blindly ask for a higher mean value or smaller coefficient of variation unless it is justified for the construction.

4.6. CONSTRUCTION TOLERANCE

The preceding discussion reviewed how the quality of construction can be affected by shifting the center of the lots (arithmetic mean) or reducing the variation within the lots (standard deviation) or both. Without this modification, there is no alternative but to lower the working capacity to a value x/μ. All these maneuverings are aimed at maintaining the confidence level in the original design. The confidence level of normally distributed events is expressed by

$$t = \frac{\bar{x} - x}{\sigma} \quad \text{or} \quad t = \frac{\bar{x} - x}{v\bar{x}} \tag{7}$$

where x is the minimum quality level and $\bar{x} - x$ is the tolerance with respect to the mean value. For achieving an identical level of confidence in design and construction, the mathematical relation between the construction tolerance and design allowance is

$$1 \pm \frac{x_c}{\bar{x}_c} = \frac{v_c}{v_D}\left(1 \pm \frac{x_D}{\bar{x}_D}\right) \tag{8}$$

where the subscripts c and D denote construction and design parameters, respectively.

The left side of the equation represents the construction tolerance, which is a function of the design allowance and the ratio of construction and design variation. The coefficient of design variation v_D consists of the actual variation of construction v_c plus the variation in design analysis, engineering judgment, and so forth. If the variation in engineering analysis is equal to the construction variation, the anticipated coefficient of design variation v_D is

$$v_D = \sqrt{2}\, v_c \qquad (9)$$

This means that the construction tolerance is approximately equal to two-thirds of the design tolerance.

Since the standard deviation is so important to the construction quality, it is necessary to review the standard deviation of multiple events. Often, multiple events occur as independent steps in the fulfillment of the construction, for instance, the quality-control test consists of independent events such as sampling, testing, dispersion of materials, and method of construction. If these events are independent components of equal importance in the process of construction, the overall variation of such construction may be mathematically written as

$$\sigma^2 = \sum_i \sigma_i^2 \qquad (10)$$

where i is an independent and discrete event in a sequential development of a universe, such as the aggregate, cement, batching, and placement of concrete.

For events performed simultaneously (as an element of a universe) as well as independently, the overall variation is

$$\sigma^2 = \frac{1}{n} \sum_{i=1}^{n} \sigma_i^2 \qquad (11)$$

In this case, the components may be the variations, such as in the slabs, beams, and columns which constitute the components of a structure performing simultaneously under a load. The overall variation of the construction is equal to the mean value of the components. In this equation, it is assumed that each performs independently, but simultaneously, with equal contributions to the system.

When the components of a structure perform not equally but at various weights of contribution, then the overall variation is

$$\sigma^2 = \sum C_i \sigma_i^2 \text{ and } \sum C_i = 1 \qquad (12)$$

where C_i is the weighting factor of each of the components. To illustrate the importance of this equation, the following examples are given. In evaluating the pavement deflection of the Newark tests, the contribution of the subgrade was found to be more than 85 percent, the remaining portion distributed among the subbase, base, and top course. The subgrade coefficient of variation was about 30 percent, and the corresponding variations for the subbase, base, and top course were 0.20, 0.15, and 0.10, respectively. The overall variation for the pavement deflection, according to Eq. (12), is equal to 0.283 (see Table 4.1). It can be seen that in terms of pavement deflection, the variation of the subgrade is primary in controlling the quality of the pavement. For pavement stress, the contribution of the subgrade is much lower than that of the top course. By using the same principle of distribution, the overall variation, in terms of pavement stress, becomes 0.142 (see Table 4.1). In this case, control of the subgrade, subbase, and base is important to the stress-level performance of the pavement. Where thermal stress in the pavement is concerned, the top course is a very significant contributor and the subgrade has very little effect. The overall variation is 0.120. In this case, the quality of the top course is a significant factor in temperature stress and wearing quality of the pavement surface.

As pavement construction involves large quantities of material, the coefficient of variation and its effect on the cost of material should

TABLE 4.1   Coefficient of Variation and Its Contribution to Pavement Design

| Pavement components | Coefficient of variation | Contribution to: | | | Unit cost, $/cu yd | Economical distribution of pavement | |
		Surface deflection	Stress in pavement	Thermal stress		Cost, %	Thickness, in.
Subgrade ..	0.30	0.85	0.05	0.02	1.00	10	24
Subbase ..	0.20	0.05	0.10	0.05	4.00	25	16
Base	0.15	0.05	0.25	0.13	8.00	25	8
Top course .	0.10	0.05	0.60	0.80	25.00	40	4
Overall variation		0.283	0.142	0.120			

be studied. In Table 4.1, the variation and unit price are given for the subgrade, subbase, base, and top-course materials of the pavement. In terms of the percentage of contribution of each material to the total cost of the pavement construction, it can be seen that the cost of the subgrade and subbase represents the least investment in pavement construction. For the most efficient quality-control program in pavement design and construction, the improvement of the subgrade should be emphasized. For many pavement constructions, the coefficient of variation of the in-place density of the subgrade may be in the range of 0.03 to 0.05. However, for the physical-strength values of the subgrade, such as are represented by CBR, k value, E value, or compressive strength, the coefficient of variation of the subbase may be 0.03 for density, whereas its compressive strength may have a coefficient of variation as high as 0.20, as shown in Fig. 4.7. A small decrease in density results in a much larger decrease in strength.

In this analysis, it can be seen that the use of density or dry weight as a parameter of quality control can be very misleading. The pavement performance, indicated by its deflection, is a function of the E value of the subgrade and the compressive strength of pavement material. Therefore, in a quality-control program, the variations of the subgrade modulus and material strength should be used as the prime parameters.

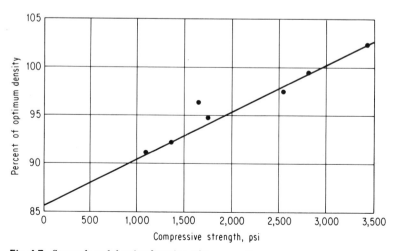

Fig. 4.7 Strength and density function of pavement base.

SUMMARY

In summarizing the above discussion, it can be stated that the lower the coefficient of variation, the better the quality of construction. However, a lower coefficient of variation also means higher construction costs. To achieve a better and more economical design, the effect of the variation of material and labor on the overall performance of a structure should be taken into consideration during the design. The anticipated coefficient of variation in the design should be governed by the practical construction conditions and the degree of confidence to be designed into the structure. In the specifications, the construction tolerance and the minimum quality level should be specified to conform with the value assumed in the original design. The concept of the "end result" used in the past should be improved by the adoption of the quality-assurance program. When statistical concepts are introduced in the original design as well as in the subsequent construction, a more reliable structural performance can be anticipated.

REFERENCE

1. E. C. Granley, Quality Assurance in Highway Construction, *Public Roads,* vol. 35, no. 11, pp. 257–260, December, 1969.

CHAPTER FIVE

Material Concept of Pavement Construction

In current pavement construction, the selection of paving material seems to be confined to three major catagories: portland-cement concrete, asphalt concrete, and aggregate. There has been some discussion about soil stabilization. Frankly, many engineers consider this to be of dubious and inferior quality. There are several reasons why the selection of pavement materials is so limited. The wide acceptance of the above-mentioned materials should be credited to organized promotion by materials industries. Their research effort has resulted in the upgrading of pavement design methods for their own material. Consequently, current pavement design methods are classified as either *rigid* or *flexible*. Hiding behind is the sales pitch for the proprietary material.

If he were objectively to review the basic material concept, a paving engineer would find that there are many materials which can be used in pavement construction. The major fallacy of today's paving concept is that the subgrade has not been considered as a construction material for the pavement. The physical properties of the subgrade are not properly defined in design methods governing either rigid or flexible pavements.

In this chapter, a review will be made of the basic concept of material performance and, it is hoped, new light will be shed on the use of material in pavement construction.

5.1. COST OF MATERIAL IN PAVEMENT CONSTRUCTION

In today's design procedure, the engineer selects the type of pavement material and then computes the thickness of pavement construction. On some occasions, the material selection is greatly influenced by political lobbying and sales promotion. The interaction and compatibility of pavement material and subgrade have been neglected. As a consequence, subgrade deformation and pavement overstress can be anticipated when the subgrade yields more than the pavement structure. Furthermore, in today's design practice, there is a general lack of economic study to compare objectively the use of one material against others. The competition among materials industries does not encourage open discussion on the use of alternate materials. If the funding policy dictates the pavement construction program, the thickness of pavement may be reduced in order to comply with the fiscal constraint. The pavement design, in reality, reflects the manipulation of administrators instead of scientific analysis by paving engineers.

For modern pavement construction, high-speed roads, or airports, the pavement structure is usually very heavy. A large volume of material will be required for the construction. In the past, there was an abundance of borrow pits and good aggregate. It was possible to utilize such material for the pavement construction. In recent years, with the rapid development of land use plus rigid zoning and environmental regulations, the supply of good aggregate is depleting, and in many cases, it is impossible to obtain the specified material. Engineers must develop a method of utilizing marginal or sub-standard materials.

In order to understand the magnitude of the problem, it is necessary to review the cost of pavement construction and to pinpoint the problems. Table 5.1 shows the types of paving material and costs of delivery and finished pavement in the New York area in 1971.

TABLE 5.1   Cost of Material in Pavement Construction

Type of material	Cost, delivered at site, $	Cost, finished in pavement, $	Coefficient of variation
Portland-cement concrete	14–18/cu yd	28–36/cu yd	0.08–0.15
Asphalt concrete	8–10/ton	11–15/ton	0.10–0.15
Crush stone	3–5 /ton	5–8 /ton	0.15–0.20
Aggregate base	2–4 /ton	3–6 /ton	0.18–0.25
Fill (borrow)	1–2 /cu yd	2–3 /cu yd	0.20–0.30

Cost of material delivered at site   = 65% of finished pavement
Cost of materials handling at paving = 35% of finished pavement

It can be seen that the cost of material in pavement construction is about two-thirds of the total cost of pavement construction. As the economy of a design is a primary consideration in engineering endeavor, an appropriate material concept in pavement design is as important as structural analysis. In the past, the effort in evaluating materials has not been adequate. The selection of material has been determined by the engineer's intuition or by consultation with the materials industry. Once politics and proprietary materials are involved, there will be no truth in evaluating the efficiency of material used in the pavement construction and no cost-benefit study will be meaningful.

For modern pavement construction, an independent material concept should be introduced in the design system. Instead of designing pavement for a given type of material, it is necessary to design materials based on the structural requirements of a pavement. This is a diametrical departure from the existing pavement design procedure. The following sections deal with the basic philosophy involved in developing an appropriate material concept.

5.2. NATURAL PROCESS OF MATERIALS

Since pavements are exposed to environmental processes, more pavements are destroyed by natural forces than by the application of vehicle load. For instance, the busiest runway, such as the one at Kennedy Airport in New York, has approximately 500 aircraft movements a day. Each movement takes less than 0.01 sec for an aircraft to pass over a given location on the runway. The total loading time does not exceed 5 sec/day. This means that for 0.99994 part of each day, the pavement is exposed to the deteriorating effects of the natural environment. Therefore, a good pavement must be designed to resist natural destructive forces, such as temperature and moisture.

In order to develop an understanding of the processes of the natural environment, a review of the geological formation of natural material is given. To simplify the analysis, let us assume that the material starts as igneous rock, which is formed under high intensity of heat and pressure. (For man-made materials, it is not practical to reproduce material like igneous rock. The cost of producing such "igneous" material under high intensity of pressure and heat would be prohibitive.) As the weathering process takes place, the igneous rock is eventually disintegrated and decomposed. The mechanical as well as chemical process produces materials of various particle size and mineral content. When the decomposed rock is carried away by water, the current force becomes the natural process in sorting the material mechanically. When the transporting force is strong, that is, when the current velocity is high, the deposit is coarse-grained soil, such as boulders, gravel, and sand. In slow-flowing water, the sediments consist primarily of fine-grained soil, such as silt, clay, and colloidal-size particles.

After sedimentation, the deposit consolidates gradually under its overburden pressure and gains strength due to compaction. In the meantime, the intrusion of cementitious compounds, such as those derived from calcium solution and clay minerals, bonds soil particles into a cemented material. When external pressure exists, the unconsolidated clay or silt deposit is transformed into a hardened shale and the sand-gravel into sandstone or conglomerate. When these soft sedimentary rocks are subsequently subjected to high heat and pressure, the deposit is transformed into a hard metamorphic rock, which is stronger and more durable in resisting weathering. In the natural process of the material, the destructive force is weathering and the solidifying forces are chemical bond, heat, and pressure. The strength and resistance of a transformed material depend on the level of solidifying forces. Hence, in developing an appropriate material concept, the elements of chemical bond, heat, and pressure should be carefully scrutinized.

For man-made materials, the crushing and sorting of aggregate are very similar to the natural processes of weathering and sedimentation. The result is the supply of graded fine or coarse aggregate. The introduction of portland cement or asphalt in paving aggregate resembles the natural process of cementation in the formation of sedimentary rock. The metamorphic process of man-made materials can be illustrated by the production of glass, brick, steel, and aluminum. These materials are produced by fusion under high heat. Because of the lack of adequate pressure, intensity, and time, the man-made material is not exactly like its natural counterpart, metamorphic rock. Insofar as pavement construction is concerned, the heat-processed material is rather costly. An economical paving material is similar to sedimentary rock, such as sandstone or conglomerate. The material strength is derived from chemical bond and mechanical compaction. However, sedimentary rocks are not strong in resisting weathering and wearing, but are good for bearing. A realistic material concept reflects the geological principle of material formation as well as the material's inherent limitations when establishing the maximum cost-benefit performance.

In the natural process, the gradation of sedimentation is random and so is its strength. If a reasonably random strength can be tolerated in the material formulation, there will be an abundant supply of low-cost aggregate for developing an economical construction material. The new material could be designed to be strong enough to resist deformation under load, volumetric change due to temperature, and durability against weathering.

The difference between an unconsolidated sand deposit and consolidated sandstone lies in its shearing strength to resist external load. In general, the shearing strength of material can be expressed by

$$s = c + (p - u) \tan \phi \qquad (1)$$

where s = shearing strength of material
   c = cohesive bond of material
   p = normal pressure on the deposit
   u = pore water pressure
   ϕ = internal frictional angle of the deposit

The difference between sandstone and sand deposit is in the cohesive bond c. The angle of internal friction varies within a very narrow range, from $28°$ for very loose sand to $42°$ for a very compacted deposit. There are two ways to obtain a significant improvement in the strength of a material: either increasing the cohesive bond of the material or increasing the normal pressure of the deposit. If the sand deposit is located above the water table, the pore water pressure u is equal to zero and the normal pressure is governed by external overburden pressure. For man-made process, the external overburden pressure can be increased by the placement of surcharge or mechanical compaction. The cohesive bond, in the natural process, is achieved by the intrusion of lime solution. For man-made material, the cohesive bond can be obtained by the use of portland cement, asphalt, hydrated lime, or other chemicals. Just as all natural deposits have their limitations in the physical environment, no man-made material is perfect in performance.

The adoption of geological principles in evaluating pavement material reflects the macroconcept of developing new paving material. There is no ideal material for pavement construction. Any material which meets the requirements of (1) low cost, (2) desirable structural performance, and (3) reasonable quality variation should be considered in pavement construction.

5.3. PROPERTIES OF MODERN CONSTRUCTION MATERIAL

The above analysis gives an overall view of the material concept. In this section, we shall begin to look into the microconcept of the material properties relating to the performance of pavement construction.

Compressive Strength

The first property to be considered in pavement construction is its strength against crushing by wheel load. This property is known as the *compressive strength* of the material, which can also be explained by Eq. (1). For unconsolidated coarse-grained sediments, such as sand and gravel in their natural deposit, the cohesive bond is insignificant and the stability of the deposit depends on its confining pressure. For clay soils, the high cohesive bond governs its strength and the confining pressure is relatively insignificant. For cemented material, the compressive strength is usually expressed as unconfined compressive strength; that is, the test specimen is compressed to

failure without confining pressure ($p = 0$). Therefore, the so-called
compressive strength is actually the cohesive bond of the material. As
the development of cohesive bond depends largely on the time
interval and temperature range of the test, it is necessary to include
such variations in formulating the material concept. With respect to
compressive strength, the higher the strength, the more adequate will
be the load-carrying capacity. On the other hand, high-strength
material always costs more in construction. In considering the overall
performance of a material, its compressive strength is not necessarily
the best indication of its quality. However, because of its simplicity
in testing, many engineers have related the compressive strength to
the E modulus, shear strength, and bending resistance of the
material. For more objective study, other physical properties should
be reviewed as well.

Deformation under Load

This is referred to as the *stress-strain property* of a material. For
ordinary construction materials, there are three distinctive stress-
strain relationships. At the very beginning of load application, the
deformation of material is proportionate to the intensity of load
application. The stress-strain function at this stage is said to be *linear
elastic.* At the second stage, the strain increases faster than the
stress. The rate of excessive strain is significantly influenced by the
duration and intensity of load application. It is said that the material
is in a *plastic state of equilibrium.* The third stage occurs prior to the
failure of the material, when the rate of stress increases faster than
the corresponding strain. This is known as the *stress-hardening stage.*

In modern pavement construction, the primary concern is to
maintain a longer service life and a smooth surface after repeated
loadings. Therefore, the linear stress-strain condition is the one most
pertinent to design problems. The material should be able to rebound
after the removal of load. The linear stress-strain relation, known as
the *modulus of elasticity,* can be expressed by

$$E = \frac{\partial \sigma}{\partial \epsilon} \tag{2}$$

It is very simple in mathematical definition but rather difficult in
actual testing to measure the stress and strain of a specimen. Because

the stress is not a real function, there is no instrument to measure it. In the real world, it is the external load and its corresponding deformation. The stress-strain tests are based on the assumptions that (1) the stress distribution in a sample is uniform and is equal to the external load divided by the cross-sectional area of the sample, and (2) the measured deformation is also uniformly distributed over the entire length of the sample. Consequently, the tensile elongation of a material is not necessarily equal to its compressive strain. The determination of the E value inherits errors due to testing procedure.

In practical pavement construction, the higher E value not only costs more for the material but also induces higher bending and brittle stress in the pavement where the ground is settling. Therefore, the higher E value is not necessarily a desirable feature of pavement material. The E value of an appropriate pavement material should be designed to conform with the physical properties of the subgrade.

Tensile Strength

The third parameter is the property against stretching. That is the tensile property of material which is governed by Eq. (1) under the condition that $p - u$ is a minus value. Therefore, for all materials, the tensile strength is always lower than the corresponding compressive strength. In structural analysis, the tensile member is usually the most efficient component. When structural members are designed for compression or bending, the material strength is not fully developed. For members subject to bending, less than one-half of the material is utilized. For a compressive member, the allowable stress is governed by its elastic stability or buckling. The allowable stress of a tensile member is not governed by the geometric shape of a structural member, and full utilization of the material strength can be accomplished. The drawback of utilizing tensile members in a structural framing is that a large amount of deformation may be encountered under the load. In pavement construction, because its subgrade support is massive, it is not economical to develop a tensile membrane. However, the tensile stress is very critical for the top course of a pavement structure. The top-course material should have sufficient tensile strength to overcome temperature- or settlement-induced stresses. At many airports, tensile-stress cracks have been encountered outside the wheel path in areas where a thin layer of top course was placed on a compacted stone base. In the Newark pavement test, tensile stress was monitored about 20 ft in front of a moving load. In the wheel path, the stress pattern changed from tensile to compressive and then to tensile again when the load moved over and passed through the observing point. In the area outside the wheel path, the stress changed from zero to tensile and then back to zero again. The pavement was constantly under a tensile stress.

On warm days, the lower side of a pavement is cooler than the upper side. The top surface of the pavement is thus under tension. The combination of temperature and a moving load may result in a critical tensile stress in the pavement, causing cracks. Therefore, the tensile strength of the top course is one of the important factors in controlling cracks in a pavement. As the development of the tensile strength is closely related to the microscopic cracks produced in the material, it is necessary to design a material which has fewer initial shrinkage cracks and, subsequently, is subject to less deflection and sharp bending.

Brittle Fracture

All materials elongate under tension or at elevated temperatures and shrink under compression or at lower temperatures. Materials must be designed to take such movement. The ability of a material to elongate is known as its *ductility*.

In ductile materials, there are two distinct types of stress: yield stress, which causes yielding through a shear-type deformation, and brittle stress, which causes cleavage by a sudden break without noticeable deformation. Although the behavior of brittle fracture depends on many material properties, such as yield strength, chemical content, grain texture, and stress concentration, the most important factor is the temperature change. As all materials have two types of strength, yield and brittle, the reactions of these strengths to temperature changes are significantly different. The relation between yield strength and temperature change can be expressed by

$$\sigma_t = \sigma_y - \frac{\sigma_y - \sigma_{et}}{1 + 1/C_t} \tag{3}$$

where σ_t = yield strength at temperature t
σ_y = yield strength at standard test temperature
σ_{et} = residual strength of plastic flow

C_t = strength coefficient, a character of material and usually an exponential function of temperature

At elevated temperatures, the yield strength of material tends to decrease and the material creeps under a sustained loading. The brittle cleavage strength is also a function of temperature, but at a different rate of fluctuation as compared with the yield strength. The intersection of these two functions (t_{cr} in Fig. 5.1) represents the critical temperature of brittle fracture. At any temperature above t_{cr}, the stress in the material will reach the yield strength before it can attain the brittle strength. The fracture of material is always ductile. On the other hand, when the temperature is lower than t_{cr}, the brittle cleavage strength will be reached first, prior to the development of yield strength. The fracture is brittle, and failure takes place suddenly without any sign or warning. The material is more ductile when the yield strength is lower, the grain size is finer, and the modulus of elasticity is lower too. Brittle fractures can occur at a stress level considerably lower than the yield strength of material.

In normal construction practice, the rule of thumb in measuring the ductility of a material is the percentage of elongation under tensile stress. For instance, a material having less than 10 percent tensile elongation is usually brittle for pavement construction. The ductility of a material can be improved by increasing its content of fines or by lowering its elastic modulus of yield strength. The lower-modulus material does not offer a higher load-bearing capacity. Therefore, an appropriate material concept should reflect the balance of E modulus and ductility of the material.

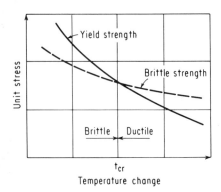

Fig. 5.1 Brittle strength of material.

Fatigue Strength

For modern pavement construction, it is necessary to design for repeated load applications during the anticipated service life. The ability of a material to resist repeated load applications is known as its *fatigue strength*. For ordinary construction materials, the fatigue strength is not too much different from the yield strength when the number of repetitions does not exceed 1,000 cycles. When the load repetitions exceed 10^6 or 10^7 cycles, the fatigue strength becomes independent of the number of repetitions. The basic strength for an infinite number of load cycles is known as the *endurance strength* of material. For a finite number of cycles, the fatigue limit is governed by

$$\sigma_n = \sigma_y - \frac{\sigma_y - \sigma_e}{1 + 1/C_N} \tag{4}$$

where σ_n = fatigue limit at N cycles of repetitions
$\quad\quad\sigma_y$ = yield strength
$\quad\quad\sigma_e$ = endurance strength
$\quad\quad C_N$ = Woehler coefficient, a function of the number of loading cycles N

The function of Eq. (4) is shown in Fig. 5.2. In actual engineering application, there are many factors which will affect the fatigue limit of the material. The most important factor is the range of stress fluctuation. Generally, the stress fluctuation can be expressed by

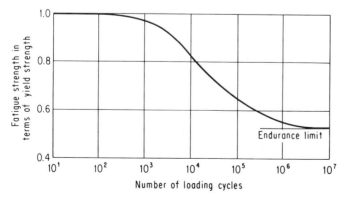

Fig. 5.2   Fatigue strength of material.

$$\left.\begin{array}{c} \sigma_{max} \\[1.2em] \sigma_{min} \end{array}\right\} = \sigma_m \pm \Delta\sigma \qquad\qquad (5a)$$

where σ_m is the mean value of stress fluctuation and $\Delta\sigma$ is the amplitude of cyclic stress. For materials with no creep effect, the maximum stress at a finite number N of stress cycles to failure is

$$\sigma_{max} = \sigma_n + \left[\frac{\sigma_{max} - \sigma_n}{\sigma_y - \sigma_n}\right](\sigma_y - \sigma_n) \qquad\qquad (5b)$$

The ratio $\sigma_{min}/\sigma_{max} = -1.0$ represents the complete reverse of cyclic loading as defined for the fatigue limit σ_n. For the ratio $\sigma_{min}/\sigma_{max} = +1.0$, there is no stress fluctuation. The maximum stress limit is equal to the yield strength of material. For the ratio $\sigma_{min}/\sigma_{max} =$ between $+1.0$, and -1.0, the maximum fatigue strength is as shown in Fig. 5.3, in which a parameter $(\sigma_{max} - \sigma_n)/(\sigma_y - \sigma_n)$ has been introduced for the curve plotting.

When the cyclic stress appears in a random magnitude, the fatigue life is usually governed by the number of occurrences in high stress.

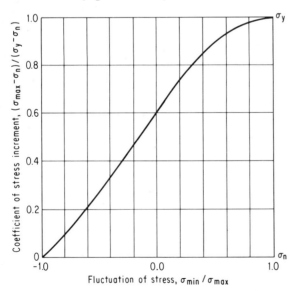

Fig. 5.3 Fatigue strength relating to fluctuation of cyclic stress.

In many cases, microscopic cracks may be formed at the high stress level and continue to propagate during the subsequent low stress level. If a material is once loaded to its yield strength, the fatigue life of that material will be abruptly reduced, no matter how low the subsequent load cycles are. If the maximum stress level is far below the yield strength of the material, the fatigue life can be estimated by the geometric means of load cycles, as given by the equation

$$N' = \frac{\sum n}{\sum (n/N)} \tag{6}$$

where N' = weighted fatigue life for the mixed range of stress

n = actual number of cycles anticipated within a specific range of stress

N = fatigue life as given by Eq. (4) for the identical range of stress

The finer the division of stress range, the more accurate will be the weighted fatigue life.

For practical design analysis, the endurance limit is the most important parameter insofar as the fatigue strength is concerned. The endurance limit for complete reverse of cyclic loadings ranges from 0.40 to 0.60 for cemented materials and from 0.50 to 0.65 for metals. The endurance strength is expressed by the yield strength of the material, and the upper limit of load applications is about 10^6 to 10^7 cycles. In order to improve the fatigue life of a structure, it is necessary (1) to reduce the level of working stress, (2) to avoid stress reversal in the system, (3) to minimize the range of stress fluctuation, and (4) to prevent occasional overstress. A 10 percent increase in overstress may mean a more than 10 percent reduction of service life. If it is anticipated that the pavement will take extra load during its service life, the pavement material should be designed with higher endurance limit or a lower working stress to improve its resistance against fatigue failure. For thin pavement construction, the change of stress in the pavement can vary from tension to compression. The fatigue strength of the top pavement layer may be a critical element in controlling tensile cracks in the pavement surface.

Volumetric Change and Durability

This refers to the material property that is influenced by changes of temperature and moisture. Because pavement is always exposed to changes of environment and weather conditions, a volumetric change of material will result in the expansion or shrinkage of a pavement structure. The pavement may buckle during the summer and shrink during the winter. It is desirable to develop a pavement material with a small coefficient of volumetric change due to both temperature and moisture.

Changes in moisture also affect the durability of a material in maintaining its physical strength. Ambiguity in testing leaves many questionable features in the method of evaluation. In designing pavement material, the result of laboratory tests must be correlated with experience in the field. Durability may not be as critical in a deep pavement as in a thin-layered pavement.

Progressive Deformation

As mentioned previously, for 99.994 percent of a day, the pavement is under the influence of its own weight. The constant strain under its own weight, plus the short-period wheel load, causes progressive deformation of the pavement surface. This is known as the *plastic adjustment* of material under long-term loading. The stress level affects the magnitude of adjustment, and the loading period affects its rate of deformation. Although plastic deformation is not a desirable feature in pavement construction, it can be a desirable feature in damping the vibration of a dynamic force. However, there are many practical problems in utilizing such advantage. For instance, if the plastic layer is under a rigid pavement, higher stress in the rigid layer will be encountered and damping of the plastic material will be realized only when the rigid layer is stressed to yield condition. This means that the rigid layer will crack more readily. Moreover, rigidity in the upper layer will prevent, or at least reduce, the transmission of vibration to the lower layer. The lower layer will then not provide a significant damping advantage for the entire pavement system. If the plastic layer is on top of a rigid base, excessive deformation will be encountered in the plastic layer. In general, plastic adjustment under long-term loading is not a desirable feature of pavement material.

Wearing and Tractive Resistance

This is an important property for the top course of a pavement. All top-course pavement materials must be designed to resist the wear and tear of traffic and to develop sufficient tractive resistance for the safe operation of vehicles. If the material is sensitive to polishing, its function as a pavement material will be reduced.

Bond Strength

For multiple-layered construction, the bond strength will affect the stability of pavement against bending and horizontal shear. A good mechanical bond can be achieved by providing a clean rough surface. The pavement material should be able to bond to itself or to another compatible surface. Bond strength seems to be a secondary consideration in pavement design, but it results in many surface irregularities and local distresses. In today's airfield pavement construction, the maximum horizontal shear is encountered at a depth of less than 6 in. below the pavement surface. The top course of pavement structure should be properly designed to have sufficient bond and shear strengths.

Workability

This is a very important parameter in the cost of pavement construction. For portland-cement or asphalt concrete, the workable duration is about 2 to 4 hr, that is, within 2 to 4 hr the pavement construction has to be completed. The workability of aggregate-base course is determined by the moisture content of the material. If the aggregate is wet, it cannot be compacted. A short period of workability means that a large labor force must be available to finish the paving operation within the workable duration. When the cost of labor is the most significant item in pavement construction, a short workable duration is equivalent to a high cost of pavement construction. If the material has a workability of 24 hr, it can be rehandled next day and the contractor does not need to finish the job in the same day. Therefore, manpower and machines can be conserved and a prolonged construction schedule can be planned. This will reflect in the saving of construction cost.

Workability is influenced by weather and season of construction. For instance, portland-cement concrete has a shorter workable period in summer, whereas asphalt concrete has a shorter workable period during cold weather. A heavier asphalt layer may be more desirable for cold-weather construction. In developing a new pavement material, the ideal workable period should be longer than 4 hr but should not exceed 2 days. If material is not set in 2 days, weather and environmental conditions may affect its quality.

Shrinkage

This refers to the initial stage of volumetric change during the hardening of cemented material. The associated problem is the development of incipient cracks which will affect the tensile strength as well as the durability of material. During the hydration of portland cement or in the cooling period of asphalt concrete, the differential above ambient temperature represents the thermo-gradient which causes the potential shrinkage cracks. The slower the rate of cooling, the fewer the cracks. On many occasions, a smaller size of aggregate will cause more shrinkage cracks. Fine aggregate having claylike particles will produce high shrinkage. Soft rock, such as slate and limestone, is more sensitive to shrinkage than quartz, dolomite, and traprock. The moisture content is also an important factor in controlling shrinkage. A high moisture content in a mixture may cause excessive shrinkage during drying. On the other hand, protective action should be taken to prevent premature drying, which will accelerate the formation of cracks. In designing a low-shrinkage material, the most important consideration is to produce a dense aggregate matrix, utilizing a minimum amount of cementitious compound to fill up the void. The cementitious compound should not emit excessive heat during its chemical hardening.

Temperature Consistency

This problem arises mainly from the use of asphalt as binder. Liquid asphalt is introduced at elevated temperatures in aggregate mixtures to improve the bonding efficiency of the intersurface. During the cooling cycle of the mixture, the emitting of solvent results in a thin film of hardened residue which cements the mixture into a structural component. In the process from construction to service, the asphalt is transformed from a liquid to a plastic to a hardened state. There are three requirements for an asphalt cement:

1. Its viscosity-temperature characteristics must be within the range for good mixing and adequate compaction without delay.

2. Its viscosity must be such that a finished pavement will have a stiffness during summer to bear the traffic load without deformation.

3. The hardened asphalt residue must retain sufficient flow properties—ductility—to minimize crackings under falling winter temperatures.

As shown in Fig. 5.4, the viscosity of asphalt varies with temperature. It is obvious that a single temperature consistency cannot ensure an asphalt product meeting all three of these requirements. The ideal combination is to obtain (1) maximum stability for service condition in summer and (2) minimum ductility limits on the residue for service condition in winter. A viscosity grading system should be introduced in the mix design. If the binder consistency is correct, mixing and compaction will be improved. Although the critical mixture consistency depends on the type and gradation of aggregate, the type and amount of mineral filler, the asphalt content of the mixture, the temperature during service, and traffic conditions, the temperature susceptibility of the asphalt

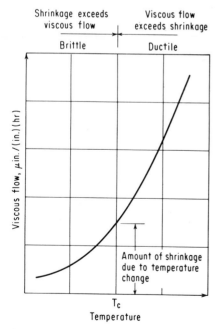

Fig. 5.4 Temperature consistency of asphalt cement.

binder could result in additional complications in the performance of an asphalt-stabilized material. It is important to point out that AASHO has adopted a viscosity-grading specification (1971) for improving the mixture design as well as pavement performance.

Criteria of Working Stress

In order to determine the safe capacity of a structure, the theory of probability is commonly used. The main factors governing the safety of a structure are (1) the actual strength of the construction material, (2) the actual dimensions and tolerance in the geometry of the structure, (3) the actual loading arising from any cause, and (4) the degree of approximation adopted in the design computations. Since all the data necessary for a rigorous probability approach to the treatment of structure safety are not available, it is convenient to utilize a "characteristic value" of the strength which defines the mechanical properties of the materials based upon a fixed probability that the actual value will be less than the value selected. Thus the material strengths, as given by appropriate tests, are used to define the characteristic strength σ_k:

$$\sigma_k = \sigma_m - k\sigma_s \tag{7}$$

where σ_m = arithmetic mean of test results
σ_s = standard deviation of test results
k = coefficient depending on reliability of performance
The design strengths σ of the material are given by

$$\sigma = \frac{\sigma_k}{\gamma_m} \tag{8}$$

where γ_m is the strength-reduction coefficient which takes account of the reduction in strength in actual construction as well as the possible local reductions in strength due to other causes. By applying the calculus of probability and the theory of statistics, Freudenthal and Julian [1,2] developed a set of factors of safety (or serviceability) corresponding to the probability of failure. The factors of safety due to material variations are shown in Fig. 5.5 for a probability of failure of 10^{-4} (1 failure out of 10^4 cases).

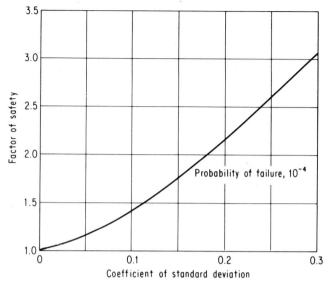

Fig. 5.5   Factor of safety for the variation-of-material testings.

5.4. CONCEPT OF MATERIAL STABILIZATION

In the preceding sections we have discussed the diminishing supply of natural material, the high cost of construction, and the urgent need for utilizing marginal or substandard materials for pavement construction. However, from the point of view of engineering application, today's material concept is not adequate for modern pavement construction. Portland-cement and asphalt concrete may not be economical for heavy construction. Engineers must develop new materials which are economical and suitable (1) to support heavy wheel loads, (2) to transfer loads onto a much weaker subgrade, and (3) to maintain structural integrity under environmental destructive forces. Because the pavement is acting as a transfer function between the subgrade and wheel load, the physical properties of pavement material should be between those of the subgrade and the landing gear structure. If the pavement material is too rigid, the pavement may be strong enough to resist the bending moment due to wheel load but not thick enough to protect the deformation of the subgrade (see Fig. 5.6). The pavement construction may then be costly but unstable. On the other hand, if the pavement material is too flexible, particularly during elevated

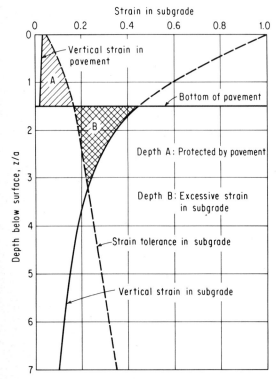

Fig. 5.6 High-modulus ($>20\,E_s$) material laid directly on subgrade.

temperatures, the pavement may not be adequate to transfer the wheel load onto the subgrade and rutting along the wheel path will result. An ideal pavement composition consists of a top-surface material which is strong enough to carry the wheel load directly and a bottom layer which is flexible enough to be compatible with the subgrade (see Fig. 5.7). With this concept in mind, a new group of pavement materials can be developed.

Gradation of Natural Deposit

As noted in Sec. 5.2, when a deposit is (1) well graded, (2) intruded by cementitious compound, and (3) under a heavy overburden pressure, the material is stable, cemented, and hard. The first step in studying the feasibility of material stabilization is to determine the mechanical gradation of the material. For deposits consisting of fine-grained particles, such as silt or clay, the method of stabilization is quite different from that with coarse-grained particles, such as sand and gravel. If the soil gradation is confined within a very narrow range of particle size, such as beach sand or hydraulic fill, the deposit lacks fines and is not stable enough to support heavy loads. The gradation of soil particles is also an indicator of the permeability of the deposit. For fine-grained soils, chemical stabilization or grouting may be the only feasible method because the low permeability of the material permits the migration of only chemical solutions. For a coarse-grained material, the stabilizer can be fine-grained minerals, such as portland cement, hydrated lime, and flyash.

Mechanical Stabilization

A poorly graded soil can be modified by adding either fine or coarse particles to achieve a better interlocking of soil structure. Well-graded materials exhibit higher stability and shearing strength. The

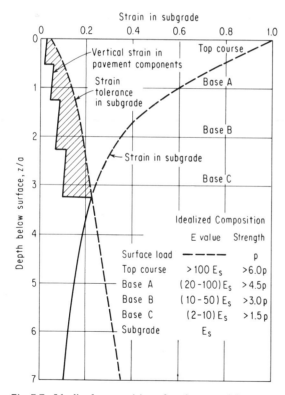

Fig. 5.7 Idealized composition of paving material.

interlocking of soil particles improves the angle of internal friction ϕ, as shown in Eq. (1). The most common mechanical stabilization in pavement construction is the use of dry or wet bound aggregate base. Graded crush stone is good for drainage but lacks sufficient stability under heavy load. When fines are added in the crush stone, the dense mix is more stable under load but less pervious for drainage. It can be seen that stabilization is not necessarily always the best answer in pavement construction. Furthermore, mechanical stabilization improves on the internal friction, ϕ value. As mentioned earlier, for practically all soil material, the ϕ value varies within a very narrow range—between 28 and 42°. The improved strength of a mechanically stabilized soil cannot exceed two times its original strength.

Chemical Stabilization

If higher strength is required for stabilization, it is necessary to seek improvements in the c value of Eq. (1), the cohesive bond of material. Chemical stabilizers may be used to adhere the interface of the particles. The development of a cohesive bond depends on the strength of the stabilizer and the surface area of the particles. A mechanically stabilized soil or well-graded dense mix will develop a higher bond strength. Thus the strength of a chemically stabilized material can be designed to meet a wide range of strength requirements for pavement construction. There are many bonding materials which can be used for construction. The common ones are portland cement, hydrated lime, and asphalt. Portland cement is a derivative of limestone. When water is added, the cement paste occupies the voids between solid particles. During the hydration of cement, the paste bond aggregates into a mass. The strength of portland-cement concrete depends on (1) the amount of water and cement used in the mix and (2) the mechanical properties of the aggregate. Because of the heat released during its hydration, portland-cement paste will cause some volumetric changes in the concrete mass. Early shrinkage cracks and short workable duration are the inherent problems in cement stabilization.

The next cementitious compound is hydrated lime. In a moist environment, hydrated lime releases its calcium oxide slowly. The process is known as *lime absorption*, which is a function of temperature and is independent of the amount of water in the mix. Therefore, lime stabilization can be applied to either coarse- or fine-grained soil at the optimum moisture content for compaction. If the clay minerals are pozzolanic reactive, lime can be a good stabilizer for clay soils. However, the excessively long time required for the migration of lime and the high temperature required for the chemical reaction may result in many difficulties in actual construction. In many cases, it takes several years after construction to realize the full benefit of the lime stabilization. This is one of the most difficult problems in modern construction.

The other common stabilizer is asphalt. The asphalt can be either emulsified or cut back. The principle of asphalt stabilization is that the asphalt will form a coating on the particles and the compacted mass will develop an effective bond strength. The effectiveness of asphalt stabilization depends on (1) cleanness, (2) gradation, (3) moisture, (4) temperature, and (5) compaction of the mass. If the natural deposit is either too fine or too coarse in particle size, the efficiency of asphalt or other stabilization may be greatly reduced. Moreover, if emulsified asphalt is used, the fluctuation of ground moisture can be a very important factor in the integrity of asphalt-stabilized soil. Furthermore, asphalt stabilization inherits two problems: (1) asphalt is a temperature-sensitive material, and (2) asphalt exhibits a time-dependent viscosity. Therefore, the deformation of an asphalt-stabilized layer is subject to the variations of time and temperature. This may be a desirable feature in some pavement construction—resulting, for example, in the formation of a resilient base for damping vibratory force—and may be a drawback in others, causing deep ruttings to develop in the wheel path.

There are several special stabilizers not widely used in pavement construction. The most common ones are petroleum derivatives. The material can be used in part or in whole to replace asphalt. However, the high cost of the chemicals does not justify their popular application except for special construction, such as emergency landing pads. Another stabilizer is natural pozzolanic material. Volcanic ash is very sensitive to lime reaction. By using volcanic ash and hydrated lime, the Romans built many enduring structures, such as aqueducts and the Coliseum. The modern counterpart of volcanic ash is flyash, derived from the burning of pulverized coal. Flyash has the same chemical properties, which make it reactive with lime. Insofar as quality is concerned, flyash is far superior in particle grade and chemical content to volcanic ash.

Strength Requirement

The purpose of the stabilization is to develop materials for different functions in the pavement construction. There should be three distinct types of material used in the pavement construction: (1) the top course, in direct contact with the wheel load; (2) the material laid directly on subgrade; and (3) the layers between. A pavement structure with these materials in its composition will have a better interacting performance between wheel load and subgrade (see Fig. 5.7). For the top-layer pavement material, the basic requirement is that it resist the wear and tear of the wheel load and survive the destruction of environmental forces. The material has to be strong enough to withstand the tire pressure. For all practical purposes, the minimum compressive strength of top material should be more than six times the unit tire pressure of the wheel load. In modern airplanes, the tire pressure of the landing wheels varies from 120 to 300 psi. Therefore, the minimum compressive strength of the top material should exceed 1,000 psi for balloon-tire operations and 2,000 psi for high tire pressure. The factor of 6 is based on considerations of construction tolerance, environmental forces, endurance limit, overload factor, and quality variation. If the endurance limit is less than 0.50 and the coefficient of variation exceeds 0.25, the factor between compressive strength and tire pressure should be appropriately increased.

For material directly on subgrade, the basic requirement is that the elastic property of the base material be compatible with the subgrade. For normal pavement construction, the E modulus of subgrade ranges from 2,000 to 10,000 psi. In considering the effect of repeated loadings and the coefficient of variation, the most desirable range of E value is between 20,000 and 100,000 psi for base material directly on subgrade. A low-E-modulus base material can be economically developed by stabilizing the natural soil deposit, using one of the stabilizers discussed previously. In theory, hydrated lime and asphalt are more desirable for low-strength stabilization. The material directly under the wheel load should have a minimum compressive strength of about 1,000 to 2,000 psi, depending on the type of operational vehicle. This represents an E modulus in the neighborhood of 500,000 to 2 million psi. For a gradual transition, the layers in between should have an E modulus in the neighborhood of 200,000 to 500,000 psi.

The concept of developing material strength according to its stress-strain position in a pavement construction is new. The economic advantages of such a concept require a rigid engineering discipline for its implementation.

Designing the Coefficient of Variation

The primary purpose of stabilization is to modify a low-cost material which can be refitted in the overall performance of a pavement. The structural performance of a stabilized material should, therefore, be more important than its cost advantage. A quality-control program should always be associated with material stabilization. As discussed previously, the smaller the coefficient of variation, the higher the cost of material stabilization. Therefore, a cost-benefit study should be conducted for all stabilization processes. As a guide, the low-strength material should be placed at the bottom of pavement construction. The coefficient of variation should be gradually reduced during the buildup of the pavement structure. A desirable coefficient of variation for the top layer will vary from 0.12 to 0.15. In no case should the coefficient of variation of an upper layer be greater than that of the lower layer. The coefficient of variation should be designed for each stabilized material, and its quality should be rigidly controlled during the construction. In the past, very little quality control has been exercised for stabilized material. The idea of "cheap" construction should not be accomplished by neglecting quality control. The real economy of using stabilized material is derived from (1) efficient utilization of material strength according to its function in the pavement structure and (2) the possibility of utilizing low-transportation, substandard local material. The formulation of stabilized material should be deduced from extensive laboratory and field tests. Rigid inspection procedure should be enforced during the process and placement of stabilized material.

Example of Soil Stabilization—LCF Mix

The most important considerations in the development of a new construction material are its simplicity of application and low cost in construction. The physical properties of stabilized material may have to be compromised in order to obtain an optimum result. In the development of lime-cement-flyash (LCF) stabilization, equal emphasis has been given to its cost and its physical performance. The proportioning and mixing processes are designed to be as simple as possible. The stabilized material is of low strength in comparison with portland-cement concrete but is far stronger than aggregate base. Consequently, the performance of LCF stabilization does not resemble that of concrete or of aggregate.

In American engineering practice, hydrated lime and flyash are commonly used in stabilization. Flyash itself is not a cementitious compound. It consists of approximately 70 percent by weight silicon dioxide and aluminum oxide of fine spherical particles. It can be used as a filler to improve the mechanical property of an open-graded aggregate, such as hydraulic fill. By mixing flyash with hydrated lime, there is a chemical reaction in the form

$$SiO_2 \text{ and } Al_2O_3 + 3Ca(OH)_2 = CaO \cdot SiO_2 \text{ and } (CaO)_2Al_2O_3 + 3H_2O \tag{9a}$$

$$Flyash + hydrated\ lime = nonsoluble\ cementing\ compound \tag{9b}$$

The chemical exchange is known as a *pozzolanic reaction*, which is usually a long and slow process. If powder lime is used in the mix, the active calcium oxide is absorbed prior to the pozzolanic reaction in the form

$$CaO + H_2O \rightarrow Ca(OH)_2 \tag{10}$$

Thus the rate of compressive strength developed in the lime-flyash mix is directly related to the rate of lime absorption. The faster the rate of lime absorption, the greater will be the development of compressive strength. The rate of strength growth is approximately as given in Table 5.2.

In terms of material utilization, the lime-flyash process is the most efficient method in utilizing stabilizers. One part of hydrated lime plus an equal part of flyash will result in two parts of cementitious compound similar to portland cement. Any flyash in excess of that needed for chemical reaction is used as filler to modify the mechanical gradation of sand.

In France, flyash has been widely used in road construction. In order to dispose of a large quantity of soft-coal flyash, French engineers are using 25 to 35 percent flyash in their sintering process of cement. The amount of cement used in the road-base stabilization

TABLE 5.2   Aging of Hydraulic Cement Mix

Age of sample (70° F)	Lime-flyash-sand (4%:20%:76%)	Portland-cement concrete (5½-sack mix)
1 day	0.10	0.34
7 days	0.12	0.60
14 days	0.21	0.72
28 days	0.33	0.85
2 months	0.65	0.92
6 months	0.90	0.98
1 year	1.00	1.00
5 years (projected)	2.00	1.10

ranges from 6 to 10 percent by weight, and of flyash ranges from 10 to 20 percent. The residual is dune sand. High early strength is always reported in the French tests.

In the United States, the price of portland cement and hydrated lime are about the same and both materials are available in large quantities. Therefore, the cost of stabilizer in the flyash-cement mix is about twice as much as in the lime-flyash mix. The merit of the French experience is the high early strength of the mix; the American lime-flyash process is known for efficiency and economy.

Prior to the construction of test pavements at Newark Airport, experiments were conducted in the laboratory utilizing mixtures of portland cement, hydrated lime, and flyash. The test results are plotted in Fig. 5.8. For the mix containing one portion of portland cement and four portions of hydrated lime, the accelerated compressive strength is about 50 percent greater than in the mix containing straight hydrated lime. A diminishing strength gain is observed when the amount of portland cement exceeds the 1:4 proportion.

During early laboratory testing and subsequent field experiments, LCF-stabilized sand exhibited the ability to be rehandled, regraded, and recompacted during a 3-day period after its initial mixing when the ambient temperature was below 70°F. This workability allows a contractor to hire a smaller crew than would be necessary if he were placing soil-cement or concrete pavement. In the laboratory test, samples were remolded and recompacted 24 hr after the initial mixing and compaction. The remolded strength of various mixes is also shown in Fig. 5.8. It can be seen that the introduction of small amounts of portland cement improves the ultimate strength after

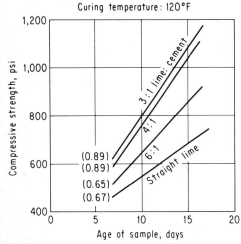

Lime and cement: 4% by dry weight
Flyash: 12%
In-place sand: 84%
Curing temperature: 120°F

Fig. 5.8 Compressive strength of LCF-stabilized sand. Numerals in parentheses denote the strength of remolded and recompacted samples.

remolding. The LCF mix is a hybrid formula which represents the best combination of French-American engineering practice. The physical strengths of the LCF mixes are designed for their function in the pavement construction. They are given in Table 5.3.

The wide range of compressive strength is designed for economical construction. When the designed coefficient of variation is high, the tolerance in material processing is less rigid and the cost of production can be reduced. However, the large coefficient of variation will affect the reliability of material performance. The strength range given in the table seems to represent a good compromise between cost and quality of LCF material.

As the unconfined compression test is the simplest for quality control, a larger number of tests were performed in the laboratory to establish the guidelines for field control. In actual pavement

TABLE 5.3   Mixing Formula for LCF-stabilized Sand

Type of pavement component	Proportion of mix, % (L-C:flyash:stone:sand)	Compressive strength, 90 days @ 70°F	Designed coefficient of variation
Subbase	3½:12: 0:84½	600–1,000 psi	0.20–0.25
Base	4 :12: 0:84	800–1,200 psi	0.15–0.20
High-grade base .	4½:10:35:50½	1,200–1,600 psi	0.12–0.15

construction, the quality of LCF mixes will be greatly influenced by the manner of placement and the degree of compaction of the in-place material. A typical test result is shown in Fig. 5.9. The other test results are shown in Fig. 4.7. In addition to compressive strength, the LCF mix has been tested for its modulus of elasticity, fatigue, strength, durability, volumetric change, and other pertinent physical properties to be used in pavement design.

5.5. PROGRAM FOR UTILIZING LOCAL MATERIALS

In the preceding section, it was indicated that 65 percent of pavement cost is for material delivered to the site. Of this figure, the

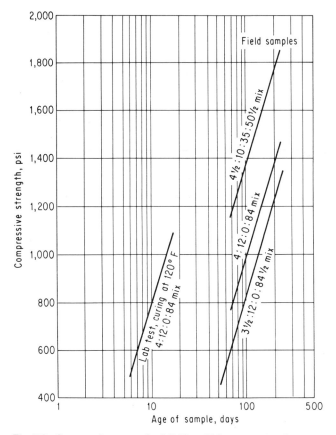

Fig. 5.9 Compressive strength of field and laboratory samples.

costs of handling the material at the source of supply, processing at the plant, and transporting to the site may constitute as much as 90 percent for construction aggregate and 70 percent for portland-cement or asphalt concrete. In the overall cost of the pavement construction, the cost of handling and transporting material may range from 40 to 60 percent of the total pavement cost. If local material can be utilized for the pavement construction, the handling and transportation cost can be significantly reduced and the pavement can be economically constructed. However, not all local material can be directly used in pavement construction. Many local materials require extensive modification to improve their physical performance and to meet the requirements for pavement construction, Therefore, a research program should be conducted to determine the feasibility of material stabilization prior to the design of pavement structure. A step-by-step procedure is outlined herein.

Survey of Local Materials

The survey should include the supply of all materials pertaining to the pavement construction, such as portland cement, hydrated lime, asphalt, sand, gravel, crush stone, natural pozzolanic deposits, and local industrial by-products. The survey should also include the demand, supply, and economic stability of the materials market. Sources of information are local materials suppliers, the agriculture department, highway engineers, and geologists. A local geological map can be of value in determining the formation of sand and gravel and even silt and clay.

Cost of Transportation

During the survey of the materials supply, it is necessary to evaluate the cost of obtaining and transporting such materials to the construction site. In many cases, the materials are available but transportation is difficult. Therefore, the cost of material should include the cost of obtaining the material at its source, the cost of processing, and the cost of transportation. Water transportation is cheaper than truck transportation, but the hauling distance may change the economic picture. As pavement construction involves large volumes of material, the cost of transportation is one of the important items in developing an economical pavement material.

Sampling of Material

The next step in the development program is the sampling of materials in sufficient quantity to determine mechanical gradation, density, moisture content, specific gravity, and mineral contents. For the mechanical gradation, there are two important particle sizes, namely, the No. 200 and No. 4 sieves. Materials retained on the No. 4 sieve represent pea gravel, and those passing the No. 200 sieve represent fined-grained soil, such as silt and clay. With materials having more than 20 percent passing the No. 200 sieve, a high-quality stabilization may be difficult to achieve.

Optimizing Mechanical Stabilization

After determining the gradation of different materials, a trial mix should be developed to determine the maximum density of the mix by varying the proportion of different particle sizes. The maximum density is determined by compaction at optimum moisture content. By varying the percentage of the coarse and fine materials, a very dense mix can be developed. If this dense mix contains a high percentage of fines passing the No. 200 sieve, the material can be of low durability resistance. On the other hand, many coarse-grained materials require the addition of material passing the No. 200 sieve to achieve optimum mechanical stability.

Chemical Stabilization

After trial runs on mechanical stabilization, several mix formulas should be selected for chemical stabilization by introducing portland cement, lime-cement-flyash, asphalt, or other chemical stabilizers. The selection of the chemical stabilizer should also reflect the cost and supply in the local materials market. For instance, in countries where the supply of portland cement is plentiful but hydrated lime has to be imported, there is no reason to use lime-flyash stabilization and portland cement can be used to its best advantage. In the development of a mixing formula, all stabilizers should be treated as equals and evaluated equally; the optimum chemical stabilizer should be determined by its cost against the benefit derived from the material utilization. At this stage of study, program engineers should not be influenced by outside interests that promote the use of proprietary material. After a series of initial tests, such as for compressive strength, durability, and density, selection of the optimum stabilizer should be reduced to a choice of one or two types of material.

Comprehensive Testing Programs

After the trial mix has been narrowed down to two or three different types of gradation and one or two types of chemical stabilizers, a comprehensive test program should be conducted to determine the durability, compressive strength, workability, volumetric change, temperature effect, and elastic and plastic properties.

Construction Practice

In the process of finalizing the material formulation, it is necessary to review the method of construction and material process, such as the efficiency of mass production, the type of mixing plant, the capacity of the compactor, and the performance of the paving equipment. In many cases, a sophisticated and expensive mixing plant may not be the right answer for stabilization. An appropriate construction program reflects local construction practice, efficiency of manpower, and machine utilization. The final mix formula should be simple and the designed coefficient of variation and construction tolerance should reflect the reality of local construction practice. A small tolerance means higher costs in producing the material and in pavement construction. The material formulation should be simple and explicit, so that a superintendent can understand it clearly. The communication between engineer and superintendent is different from that between the engineer and the professors. Good quality of construction is achieved by the superintendent.

Designing Material Strength

After developing a practical mixing formula, a design analysis should be made to determine the strength requirements for various components of various thickness. An ideal pavement construction consists of at least three distinct layers. So there should be at least three formulations of pavement material.

Cost-Benefit Study

The study should reveal the cost of the total construction when one type of material is utilized as compared with costs using other types of material. It should also review the cost of pavement construction using different thicknesses in trading with the strength of material. The subgrade is a natural construction material. Its inherent strength variation will dictate the pavement design and the development of economical pavement material.

Designing Quality Variation

This is the final consideration in formulating the material concept. Natural materials are not perfect, and their quality varies. In the material formulation, it is necessary to design the tolerance of quality variance. Consequently, the pavement structure should be designed to reflect the reliability of material performance and saving in construction cost. A tradeoff should be established between the degree of quality control and the physical requirements of pavement components. The coefficient of variation shown in Table 5.3 represents the best compromise between cost and performance in today's construction practice.

REFERENCES

1. A. M. Freudenthal, Safety and the Probability of Structural Failure, *ASCE Trans.,* vol. 122, 1956.
2. O. G. Julian, Synopsis of First Progress Report of Committee on Factor of Safety, *Proc. ASCE,* vol. 83, no. ST4, July, 1957.
3. F. Stuessi, Theory and Test Results on Fatigue of Metals, *ASCE Trans.,* vol. 128, pp. 896-922, 1963.
4. B. Tremper and D. L. Spellman, "Shrinkage of Concrete, Comparison of Laboratory and Field Performance," paper presented at the 42nd Annual Meeting of the Highway Research Board, 1963.
5. G. Winter, Properties of Steel and Concrete and the Behavior of Structures, *ASCE Trans.,* vol. 126, 1961.

CHAPTER SIX
Environmental Effects on Pavement Systems

It is important to recognize that pavement systems are exposed to many environmental factors which could cause the distress and failure of pavements without the application of wheel load. Cracks and disintegrations may be encountered outside the traffic wheel paths of runway and taxiway pavements. A pavement is a dynamic, or changing, system which must survive the never-ceasing natural destructive forces. The general mechanisms by which the environment influences pavement behavior and performance are (1) the effect on engineering properties of component materials, such as physical strength and tractive resistance; (2) the effect on the integrity of materials, such as durability and physicochemical disintegration; and (3) the effect on volumetric change and the resulting internal stress equilibrium in the pavement system.

In classic pavement design, these mechanisms have been closely related to climatic factors, i.e., temperature and moisture, in performance analyses. Many researchers have characterized the moisture and temperature in the pavement system as functions of space and time. A comprehensive report on this subject has been prepared by M. R. Thompson [5]. In the construction of modern airports, there are several other environmental factors which have caused considerable concern: (1) the volumetric change due to consolidation of subgrade, (2) excessive reduction of tractive resistance at normal landing and takeoff speeds, (3) the stability of surface material in resisting the hot jet exhaust, (4) oil and chemical corrosion on the pavement surface and its environment, and (5) the cleanness of pavement surface and its effect on engine ingestion of aircraft.

In this chapter, we shall review the nature of each environmental factor to assess its effect on the basic equilibrium and stress condition of a pavement system. Finally, we shall introduce practical solutions for airfield pavement construction.

A. Temperature of Pavement System

6.1. FACTORS INFLUENCING TEMPERATURE

Since methods for predicting temperatures in pavement systems are complex, no simple system can be effectively used to describe the nature of temperature. For practical purposes, temperature influence factors can be divided into extrinsic and intrinsic categories. The extrinsic factors are usually weather conditions, such as air temperature, solar radiation, wind, precipitation, evaporation, and condensation. It is evident that geographical location will have a strong influence on the climate of a region. The intrinsic factors generally refer to the emission of long-wave radiation from the ground and thermal properties, which include the thermal conductivity, heat capacity, and latent heat of fusion of the pavement materials and subgrade. The geological features of a region have a significant effect on the contribution of intrinsic factors. In a dynamic environment, the intrinsic factors are relatively constant as compared with the fluctuation of the extrinsic factors. Among all extrinsic factors, the ambient temperature is the most important factor in the thermal equilibrium of a pavement. Straub, Schenck, and Przybycieu [5] observed the hourly fluctuation of ambient temperature as well as the pavement temperatures at various depths. The results are reproduced in Fig. 6.1a. It can be seen that solar radiation and ambient temperature had a significant effect on the pavement temperature. The peak of pavement temperature lagged at least 1 hr behind the peak of solar radiation. Although the pavement temperature assumed a cyclic fluctuation at various depths, the amplitude of fluctuation decreased with increasing depth below the surface. A typical thermal profile is shown in Fig. 6.1b for the measurements at 4 A.M. and 2 P.M. At 4 A.M., the long-wave radiation from the earth kept the pavement temperature warmer than the night air. At 2 P.M., corresponding to the peak of air temperature, the solar radiation absorbed by the pavement caused the rise of pavement temperature. The rate of heat penetration at various depths resulted in a thermal gradient, as shown in Fig. 6.1b.

As all materials are sensitive to the volumetric change due to thermal fluctuation, the effect of thermal variation will be reflected by the volumetric expansion or shrinkage of the pavement system. If the temperature variation is uniform throughout the entire depth of the pavement and if the bottom of the pavement system is free to move without any restraint, the pavement system experiences a linear movement without its internal equilibrium being affected. In reality, the thermal conductivity and heat capacity of pavement material will affect the rate of heat flow in the pavement and will result in a thermal gradient across the vertical axis of the pavement

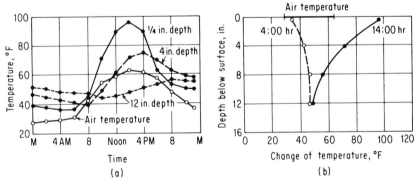

Fig. 6.1 Hourly temperature change in asphalt pavement.

system. The magnitude of thermal expansion is not a constant. The edge of a pavement system tends to bend upward during the cool night and downward when the air temperature is higher than the ground. An internal thermal stress therefore results in the component materials of a pavement. When the bending stress exceeds the strength of material, cracks develop, whether or not the pavement system is subject to wheel load. This explains why some concrete airfield pavements experience fine cracks outside the normal aircraft operational area.

Many researchers have attempted to relate the air temperature to the change of pavement temperature. By using meteorological parameters, they have reviewed the solar radiation, convective heat transfer, and latent heat of fusion. The complexity of the meteorological parameters has resulted in the use of an empirical factor to convert air temperature to pavement temperature. In other words, all researches are still far from scientific analysis. The state of the art is that "the latitudinal extent is the most important cause of temperature variations. Temperature isotherms vary from north to south. However, wind, storms and predominant air mass can skew some of the isotherms at times."

The thermal conductivity of a material is defined by the quantity of heat which flows normally across a surface of unit area per unit time under a unit thermal gradient. For a given amount of heat input across a given area within a given time period, the higher the thermal conductivity of the material, the less the thermal gradient will be. Consequently, a lower temperature stress will be encountered in the material.

The thermal gradient is also affected by the *heat capacity* of the material. This is defined by the amount of thermal energy necessary to cause a unit temperature change in a unit mass of substance. The higher the mass heat capacity is, the less the thermal gradient will be. Table 6.1 shows the thermal properties of some pavement surface materials.

There have been many attempts to relate the intrinsic and extrinsic factors in predicting temperature fluctuation in pavement systems. The complexity of the parameters involved prevents the development of a reliable solution. The advent of the digital computer has created interest in a numerical method for solving transient heat flow in pavement systems. The heat-transfer model

TABLE 6.1 Thermal Properties of Paving Materials [5]

Surface materials	Solar absorptivity, %	Thermal conductivity, Btu/(hr)(ft^2)($^\circ$F)	Heat capacity, Btu/(lb)($^\circ$F)
Asphalt concrete	85–90	0.70–1.8	0.20–0.22
Portland-cement concrete .	60–70	0.54–2.0	0.20–0.25
Wet sandy surface	80–90	1.0 –1.3	0.20–0.25
Dry sandy surface	60–80	0.5 –1.0	0.16–0.19
White paint	40	- - - -	- - - -
Water	- - - -	- - - -	1.0

developed by Thompson and Dumpsey utilizes the statistical data of a 30-year climatic record to generate the time-temperature regime in a pavement system. It is apparent that the finite-difference method does offer a meaningful analysis in predicting the regional variation of pavement temperatures if the input parameters and boundary conditions are properly defined.

For practical engineering application, the most reliable information on temperature variation and thermal gradient can be obtained from actual field observations. The reliability of observed data depends on the technique, scope, and time span of the observation. Straub and Schenck observed the hourly fluctuations of pavement temperature during various days of the year. One of the observations is shown in Fig. 6.1a. During the pavement test at Newark Airport, the temperature observation was carried out from August to February, the hottest and the coldest months of the region. Thus, the maximum seasonal variation of pavement temperature was recorded for that year. The results are shown in Fig. 6.2. It should be pointed out that the drift of electric instruments from one season to the next may have had some effect on the reliability of the observed data and that the lack of appropriate recording technique prevented the simultaneous monitoring of the pavement surface and ambient temperature. The original purposes of the temperature observation were:

1. To monitor the daily and seasonal temperature variation to be used in evaluating the change of physical properties of materials as well as the volumetric expansion and shrinkage of the pavement system.

2. To establish the daily and seasonal thermal gradient in order to estimate the internal stress equilibrium of a pavement system at various times of the year.

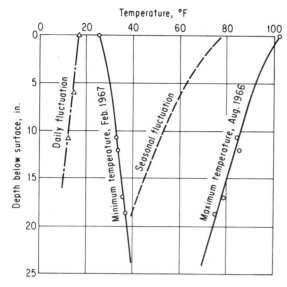

Fig. 6.2 Fluctuation of temperature in pavements.

6.2. CHANGE OF PHYSICAL PROPERTIES OF MATERIAL

The most common physical property used to measure the effect of temperature is the compressive strength or modulus of elasticity of the material. It is chosen not necessarily for theoretical analysis but for the convenience of engineering design. In Fig. 6.3, a test result is given showing the effect of temperature on the strength of portland-cement concrete. The standard temperature at testing was 70°F. Within a temperature range of 40 to 100°F, the relative compressive or flexural strength decreased from 1.25 to 0.80. The concrete strength decreased at the high temperature. Similar, but much more pronounced, temperature influence is experienced for asphalt concrete. Within the same temperature range, the stability of asphalt concrete may decrease from 15,000 lb at 40°F to approximately 3,000 lb at 100°F. The viscous properties of the asphaltic binder take a heavy toll on the elastic properties of the asphaltic concrete. The modulus of elasticity may decrease from 500,000 psi at 40°F to less than 4,000 psi at 100°F (see Fig. 6.4).

The effect of temperature on the physical properties of the subgrade is complicated by the ground moisture. For a given

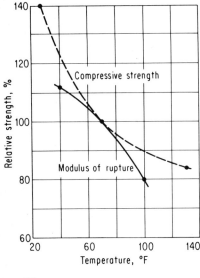

Fig. 6.3   Effect of testing temperature on strength of concrete.

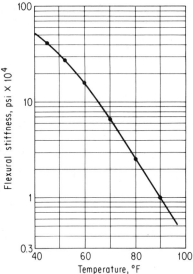

Fig. 6.4   Effect of temperature on strength of asphalt concrete.

moisture content, the strength of the subgrade tends to decrease at an elevated temperature. Murayama has reported that the elastic modulus of a clay soil may decrease from 1.00 to 0.70 if the temperature of the clay increases by 60°F (see Fig. 6.5). This change of physical strength of clay soil is practically identical with that in portland-cement concrete. It is likely that the similar rate of change may be encountered in sandy subgrade.

Another important physical property affected by the temperature variation is the volumetric change of pavement material. For the common paving materials, such as steel, concrete, lime-cement stabilization, and compacted subgrade, the coefficient of expansion is normally in the range of 2 to 8 x 10^{-6} in./(in.)($^\circ$F). Within the temperature range of 0 to 150°F, the thermal expansion is assumed to be a linear function of temperature change. Thus the change of pavement length is expressed as the product of the coefficient of expansion ϵ, the length of pavement l, and the temperature range t.

The thermal expansion of asphaltic concrete is primarily governed by the viscous properties of the asphaltic binder and the physical properties of the aggregates. The viscous properties are time-dependent functions, and the aggregate properties are normally influenced by the internal equilibrium of the aggregate structure. There is no coefficient of expansion for asphaltic concrete, as there is for steel and concrete materials. The volumetric change of asphaltic concrete has to be evaluated for each mix formula within each given range of time-temperature domain. In engineering practice, the effort of such testings and evaluations can be further complicated by the construction variations and environmental factors. Knowledge of the thermal expansion of asphaltic concrete is more of an art than a science.

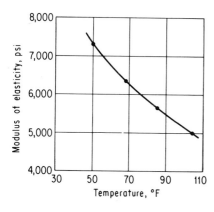

Fig. 6.5 Temperature effect on strength of clay soil.

6.3. STRESS INDUCED BY TEMPERATURE

If a linear thermal expansion exists for a given pavement material (by implication, this assumption does not apply to asphaltic material), the unit linear expansion is ϵt and the unit volumetric expansion is approximately $3\epsilon t$. If the stress-strain property of the material is linear elastic, the external force p required to restrain the thermal expansion is

$$f = \epsilon t E \tag{1}$$

As mentioned above, the coefficient of expansion varies within a very narrow range for many pavement materials; the thermal stress in a pavement is practically governed by the elastic modulus of the material. The higher the modulus of elasticity, the more pronounced the thermal stress will be. Therefore, low-strength lean concrete or lime-cement stabilization is not necessarily less desirable than high-strength paving materials.

The change of temperature with depth known as the thermal gradient and expressed by $\Delta t / \Delta z$ will result in warping of the pavement. The radius of curvature R of the warping and the resulting internal moment M in the pavement components are

$$\frac{1}{R} = \epsilon \frac{\Delta t}{\Delta z} \tag{2}$$

and

$$M = \frac{EI}{R} \tag{3}$$

For a homogeneous pavement component, the bending stress σ_t is given by

$$\sigma_t = \frac{1}{2} \epsilon E h \frac{\Delta t}{\Delta z} \tag{4}$$

It can be seen in the above equation that for a given pavement material in a given temperature environment, the temperature stress increases with increasing pavement thickness. Experience has indicated that a 16-in.-thick concrete pavement may have more surface cracks than a 12-in. concrete pavement.

The E value used in Eq. (4) depends on the selection of temperature variation. If the seasonal temperature variation is used in determining the thermal gradient, the warping of pavement will take place in a prolonged time interval. The plastic adjustment of the material will result in a small modulus of elasticity. The long-term modulus of elasticity of a regular portland-cement concrete is about one-third to one-half of its normal elastic modulus. For short-term temperature variation, say in a day or a week, the normal elastic modulus should be used to evaluate the daily or short-term variation of thermal gradient.

In analyzing the equilibrium of a pavement system, there are three conditions to be satisfied: (1) the linear movement at the surface, (2) the linear movement at the bottom of the pavement, and (3) the resulting bending stress due to the unequal movement of (1) and (2). If the temperature change at the bottom of a pavement is equal to zero, the temperature cracks in the pavement will depend on the strength of the material to resist the bending stress induced by the thermal gradient, as given by Eq. (4). If the strength of the material exceeds the temperature stress of Eq. (4), the formation of pavement cracks will depend on the magnitude of temperature change at the bottom of the pavement. The total temperature stresses consist of the bending stress by Eq. (4) and the direct axial stress by Eq. (1). The combined thermal stress becomes

$$\sigma_t = \epsilon E \left(t_b \pm \frac{1}{2} h \frac{\Delta t}{\Delta z} \right) \tag{5}$$

where t_b is the temperature range at the bottom of a pavement. For a given pavement material, the E and ϵ values are constant and the formation of temperature cracks will depend on the thickness of pavement, the thermal gradient in the pavement, and the temperature variation at the bottom of the pavement. If a linear thermal gradient exists in the pavement system, the zero temperature variation is encountered at a depth d below the surface, and the above equation can be rewritten for the maximum combined stress as

$$\sigma_t = \epsilon E \left(d - \frac{h}{2} \right) \frac{\Delta t}{\Delta z} \qquad \text{for } d > h \qquad\qquad (6)$$

It can be seen that the thermal stress σ_t can be reduced to a minimum if the temperature range at the bottom of the pavement is equal to zero, i.e., if the thickness of pavement is equal to or exceeds the depth where a constant temperature is maintained. This explains why the crack mechanism of a thin pavement is most likely to be governed by the temperature variation at the bottom of the pavement whereas the crack mechanism of a thick pavement apparently results from the thermal gradient in the pavement components.

6.4. EFFECT ON DESIGN AND PERFORMANCE OF PAVEMENT

If the mechanics of thermal expansion can be defined by the boundary conditions outlined under the previous heading, the design process can be narrowed down to the determination of (1) the physical strength of the material to resist the thermal stress and (2) the detail of expansion joint to accommodate the movement, either on a restricted or a free bottom.

The thermal stress given by Eq. (4) is applicable for a free-moving pavement bottom, and Eq. (5) is valid for a fully restricted base against horizontal expansion. The fully restricted condition satisfies the following relation:

$$\sigma'_t > \mu \gamma \frac{L}{2} \quad \text{and} \quad \sigma'_t > E \epsilon t_b \qquad\qquad (7)$$

where σ'_t = ultimate tensile strength of pavement material

μ = coefficient of friction, ranging from 0.6 to 0.7 for granular base

γ = unit weight of pavement material, ranging from 0.07 to 0.08 lb/cu in.

L = maximum spacing of expansion joint or pavement cracks, in.

If the temperature fluctuation assumes the same trend as that shown in Fig. 6.2, the average temperature stress due to seasonal variation in a 20-in. pavement will be

$$\sigma_t = 600,000 \times 6 \times 10^{-6} \times 38 = 137 \text{ psi}$$

If the ultimate tensile strength of the pavement material is 70 psi, the maximum spacing of expansion joint is, therefore,

$$L = 2 \times 70/0.7 \times 0.08 \times 12 = 208 \text{ ft}$$

If the joint spacing is designed to be 150 ft, the maximum movement of joint at the bottom of the pavement is

$$6 \times 10^{-6} \times 38 \times 1,800 \times 257/407 = 0.26 \text{ in.}$$

The relative movement between the bottom and top surfaces of the pavement is

$$6 \times 10^{-6} \times 40 \times 1,800 = 0.44 \text{ in.}$$

The total movement at the top of the expansion joint is 0.70 in. If the average temperature for laying the expansion joint is 65°F, the movement of the joint is approximately 0.35 in. in expansion and 0.35 in. in contraction. If the joint filler is designed for 35 percent compression, the width of the expansion joint will be 1 in. The minimum elongation of the joint sealer will be 35 percent.

The effect of temperature variation on the bending stress of pavement components is usually not as significant as the effect of wheel load. The most serious combination of wheel load and thermal stress is encountered during the winter months or cool nights, when the surface temperature is far cooler than the base of the pavement. It is likely that the thermal gradient is less than 0.5 and 1.5°F/in. for the daily and seasonal variations, respectively.

The thermal properties of asphalt concrete are radically different from those in the above analysis. First of all, the volumetric expansion is a nonlinear and time-dependent function. The drastic decrease of elastic modulus at elevated temperature (see Fig. 6.5) invalidates the meaning of Eqs. (1) to (7). The stability of asphalt concrete depends on the viscous cementing properties of asphaltic binder and the mechanical equilibrium of aggregates. In other words, the stability of asphaltic concrete can vary from batch to batch if quality control is not reasonably enforced. As indicated in Table 6.1, the solar absorptivity and the thermal conductivity of asphalt concrete are much higher than in most pavement materials; asphalt concrete can absorb more heat and can tolerate a much greater thermal gradient. As a consequence, the pavement base with asphalt cover can be somewhat cooler than with other types of pavement surface material. A significant reduction in temperature cracks will be encountered in base courses having asphalt concrete as surface.

The major drawbacks of asphalt concrete in airfield pavement are the development of surface indentation and horizontal shoving under the influence of gear wheels, particularly during warm weather. There seems no effective method for eliminating these problems as long as asphalt concrete is used in the pavement construction. The introduction of low-viscous asphalt binder and the increase of aggregate size and density may improve the stability of asphalt concrete at elevated temperatures but may result in shrinkage cracks during cold months. Although many research works have been carried out in this field, the design of asphalt concrete, from the point of view of engineering practice, remains an art.

B. Moisture of Pavement System

6.5. FACTORS INFLUENCING MOISTURE

Moisture is a fundamental variable in all problems of soil stability. It has special significance in pavement systems since the shallow depth of pavement construction is usually subject to large variations in moisture content. The principal conditions in which moisture-content changes can occur in a pavement system are shown in Fig. 6.6. These are:

1. Seepage of water into the pavement from higher adjacent ground
2. Rise and fall of the water table
3. Percolation of water through the pavement surface
4. Transfer of moisture either to or from the shoulder areas

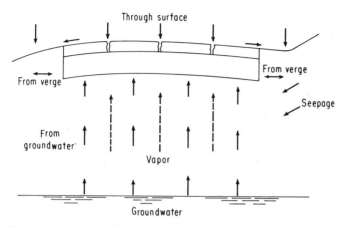

Fig. 6.6 Conditions affecting the change of moisture.

5. Transfer of moisture either to or from the subsoil
6. Transfer of moisture by vapor from the subsoil

The influence of seepage water depends mainly on the hydraulic gradient, which is influenced by the position of the adjacent water table and the coefficient of permeability of pavement materials as well as the subgrade soils. In a saturated field, the rate of flow can be assumed to follow Darcy's law:

$$Q = kiA \tag{8}$$

where Q = rate of flow, cu ft/sec
   k = coefficient of permeability, fps
   i = hydraulic gradient
   A = cross-sectional area of soil mass normal to seepage flow, sq ft

The utilization of flow net and the finite-element method for continuous flow would appreciably increase the reliability of seepage analysis. However, the major drawbacks of the analysis are (1) that Darcy's law is not valid for laminar flow if entrapped air is encountered in the soil and (2) that the coefficient of permeability is not a constant in both the vertical and horizontal directions. In analyzing the movement of water in unsaturated systems, Phillip has indicated that moisture suction or moisture tension generally dominate over the gravitational flow [5].

The rise and fall of the water table are closely related to climatic factors. A high water table may be encountered after an extended rainfall. The degree of fluctuation is largely governed by the permeability of the ground. In granula base, subbase, and subgrade, the elevation of the water table is more dependent upon the drainage characteristics of the pavement system and the site than upon precipitation.

Percolation of water through the pavement surface is one of the major causes of the fluctuation of pavement moisture. It is obvious that the intensity and duration of precipitation as well as the perviousness of the pavement surface have a significant influence on the rate of change of pavement moisture. The percolation of water has a detrimental effect on pavement performance and results in frost heaval if the subgrade is frozen and impervious.

In many airport pavements, the moisture condition may go up and down following construction and appear to stabilize after a few seasonal changes. However, the moisture contents at pavement edges remain unstable and are generally higher than those at interior locations. Similar conditions have been experienced for many highway constructions. There seems to be a continuous migration of moisture, from the edge toward the interior if side drainage is not provided to intercept the moisture movement.

The transfer of moisture either to or from the subgrade is the most important factor in stabilizing the moisture content in pavement. If the groundwater table is low, the downward movement of water percolating from the pavement surface is a gravitational flow. The rate of equalization depends on the permeability of the pavement material and the subgrade, as discussed above for seepage flow. Migration of moisture from the subgrade will be treated in detail in the following section.

The transfer of moisture by vapor from the subgrade becomes significant only when the ground is warmer than the ambient temperature. The upward movement of vapor has an important effect on the magnitude of frost heave when the upper ice layer of a frozen ground acts as a vapor barrier and entraps the moisture under the frost line. The degree of vapor migration depends on the thermal gradient in the pavement and its subgrade. Consequently, the climatic factors and geophysical conditions of the site have a significant influence on the moisture and temperature of a pavement.

6.6. TEMPERATURE-MOISTURE ENVIRONMENT

The problems of temperature and moisture are so closely associated that their common name is "environment." In a climate above the freezing point, the effect of temperature is very important and problems of moisture are usually relatively minor except when high precipitation is encountered. In temperatures below the freezing point, the moisture in the pavement has a decisive influence on the performance of a pavement. The depth of frost penetration becomes an important parameter.

The most common theoretical method for predicting temperature and the related frost penetration in pavement systems was originally developed by Stefan based on the hypothesis that the latent heat of soil moisture is the only heat that needs to be conducted to or from a point which is in the process of thawing and freezing. Heat quantities involved in temperature changes above or below the freezing point were considered to be of minor importance and, therefore, ignored. A common form of the Stefan equation for uniform soil condition is

$$D = \sqrt{\frac{48K_i F}{L}} \tag{9}$$

where D = frost depth, ft
  F = surface freezing index
  L = latent heat of fusion, Btu/ft^3
  K_i = thermal conductivity of frozen soils

Since the Stefan equation neglects the heat capacity of the soil, it will usually give a frost penetration depth that is too large. The freezing index F depends on the climate at the site and is the most difficult parameter to be assumed. Many engineers have attempted to use local experience or published accounts of maximum-frost-depth experience in their studies. Canadian engineers developed an empirical equation for frost depth D, in feet, based on field studies of frost penetration:

$$D = P + E\sqrt{F(29)} \tag{10}$$

where $F(29)$ represents the cumulative degree days below $29°F$. The values of constants P and E are dependent upon the soil type, drainage conditions, and nature of snow cover. Armstrong and Csathy found that P varied between 0.5 and 3.5 and E varied between 0.05 and 0.40, with average values of 1.3 and 0.12, respectively.

The depth of frost penetration represents the potential depth in which freeze-thaw and volumetric expansion of ice effect the integrity of a pavement. It is an important parameter, but its determination is far from scientific. The most reliable information still depends on experience and actual observation at the site.

6.7. THEORY OF MOISTURE MOVEMENT

Moisture held in soil above the water table is retained by surface tension at the points of contact of the particles or in the soil pores and capillaries. In the capillary tubes shown in Fig. 6.7, the water held above the water table will be in equilibrium when the upward component of the surface-tension force is equal to the gravitational force acting on the suspended water. The pressure or the suction in water immediately below the meniscus is lower than that at the boundary surface. The height to which the water will rise in the capillary tube is related mainly to the surface tension and the radius of the meniscus by

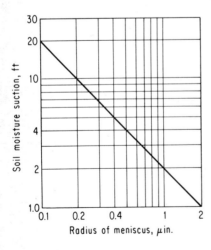

Fig. 6.7 Equilibrium of surface tension and rise in capillary tube.

$$H = \frac{2T}{\gamma r g} \tag{11}$$

where T = surface tension per unit length of boundary
  r = radius of tube or meniscus
  γ = density of water
  H = height of capillary rise

The capillary-tube analogy can be applied to soil suction. When the moisture of a soil is reduced, the water interfaces will recede into the smaller pores. The radius of curvature will decrease, and the soil suction will increase. Consequently, moisture will be sucked in from the surrounding areas to replenish the lost moisture. The most likely source of supply is the moisture from the underground water table. Since the vertical movement of moisture is affected by gravity, the upward migration of moisture cannot exceed the height at which the suction force is equal to the gravitational force.

In a soil mass, the moisture content is based on the equilibrium of soil suction. Moisture may migrate from areas of low moisture content, such as sand layers, to areas of higher moisture content, such as clay and silt layers. There is no moisture gradient which is responsible for moisture movement. The distribution of moisture is erratic and random. In general, sandy soil exhibits large pore space and low suction force. The moisture content is low, and the migration of moisture is rapid; therefore, a hydrostatic pressure is easily formed in sandy soil. For clay soils, all conditions are diametrically opposite to those in sandy soils: the pore space is small, the suction force is large, the moisture content is high, the migration of moisture is slow, and the clay layer is practically a waterproof layer. For silty soils, the range of characteristics is between those of sandy soils and those of clayey soils. Silty soils often have the worst features of sandy and clayey soils.

In addition to the soil suction, the temperature gradient in a soil can also affect the migration of moisture. This is particularly important when the pavement surface is in the cold cycles. If any soil or porous material in hydrostatic equilibrium with its surroundings is affected by local freezing, the equilibrium will be disturbed and there will be a tendency for water to move toward the freezing zone.

Although theoretical concepts have been reviewed for predicting moisture movement and moisture equilibrium, it is not practical to use these computations for actual pavement design. A carefully planned field observation, supplemented with the theoretical background, may produce a much more promising result. The concept of soil suction rather than water content appears to be more appropriate for defining the influence of water on the engineering properties of pavement materials, particularly the subgrade soils.

6.8. EFFECT ON DESIGN AND PERFORMANCE OF PAVEMENT

The so-called *environment design method* of highway pavement was advanced by the Michigan State Highway Department some forty years ago. The destructive effect of temperature and moisture was recognized to be more serious than that of wheel load. Benkelman introduced a simple straight beam in measuring the surface deflection of a pavement under the influence of a wheel load. By observing the seasonal variations of pavement deflection, considerable knowledge has been gained about the nature of environmental effects—a synonym for temperature and moisture. In the last several years, the Dyna-Flex machine has been used by many highway departments to replace the Benkelman beam in measuring the pavement deflection. A typical measurement is shown in Fig. 6.8.

The performance of the pavement can be divided into four groups:

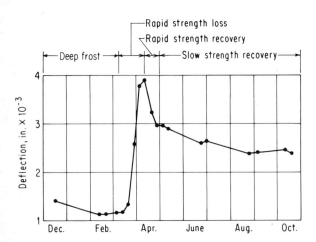

Fig. 6.8   Typical seasonal variation in pavement deflection.

1. During the months of December, January, and February, the ground of Northern states is in a period of deep frost. The moisture in the frozen ground acts as a cementitious binder. The effective thickness of pavement is equivalent to the depth of frost penetration. Consequently, the deflection measurement is small and the load capacity and service performance of the pavement are much better.

2. From the late February to early April, the weather changes rapidly from cold to warm and to cold again. The deep ground remains frozen, but the surface layer may undergo cycles of thawing and freezing. The groundwater vapor entrapped under the frost line becomes denser and denser. Meantime, during the short thawing periods, more surface water percolates into the pavement and accumulates on top of the frozen base. As the supporting capacity of any pavement material is governed by the relation

$$s = c + (p - u) \tan \phi \qquad (12)$$

where s = shearing strength of pavement material
c = cohesive bond
p = normal pressure
u = excessive pore-water pressure
ϕ = internal frictional angle

the excessive increase of moisture in the pavement during the spring thaw will greatly reduce the bearing capacity of the pavement. Consequently, a significant increase of pavement deflection results. If the freezing weather is followed by a brief thawing period, the excessive moisture accumulated on the frozen base, together with the vapor moisture entrapped at the bottom, will cause the heaval of the pavement. A major destruction, such as cracks, potholes, and disintegration, is a common phenomenon in pavement at this period.

3. During the month of April, the night temperature is persistently above the freezing point. The ground thaws completely. The migration of vapor moisture ceases. The high moisture in the pavement tends to move downward, and the pavement enters into a recovery stage in regaining its original strength.

4. During the months from May to November, the pavement functions normally and the effect of temperature-moisture is not significant.

Pavement distress during the spring thaw is environmental in origin, and remedial measures should also be environmental. Many engineers and scholars have defied this principle and searched for other, artificial means to improve the moisture-temperature condition. An example is the introduction of waterproofing seal on the pavement surface to prevent the percolation of surface water. During cold weather, this waterproof seal becomes a vapor barrier and entraps the moisture migrated from the warmer subgrade. At freezing temperature, the surface texture of the pavement may be destroyed by the blistering effect under the waterproof seal. A second example is the use of an insulating layer under the pavement. The insulating layer may prevent frost penetration below the base of insulation but also may effectively eliminate the long-wave radiation, that is, the heat flow from the ground. The temperature of an insulated pavement can be several degrees cooler than that of the area without insulation. The pavement surface can be more icy and slippery than any pavement can be. Excessive moisture from surface percolation will accumulate in the pavement base and can be more destructive to pavement structure than the wheel load.

The control of moisture in a pavement system is extremely difficult. If you cannot control the moisture, you should keep the moisture away from the pavement. There are no clear-cut rules which can be followed in the pavement design, but a few guidelines can be drawn for general reference:

1. Maintain the pavement base above the field area if possible.

2. Develop an adequate drainage system to improve the collection of surface runoff and minimize percolation.

3. Provide subdrains and an intercepting system to drain the moisture in the soil prior to its migration to the pavement area.

4. Lower the water table wherever possible.

5. Do not use silt as subgrade within frost depth and also above the water table.

6. Keep the pavement surface reasonably watertight, and reduce percolation of runoff.

C. Traction of Pavement Surface

The steadily increasing volume of traffic has increased the wearing and polishing of pavement surfaces. Such pavements may lose skid resistance, particularly when wet. The safe operation of a vehicle depends on the traction between tires and pavement surface to reduce the vehicle speed and bring the vehicle to rest within a tolerable stopping distance. Although the accident rate due to low tractive resistance is relatively low among all accidents, there is, nevertheless, a high potential risk. The operator of today's highway and airport pavement systems has faced the greater demands on the construction of pavement which have adequate skid resistance. In order to develop an understanding of the problems involved in the design of appropriate antiskidding pavement surfaces, a review of the basic mechanics of pavement friction is herein presented. A proper understanding of the problem will lead to a reliable interpretation of the data and sound engineering judgment.

6.9. FACTORS INFLUENCING TIRE TRACTION

The traction between tires and pavement surface when both are dry depends largely on the frictional resistance of the pavement surface and on the tire tread design. On wet surfaces, the frictional resistance decreases as the speed is raised. Lander [4] reported that on wet surface the risk of a skidding accident is almost twice that on dry pavement and is almost three times greater if the vehicle speed exceeds 30 mph. This observation is in general agreement with the test results on frictional resistance at NASA Wallops Station [4]. A typical test record is shown in Fig. 6.9. The friction coefficient on dry surface is about three times that on wet pavement when the vehicle speed exceeds 30 mph.

Interaction of Tire and Pavement Surface

The NASA tests also indicate that the friction coefficient is a function of load duration. For a high-speed vehicle, the loading period is shorter than for a slow-moving vehicle. Consequently, the

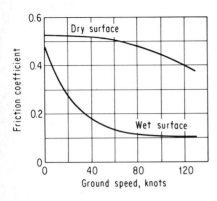

Fig. 6.9 Effect of wet surface on friction coefficient.

frictional resistance of a high-speed vehicle is much reduced (see Fig. 6.9). The rate of loss of friction largely depends on the condition encountered at the interface of tire and pavement. Water is much heavier than air and offers greater resistance at its displacement. The result is that the effective load on the interface is reduced, and consequently the friction resistance decreases drastically when the surface is wet.

In theory, the frictional resistance is proportional to the normal load, with no direct relation to the contact area of the load. However, actual tests indicate that frictional resistance increases with decreasing inflation pressure. A typical test result is shown in Fig. 6.10. If the bearing of the tire walls is assumed to be constant and the contact area is equal to the total load divided by the tire

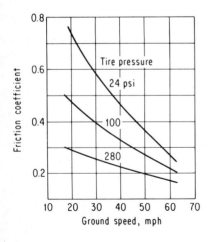

Fig. 6.10 Relationship between tire pressure and friction coefficient.

pressure, the test result suggests that the frictional resistance is proportional to the square root of the contact area.

Environmental Influences

Surface characteristics of a pavement are affected by weather. The skid resistance varies from one season to another. The mechanisms responsible for seasonal change in friction are thought to be the fluctuation of pavement temperature. Studies [1] have revealed that skid resistance tends to decrease with increased temperature. The magnitude of the loss depends on the texture of the pavement surface, as well as on the viscosity of the surface material. The effect of change in viscosity is particularly noticeable on mixes containing a predominance of sand-asphalt. A typical test result is shown in Fig. 6.11. Freezing of pavement surface is encountered when the ambient temperature is several degrees below 32°F. The presence of ice has an important effect in modifying the surface texture of the pavement. The friction coefficient ranges from 0.15 on hard ice to almost zero on wet frozen surface.

Wearing of Aggregate

Pavement surfaces, when properly designed and constructed, will initially exhibit high skid resistance. Traffic polishes as well as wears pavement. Any surface containing limestone aggregates, either coarse or fines, will polish and in time may become slippery. Loss of friction is usually most severe during the first 2 years after construction. Thereafter, the rate of polishing decreases and eventually reaches a stable level of smoothness. A typical test result is shown in Fig. 6.12.

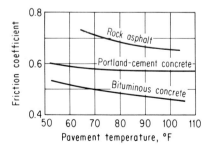

Fig. 6.11 Temperature influence on frictional resistance.

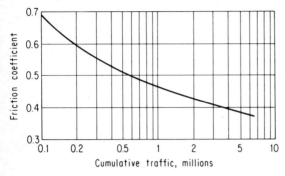

Fig. 6.12 Effect of traffic on friction coefficient of pavement surface.

Operation Requirements

A deficiency in frictional resistance is indeed a significant contributing cause in many accidents. Studies [1] of accident locations on highways indicate that 87 percent of the pavement surface exhibited a coefficient of friction of less than 0.40. This suggests that a pavement surface on a high-speed, high-traffic-volume highway with a friction coefficient of less than 0.40 may be hazardous when wet.

For airport pavements, the operation varies with the airport facility, the type of aircraft, and the pilot's maneuvering. Although there have been many studies and tests of the frictional resistance of runways and taxiways, there is no set of guidelines to suggest the minimum requirement of frictional resistance below which aircraft operation may be hazardous. Hall [4] reported that 35 percent of aircraft operational accidents can be related to the inadequacy of frictional resistance. Among these accidents, 28 percent occurred in ice and snow conditions and 42 percent were due to hydroplaning. For both conditions, the friction coefficient in accidents was probably less than 0.1. The remaining 30 percent of the accidents were encountered on wet runways when the friction coefficient was likely to be in the range of 0.1 to 0.20. This suggests that a pavement surface with a friction coefficient of less than 0.2 may be hazardous for aircraft operation when wet.

6.10. SURFACE TEXTURE OF PAVING MATERIALS

In terms of frictional resistance, the adequacy of a pavement surface must be judged on the basis of in-service performance.

Environmental weathering, as well as traffic wearing, profoundly affects frictional resistance. The basic mechanics of frictional resistance are described in the following.

Resistance on Dry Surface

The mechanism of developing frictional resistance can be illustrated by a hypothetical condition, as shown in Fig. 6.13. The spherical aggregate is embedded in cement with its top surface in contact with a rubber tire. The total vertical pressure from the tire is equal to $p(D + S)^2$, where p is the unit tire inflation pressure, D is the diameter of the aggregate, and S is the thickness of the cement bond between the aggregates. The horizontal frictional resistance is equal to $e\Delta\pi L^2/4$, where e is a coefficient related to the elastic properties of the rubber tire, Δ is the depth of indentation, and L^2 is the contact area between tire and aggregate. For spherical aggregate, the relation $L^2 = 4\Delta(D - \Delta)$ exists if the aggregate does not deform under the vehicle load. The coefficient of maximum frictional resistance becomes

$$\mu = \frac{2e}{p} \frac{\Delta^2}{D + S} \tag{13}$$

In words, the frictional resistance is profoundly affected by the

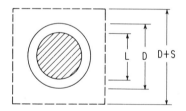

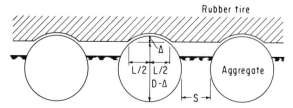

Fig. 6.13 Hypothetical condition of developing frictional resistance.

depth of indentation which is closely related to the polish resistance and sharpness of the aggregate. Crushed hard rock offers higher frictional resistance than polished gravel. The frictional resistance is significantly influenced by the size of aggregate and the cement binder. The finer the aggregate and the thinner the binder, the higher the frictional resistance will be. In designing a high-frictional-resistance surface material, particular attention should be given to the physical properties of fine aggregates and cement binders. Studies [1] have indicated that silica sand asphalt and dense graded concrete using a thinner film of binder demonstrate a higher frictional resistance.

For determining the type of cement to be used in the surface material, tests [4] have indicated that asphalt and portland-cement concrete offer an identical range of variation in frictional resistance (see Fig. 6.14). However, the breeding of excessive asphalt during the summer months may have a significant effect on the slipperiness of the pavement surface. For runway end and gate positions, cement concrete remains the best surfacing material. It is not damaged by fuel spillage, heat, or blast, and it is easier to clean. Where cement concrete cannot be provided, an asphalt surface with coal-tar pitch seal is the best alternative. It resists damage by moderate fuel and oil spillage.

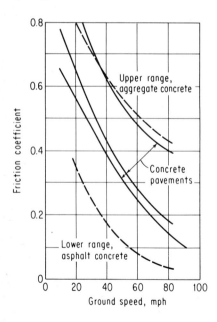

Fig. 6.14 Friction coefficients of various pavement surfaces.

Tight specifications of the shape and hardness of the aggregate, both coarse and fine, may improve the frictional-resistance value, but improving frictional resistance by being restrictive in the selection of aggregates may not be an economical solution, even if it were physically possible. Methods of roughening new surface during construction have been tried. Although many tests have shown that these methods offer good results during the early stages of in-service life, extended traffic volume tends to wear down the mechanical roughness and polished surfaces result.

In addition to mechanical methods of roughening, surface dressing of high-strength cement and hard angular chippings have also been tried. From a number of experiments [3], there are indications that surface dressing may be the simplest, most economical, and most effective method of improving the frictional resistance of existing pavement surface. To understand the concept of bonding or adhesion, one must examine the material behavior at a micro-structural level [2]. The bonding surface can be considered as an elastic sheet of which the ultimate strength depends on the energy required to produce a fractured surface. When a small microcrack is encountered, the integrity of the bonding sheet is determined by the conditions that permit the crack to propagate through the cross section. The tensile strength of the sheet is determined by the surface energy of the solid, the modulus of elasticity, and the size of microflaw of the bonding material. In composite materials, the importance of bonding surfaces is magnified because of the fact that the behavior of composites also depends on the nature of interfaces between matrix, filler, and binder. The degree of adhesion is controlled by a number of physical and chemical factors. Among the physical factors are surface area, surface roughness, degree of coating and wetting, difference in elastic properties, and thermal expansion coefficients. Among the chemical factors are difference in cohesive energies, polarities of surface energies, relative solubilities, and susceptibility to heat, oxidation, and hydrolysis. The strength of an interface depends in a very complicated fashion on a combination of these factors.

There are several drawbacks to applying surface dressings:

1. The development of frictional resistance largely depends on the wearing of chippings to be bonded on the adhesive coating. Only a small selection of chippings, such as quartz sand and crushed bauxite, can be effectively used for this purpose. It is a costly operation.

2. The high surface energy of the bonding material is always associated with a high modulus of elasticity, which means brittleness at small deflection and lower fatigue resistance under repetitive loadings.

3. Wet-dry and freezing-thawing cycles are important factors in destroying the integrity of surface dressing.

Resistance on Wet Surface

Although the frictional resistance of a pavement surface can be improved by the use of selected aggregates, mechanical surface roughness, and high-strength surface dressing, the effective frictional resistance will be, nevertheless, significantly reduced when the pavement surface is wet. The surface water encountered outside the contact area between tire and aggregate (see Fig. 6.13) produces a hydrostatic pressure, which transfers the tire pressure directly on the pavement surface and, therefore, reduces the effective contact pressure between the tire and aggregate. The reduction of effective contact pressure will result in a decrease of frictional resistance. Since the development of hydrostatic pressure is caused by the entrapment of surface water as well as the vehicle velocity (see Fig. 6.9), the frictional resistance will be significantly reduced when the surface is wet and the vehicle travels at high speed. No surface treatment for improving dry friction will be effective when the surface is wet.

One exception to this statement is found in the use of "friction courses" in the United Kingdom. The surface coarse consists of an open graded 1/4 to 3/8 in. size of crushed hard rock, such as basalt or granite. The binder is of grade-200 penetration asphalt at a rate of 5 percent by weight and mixed at 250°F. The compacted thickness of asphalt macadam is 3/4 in. on a densely graded impervious wearing course. During the rain, the water penetrates into the pavement. No surface water is in contact with rubber tires, and the surface is effectively a "dry surface" in all weather conditions. However, the friction courses can be recommended only when they are to be laid over a new or an existing pavement surface which is impervious to water and has good drainage. The water must flow quickly to the drainage channels. Any underlying surface on which water lies or which is pervious can lead only to the deterioration of the friction course itself. It is questionable that the U.K. experience can be extended to areas where wet-dry and freezing-thawing are the significant factors in the integrity of pavement surface.

6.11. HYDROPLANING AND RUNWAY GROOVINGS

The problem of aircraft skidding on wet runway surfaces becomes important when the landing speed of an aircraft is such that the distance needed for landing in wet weather is almost equal to the length of the runway and the skid-resisting force is inadequate to reduce aircraft speed without skidding, side pitching, or overshooting the runway ends.

Low frictional resistance on a wet surface is largely due to the lubricating effect and the displacement of the water film at the interface of tires and pavement. The effect becomes more pronounced when water depth and aircraft speeds increase. At high speeds, the water layer tends to build up resistance against displacement. A wedge of water is encountered at the interface. The hypothesis advanced by Martin [3] and Horne [4] is that the vertical component of the resistance force progressively lifts the tires, thereby decreasing their area in contact with the runway until ultimately they are completely clear of the surface. When this happens, the aircraft is *hydroplaning*. This process has been demonstrated by the tests at NASA Wallops Station. The Ministry of Aviation, United Kingdom, has conducted a more meaningful test to measure the friction coefficient of tire-pavement interface under the actual operation condition of an instrumented aircraft. The result is shown in Fig. 6.15. It seems that the hydroplaning may occur when

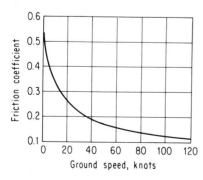

Fig. 6.15 Friction coefficient measured by an instrumented aircraft.

the speed exceeds 100 knots and the friction coefficient is approaching below 0.1.

Many accidents attributed to hydroplaning might have been avoided if the braking force after decelerating below hydroplaning speed had been able to stop the aircraft in the remaining runway length available. It is important, therefore, to have uniformly good surface with adequate friction value through the whole runway length. The loss of directional control which occurs when an aircraft is hydroplaning is just as serious as the danger of overrunning. A longer runway is not a foolproof solution to the hydroplaning problem.

The British experience [3] indicates that water depths of about 1/4 in. are probably needed for hydroplaning to be initiated. Once started, it can be sustained by considerably less water depth. The depth of water on a pavement surface depends on several factors: (1) the rate of rainfall and its duration, (2) the side slope and cross section of the runway, (3) the width of the runway, (4) the longitudinal profile of the runway, (5) the surface texture and retention of runoffs, and (6) the direction of wind and its velocity. Although many runway surfaces probably have adequate side slope and longitudinal swirl to ensure adequate surface drainage, the condition may be different during adverse high winds and storms. Even if an aircraft does not land during a storm, a period will exist after the rain during which the runway surface may be suspected of causing hydroplaning.

Normal surface treatment to increase the frictional coefficient does not prevent the development of hydroplaning. The surface water must be eliminated before hydroplaning takes place. Runway grooving is the most efficient way to get rid of the surface water. Although this method has been known for more than 10 years, there is no rational procedure for designing the geometry of the groove. According to past experience, a satisfactory groove is said to have a depth and width ranging from 1/8 to 1/4 in., with a pitch ranging from 1 to 2 in. None of the six factors cited above on the depth of water film has been related to the groove design. In Fig. 6.16a, the footprint of an aircraft tire is assumed to be of elliptic configuration, with the major axis parallel to the traffic direction. For a point 0 along the major axis of the tire contact area, the time required for the passage of a tire is equal to $2b/v$, where v is the velocity of the

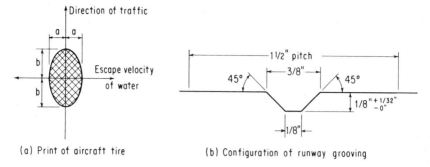

(a) Print of aircraft tire   (b) Configuration of runway grooving

Fig. 6.16   Footprint and runway grooving.

vehicle. The water particle at 0 has to escape laterally for an average distance of $\pi a/4$ during the time period of $2b/v$. The external force causing the water particle to escape is the tire pressure, and the escape velocity is equal to $cp^{1/2}$, where c is a constant and p is the tire pressure in pounds per square inch. The critical moment at the development of hydroplaning is

$$cp^{1/2} = \frac{\pi a/4}{2b/v} \tag{14}$$

This means that when the escape velocity due to external pressure is slower than the water travel sideways, there is hydroplaning. Let V_c be the critical velocity selected for preventing hydroplaning; the pitch of grooving, l, will satisfy the relation

$$\frac{l}{2b/v} = \frac{\pi a/4}{2b/V_c} \tag{15}$$

where v is the maximum velocity anticipated for landing and/or takeoff of an aircraft. If the critical velocity V_c is selected for providing a given level of friction coefficient, the same relation as given in the above equation is also valid. For a maximum speed of 140 knots, the pitch of runway grooving is as shown in Fig. 6.17.

The pitch is designed for the surface water to escape, whereas the cross-sectional area of the groove is designed to store the surface water and to prevent the development of hydroplaning. The quantity of water to be drained off the surface during the passage of a tire is equal to

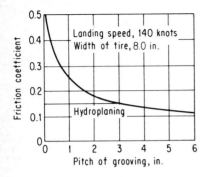

Fig. 6.17 Relation between friction coefficient and pitch of runway grooving.

$$2bc's^{1/2}t_c\,\frac{2b}{V_c} \qquad (16)$$

where
c' = coefficient of surface runoff velocity
s = side slope of runway surface
t_c = critical thickness of water film
$2b/V_c$ = passage of time at critical velocity

Within the tire footprint area, the accumulation of water on the pavement surface is equal to

$$\pi abl\,\frac{2a}{c's^{1/2}} \qquad (17)$$

where I is rainfall intensity during the period water travels a distance of $2a$ by surface flow. Immediately before the development of hydroplaning, the following relation exists:

$$2bc's^{1/2}t_c\,\frac{2b}{V_c} \geq \pi abl\,\frac{2a}{c's^{1/2}} \qquad (18)$$

The critical thickness of water film becomes

$$t_c \geq \pi \left(\frac{a}{c'}\right)^2 \frac{I}{s}\,\frac{V_c}{2b} \qquad (19)$$

For a runway cross section having a side slope s of 0.015 and a rainfall intensity of 3.6 in./hr, or 0.001 in./sec, the corresponding coefficient c' of surface runoff velocity is 1.8 fps or 22 in./sec. For a critical velocity of 100 knots, or 170 fps, the thickness of water film at hydroplaning is

$$t_c = \pi \left(\frac{8}{22}\right)^2 \frac{0.001}{0.015} \frac{170}{20} = 0.23 \text{ in.} \tag{20}$$

The cross-sectional area A of the grooving should satisfy the relation

$$A = t_c l \left(\frac{1}{a}\right)^2 \tag{21}$$

For a pitch of 2 in., the cross-sectional area of the grooving is

$$A = 0.23 \times 2(2/8)^2 = 0.029 \text{ sq in.} \tag{22}$$

The size of grooving can be 1/4 in. in average width and 1/8 in. in depth. In considering the weathering and spalling caused by wheel load and freezing, the test at Kennedy International Airport confirmed the superiority of V-shaped grooving. The actual configuration of the groove is shown in Fig. 6.16b.

D. Regional Differential Settlement

In recent years, the rapid growth of transportation systems has resulted in the increasing use of such marginal land as swampy areas and hilly uplands. This adds new dimensions and problems, which were not pronounced in previous airport and highway construction. The cost for land preparation increases rapidly, and the quality of ground support becomes more difficult to control. As swampy areas always consist of soft mud which settles due to the weight of the fill and/or the improvement of drainage of the area. The uneven distribution of natural deposit will result in an uneven settlement of the land. Therefore, any pavement constructed on such land will exhibit surface undulation or wave condition. It is a basic requirement of modern pavement engineering to design a tolerable wave condition due to uneven settlement of the ground.

6.12. FACTORS INFLUENCING SETTLEMENT

If the land subsides uniformly, the pavement structure more or less floats on the ground and the pavement surface will be even and smooth. There is nothing wrong with such pavement, and the operation of the vehicle is not hindered. In reality, however, ground settles to varying degrees from location to location, either due to the physical nature of the deposit or due to conditions of land preparation. The problem of pavement surface is directly related to the differential settlement of the area. However, the differential settlement is a function of the total settlement. In order to understand the problem of differential settlement, it is necessary to review the problem of total settlement. There are three factors in the settlement of ground: (1) the consolidation of the subsoil due to land preparation or increase of superimposed load, (2) the drawdown of the water table, causing the increase of overburden pressure (every foot of dewatering increases the effective overburden pressure about 40 to 60 psf), and (3) regional subsidence due to geological disturbances, such as mining and earthquake.

In Fig. 6.18, a typical time-settlement curve is shown. The rate of settlement decreases gradually during the period of primary consolidation and approaches a constant during secondary consolidation. As defined in soil mechanics, the primary consolidation represents the condition due to the change of pore-water pressure. While the adjustment of soil structure is considered to be the principal reason causing the secondary consolidation, the basic mechanics governing

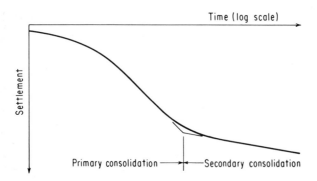

Fig. 6.18 Time-settlement curve of compressible deposit.

the primary consolidation can be expressed by

$$S = H \frac{C_c}{1 + e_0} \log \frac{p_0 + \Delta p}{p_0} \tag{23}$$

where S = magnitude of primary settlement
   H = thickness of compressible layer
   C_c = compression index
   e_0 = initial void ratio
   p_0 = initial consolidation pressure or overburden pressure
   Δp = overburden pressure causing consolidation of compressible layer

As defined in soil mechanics, the compression index C_c, the initial void ratio e_0, and the initial consolidation pressure p_0 are closely related to the initial moisture content of the compressible soil. The above equation can be simplified as

$$S = f(H, w_0, \Delta p) \tag{24}$$

This means that the total settlement of a compressible layer is a function of the thickness, moisture content, and load increment on the compressible layer. If these three parameters are uniform throughout the consolidation area, the total settlement is uniform and there should be no differential settlement. However, a natural deposit may exhibit a coefficient of variation in moisture content ranging from 0.25 to 0.35, and the thickness of the compressible layer depends largely on the geological nature of the deposit. It should not be a surprise that the high value of the moisture content can be twice the low value. Therefore, the differential settlement of a layer of uniform thickness can be anticipated to be as much as 50 percent of the total settlement. If the layer thickness is not even, the differential settlement of the region may be very erratic.

   In treating pavement problems, it is necessary to understand the moisture and thickness of compressible layers below the pavement area. The other parameter, Δp, is much easier to determine. It usually represents the weight of fill or the overburden due to dewatering. During the service life of a pavement construction, the performance of the pavement surface is governed by the rate of settlement. During primary consolidation, the time required to complete the pore-water adjustment can be expressed by

$$t = \frac{H}{k} \frac{S}{\Delta p} \qquad (25)$$

where the k value is the coefficient of permeability and H/k indicates the time required for water to seep through the compressible layer. The smaller the coefficient of permeability the larger the H/k will be (that is, the longer it will take for water to seep through layer H). In a simplified expression, the time required to complete the primary consolidation is

$$t = f\left(H, w_0, \frac{H}{k}\right) \qquad (26)$$

This means that the time required to complete the primary consolidation is a function of the thickness, moisture content, and seepage time of a compressible layer. The time required for the completion of primary consolidation is not affected by the magnitude of the overburden pressure. For sandy soils, the coefficient of permeability is in the range of 10^{-2} to 10^{-4} cm/sec. Silty soil has a k value in the range of 10^{-4} to 10^{-6} cm/sec. It takes about 100 or 1,000 times longer to drain a silty soil than a sand. Consequently, the consolidation of a silty soil is much slower than in a sand. The sandy deposit would not create a serious problem for the pavement construction insofar as settlement is concerned.

As given in the theory of soil mechanics, the characteristics of primary consolidation are very similar to those of piston-and-spring systems. The time-settlement curve approaches an inclined asymptote and continues on a gentle slope. This late stage of time settlement is known as *secondary consolidation*. It is caused by the gradual adjustment of soil structure together with the plastic deformation of soil particles. For all practical purposes, there is no single set of mechanical rules which can explain the time function of secondary consolidation. Experience indicates that the consolidation for a deep compressible layer is practically a linear function of time. As differential settlement is a function of total settlement, the pavement surface may become progressively rougher and more uneven during subsequent service years. The task of the engineer is to design the terminal condition of pavement performance for an anticipated service life.

6.13. METHODS OF IMPROVING GROUND

The factors influencing settlement are thickness, moisture content, permeability, and the load increment on the compressible deposit. Therefore, any method for improving the settlement should be directed toward minimizing the source of problems. The first approach is to reduce the thickness and unevenness of the compressible layer; in many cases, it is practical to remove all compressible soils. However, the removal of the compressible soils is not always the best answer. If the compressible soil is removed in one area and left in other areas, the difference in settlement may be aggravated and the rate of the differential settlement may become worse. Therefore, if the removal of unsuitable material or compressible soil is contemplated, the program should be carefully planned to remove all compressible materials under the pavement and its environment if necessary.

The second approach is to lower the moisture content of the compressible soil. High moisture content is a direct reflection of low consolidation pressure. The soil deposit can be highly compressible. There are several methods which can be employed to reduce the moisture content of soils, such as (1) dewatering by pumping, (2) dewatering by osmosis process, and (3) injecting chemicals to stabilize the material and reduce its moisture content. A typical example of the last method is soil stabilization utilizing hydrated lime.

The third approach is to increase the overburden pressure Δp during the period of construction. In this way, the rate of settlement will be reduced when the actual overburden pressure is reduced during the service years. This can be accomplished by adding a blanket of fill on the compressive layer. By reducing the total settlement in the service year, the differential settlement may be proportionately reduced. However, the inherent problems of this method are (1) the cost of placing and removing the surcharge and (2) the residual settlement during the service years. If the surcharge program is not associated with the improvement of subsurface drainage, the efficiency of the surcharge may be greatly reduced and the problem of differential settlement will remain.

In pavement design it is a good practice to review the geological condition of the area, to define the thickness of compressible soils and to evaluate the variation of moisture content and the compressibility of the soft layers. An effective ground-improvement program can be developed only when a meaningful cost-benefit analysis has been performed for the removal of unsuitable soils as compared with the surcharge method.

With respect to the rate of consolidation, the time function is governed by the thickness and the coefficient of permeability of the compressible soils. There is practically no way to improve the permeability of a soil as long as the soil particles remain unchanged. It is a reality that engineers can design a pavement structure to minimize the magnitude of settlement but not the time of consolidation.

6.14. EFFECT ON DESIGN AND PERFORMANCE OF PAVEMENT

As ground starts to settle, the pavement surface subsides to conform with the settlement configuration if the pavement structure is flexible enough. The riding quality of pavement surface is, therefore, governed by the equation (see Chapter Nine)

$$\frac{\Delta}{L^{1/2}} \leq K \tag{27}$$

where Δ is the vertical deviation within a wavelength of L. If the $\Delta/L^{1/2}$ value exceeds a certain value of K, it is said that the functional surface of the pavement does not meet the performance requirement for smooth and safe riding of a given vehicle at a defined speed of travel. The effect of the differential should, therefore, be evaluated by the above equation.

On the other hand, the pavement should be designed to have sufficient strength to endure the long-term settlement without cracking the pavement structure itself. By assuming the settlement contour to be a sine wave (see Fig. 6.19), the vertical deflection coordinate is

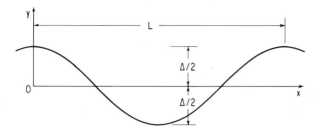

Fig. 6.19 Settlement contour of pavement surface.

$$y = \frac{1}{2} \Delta \cos \frac{2\pi x}{L} \tag{28}$$

The bending moment in the pavement layer is

$$M = EI \frac{d^2y}{dx^2} \tag{29}$$

By substituting Eq. (28) in Eq. (29), the bending moment due to differential settlement can be expressed by

$$M = 2\pi^2 EI \frac{\Delta}{L^2} \cos \frac{2\pi x}{L} \tag{30}$$

The maximum bending moment is

$$M_{max} = 2\pi^2 EI \frac{\Delta}{L^2} \tag{31}$$

when $x = 0$ or $L/2$.

  In evaluating pavement performance and design effects, it is necessary to study the vertical deviation for a given wavelength. An example of pavement evaluation is shown as follows:

Wavelength L, ft	Vertical deviation Δ, ft	Performance $\Delta/L^{1/2}$	Bending moment Δ/L^2
75	0.05	0.0058	8.9×10^{-6}
140	0.21	0.0177	10.7×10^{-6}
240	0.27	0.0174	4.7×10^{-6}
440	0.32	0.0153	1.7×10^{-6}

This table was deduced from the 10-year settlement record of runways at Newark Airport. The maximum roughness is governed by wavelengths of 140 to 240 ft, and the maximum bending moment is governed by 140-ft wavelengths. If the pavement is properly designed for differential settlement, the rigidity of the pavement will reduce the magnitude of settlement. Although the cause of differential settlement is not removed, its effect is minimized.

As the performance of the pavement and the bending stress in the pavement elements are closely related to the vertical deviation of the deflection wave, the problem of differential settlement can be effectively defined by the Δ and L parameters.

If sufficient qualitative information is available to evaluate the soil condition, the Δ and L values can be measured from the anticipated settlement profile as determined by the thickness and compressibility of the subsoil. This sounds simple, but it requires extensive subsoil survey and numerical computation to determine the rate and magnitude of settlement. In many cases, the subsoil information is not complete and the settlement computation is rather sketchy. The time and resources for conducting extensive soil exploration and laboratory tests are usually not available. It is then necessary to review the previous settlement record of the area and to predict the possible settlement for the pavement construction. For the pavement design of Newark Airport, the 10-year record of airfield settlement is used in the design of new pavement. There exists a danger that the past record may not necessarily reflect the new construction.

As the wavelength of differential settlement is closely related to the nature of the soil deposit, land preparation by the surcharge method will not alter the configuration of the settlement wave, except that the wave amplitude is reduced. On the other hand, if excavation of the compressible soil is contemplated, the best way to handle the differential settlement is to create a long transition between the backfill and the original good ground in a three-dimensional space. If the past settlement record is used in determining Δ and L values, the reliability can be improved by introducing the coefficient of variation of the observed data. The parameter Δ/L^2 should be equal to the mean value plus a desirable number of standard deviation. As differential settlement will be taking place over a long period, the E modulus in Eq. (30) should reflect the long-term deformation. The creeping modulus of cement- or lime-stabilized material is about one-third of its elastic modulus. The bending stress becomes a function:

$$\sigma = f\left(E', h, \frac{\Delta}{L^2}\right) \tag{32}$$

In words, the bending stress increases with the increase of pavement thickness, creeping modulus of material, and magnitude of settlement basin but decreases with the increase of settlement wavelength. In a subsiding region, rigidity and brittleness of pavement material will result in an increase of bending stress.

The maximum bending moment given in Eq. (31) is valid only when the pavement is flexible enough and its surface deformation coincides with the settlement configuration of the subgrade. If the rigidity of the pavement prevents such conformity, the bending moment in the pavement due to its own weight is

$$M_p = C_m \gamma h L^2 \tag{33}$$

The ratio of pavement deformation and subgrade subsidence is

$$\frac{\Delta_p}{\Delta_{max}} = \frac{Mp}{M_{max}} = \frac{6C_m}{\pi^2} \frac{\gamma}{E'} \frac{L^2}{\Delta} \left(\frac{L}{h}\right)^2 \tag{34}$$

where the subscript p represents the pavement structure and max represents the subgrade, γ is the unit weight of pavement material, and C_m is the coefficient of flexural bending and is approximately equal to 1/24 or less. In words, the rigidity of the pavement structure will minimize the effect of differential settlement of the ground. The magnitude of pavement deflection decreases with the increase of creeping modulus and thickness of pavement. Therefore, a properly designed thick pavement may equalize the differential settlement of the ground. However, if the actual wavelength of settlement configuration is such that

$$L < 2\pi \left(\frac{E'}{\gamma}\right)^{1/2} \left(\frac{\Delta}{L^2}\right)^{1/2} h \qquad (35)$$

the pavement will crack and the surface of the pavement will deflect with the subgrade. After the cracking, the performance of the pavement will be governed by $\Delta/L^{1/2}$ of the subgrade. The pavement will become rougher and may require leveling maintenance. Therefore, differential settlement in soft ground is one of the most important but most difficult parameters in pavement design. It requires careful planning and soil exploration to determine the magnitude of settlement within a significant wavelength. If the pavement is properly designed for the differential settlement, the rigidity of the pavement can be used to modify the differential settlement. The pavement will be smooth, and the full amplitude of settlement will not be reflected on the pavement surface until the termination of the anticipated service life of the pavement.

REFERENCES

1. J. L. Burchett, Jr., et al., "Pavement Slipperiness Studies," Kentucky Department of Highways Research Report PB 194 157, March, 1970.
2. A. T. DiBenedetto, General Concepts of Adhesion, *Highway Res. Record,* no. 340, 1971.
3. F. R. Martin and R. F. A. Judge, Airfield Pavements, Problems of Skidding and Aquaplaning, *Civil Eng. Public Works Rev.,* London, December, 1966.
4. "Pavement Grooving and Traction Studies," NASA SP-5073, Longley Research Center, National Aeronautics and Space Administration, November, 1968.
5. M. R. Thompson, "Environmental Factors and Pavement Systems," Construction Engineering Research Laboratory Technical Report, March, 1970.

CHAPTER SEVEN
Mathematical Models for Pavement Systems

A. Equilibrium of Pavement Systems

Mathematical models are the tools by which engineers apply scientific principles to the solution of engineering problems even without the benefit of past experience. The solution is based on the physical requirements of a structure to withstand the anticipated external loads, postulated deformations and stresses in the elements, and the mechanical behavior of materials according to the basic laws of mechanics governing motion and force. Thus, a mathematical model consists of three submodels:

1. The equilibrium of the pavement system under the influence of external loads.

2. For a given supported condition, an evaluation of the deformations and stresses in the pavement elements.

3. A characterization of the fundamental properties of pavement materials and their effect on the equilibrium and stability of the pavement structure.

The early development of mathematic models for pavement structures was confined primarily to the second submodel. Hertz [65] in 1884 proposed a mathematical method for analyzing an elastic slab supported by a liquid. In 1926 Westergaard simplified the mathematical manipulation for application to practical design problems. Pickett-Ray in 1951 introduced the use of influence charts and thus reduced the design computation to almost nothing. In the meantime, significant contributions were made by many researchers and the Portland Cement Association (PCA) on the fundamental properties of portland-cement concrete. Many good concrete pavements were designed and built. The Westergaard theory and the influence charts became a synonym of the pavement mathematical model for several decades.

The lack of equilibrium of the pavement system was subsequently brought up by Burmister in 1945. He did not emphasize the early work by Boussinesq in 1885 in solving the equilibrium equations by polynomials but plunged into the development of more refined solutions. But until the publication of the extensive listings of tabulated coefficients by Jones in 1962, no significant appreciation was developed by practicing engineers of the applications of layered theories. Meanwhile, many researchers directed their effort toward the development of a more complicated multilayer system and, in so doing, made a dwarf out of the largest modern computer. Man becomes his own enemy. He never can face the reality.

The use of layered systems alone is not sufficient to solve pavement problems. It must be supplemented with the second and third mathematical models in analyzing the conditions in pavement components. For the convenience of this presentation, the subject of mathematical models is divided into two parts: (1) the equilibrium of the system and (2) the displacement and stress in the system components. The purpose of this review of mathematical models is to provide a sound background in simplifying the mathematical manipulation. A straightforward theory can be more valuable to the engineering profession than the sophisticated one which would be appreciated only by a few researchers.

In his report to the U. S. Army Construction Engineering Research Laboratory, Barenberg summarized the state of the art in the mathematical modeling of pavement systems. Many valuable reviews have been directly reproduced herein; these are marked by ** following each subheading. Two listings of selected references prepared by Barenberg are also included at the end of this chapter. In most cases, when the text of this presentation does not adequately describe the original work, the referenced literatures should be studied.

7.1. GENERAL EQUILIBRIUM EQUATIONS

There are two kinds of external forces which may act on bodies. Forces distributed over the surface of the body, such as the pressure of one body on another or hydrostatic pressure, are called *surface forces*. Forces distributed over the volume of a body, such as gravitational forces, magnetic forces, or (in the case of a body in motion) inertia forces, are called *body forces*. In studying the equilibrium of bodies under a static loading condition, the body forces are neglected and the surface forces are usually resolved into stress components parallel to the coordinate axes. In a cylindrical coordinate system, the stresses acting on six sides of a cubic element consist of three normal stresses σ_r, σ_θ, and σ_z and six shearing stress $\tau_{r\theta}$, $\tau_{\theta r}$, τ_{rz}, τ_{zr}, $\tau_{\theta z}$, and $\tau_{z\theta}$ (see Fig. 7.1). By a simple consideration of the equilibrium of the element, the components of shearing stress on two perpendicular sides of a cubic element are equal. Thus the total number of stress components is reduced to six quantities, such as σ_r, σ_θ, σ_z, $\tau_{r\theta} = \tau_{\theta r}$, $\tau_{rz} = \tau_{zr}$, and $\tau_{\theta z} = \tau_{z\theta}$.

In studying the equilibrium of an elastic body, it is assumed that the body will not move as a rigid body, so that no displacement of

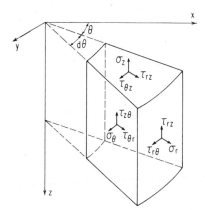

Fig. 7.1 General stress state of element in cylindrical coordinate system.

particles of the body is possible without a deformation of the body. The components of displacement of a cubic element can be denoted by u, v, and w in the radial, tangential, and z directions, respectively. The strain components of the element are

$$\epsilon_r = \frac{\partial u}{\partial r} \qquad \epsilon_\theta = \frac{u}{r} + \frac{\partial v}{r\,\partial\theta} \qquad \epsilon_z = \frac{\partial w}{\partial z} \tag{1a}$$

$$\gamma_{r\theta} = \frac{\partial u}{r\,\partial\theta} + \frac{\partial v}{\partial r} - \frac{v}{r} \qquad \gamma_{rz} = \frac{\partial u}{\partial z} + \frac{\partial w}{\partial r} \qquad \gamma_{z\theta} = \frac{\partial v}{\partial z} + \frac{\partial w}{r\,\partial\theta} \tag{1b}$$

The equilibrium of the element can be established if and only if the following differential equations of equilibrium are satisfied:

$$\frac{\partial \sigma_r}{\partial r} + \frac{1}{r}\frac{\partial \tau_{r\theta}}{\partial\theta} + \frac{\partial \tau_{rz}}{\partial z} + \frac{\sigma_r - \sigma_\theta}{r} = 0 \tag{2a}$$

$$\frac{\partial \tau_{rz}}{\partial r} + \frac{1}{r}\frac{\partial \tau_{\theta z}}{\partial\theta} + \frac{\partial \sigma_z}{\partial z} + \frac{\tau_{rz}}{r} = 0 \tag{2b}$$

$$\frac{\partial \tau_{r\theta}}{\partial r} + \frac{1}{r}\frac{\partial \sigma_\theta}{\partial\theta} + \frac{\partial \tau_{\theta z}}{\partial z} + \frac{2\tau_{r\theta}}{r} = 0 \tag{2c}$$

It is of advantage to introduce the stress functions ϕ and ψ in an attempt to solve the mathematical equations. The equations of equilibrium can be expressed in stress components:

$$\sigma_r = \frac{\partial}{\partial r}\left(\mu\nabla^2\phi - \frac{\partial^2\phi}{\partial r^2}\right) + \frac{2\partial}{r\partial\theta}\left(\frac{\partial\psi}{\partial r} - \frac{1}{r}\phi\right) \tag{3a}$$

$$\sigma_\theta = \frac{\partial}{\partial z}\left(\mu\nabla^2\phi - \frac{1}{r^2}\frac{\partial^2\phi}{\partial\theta^2} - \frac{1}{r}\frac{\partial\phi}{\partial r}\right) - \frac{2}{r}\frac{\partial}{\partial\theta}\left(\frac{\partial\psi}{\partial r} - \frac{1}{r}\psi\right) \tag{3b}$$

$$\sigma_z = \frac{\partial}{\partial z}\left[(2 - \mu)\nabla^2\phi - \frac{\partial^2\phi}{\partial z^2}\right] \tag{3c}$$

$$\tau_{r\theta} = \frac{1}{r}\frac{\partial^2}{\partial\theta\,\partial z}\left(\frac{\phi}{r} - \frac{\partial\phi}{\partial r}\right) - 2\left(\frac{\partial^2\psi}{\partial r^2} - \frac{\partial^2\psi}{\partial z^2}\right) \qquad (3d)$$

$$\tau_{\theta z} = \frac{1}{r}\frac{\partial}{\partial\theta}\left[(1 - \mu)\nabla^2\phi - \frac{\partial^2\phi}{\partial z^2}\right] - \frac{\partial^2\psi}{\partial r^2} \qquad (3e)$$

$$\tau_{rz} = \frac{\partial}{\partial r}\left[(1 - \mu)\nabla^2\phi - \frac{\partial^2\phi}{\partial z^2}\right] + \frac{1}{r}\frac{\partial^2\psi}{\partial\theta\,\partial z} \qquad (3f)$$

where μ is Poisson's ratio and the symbol ∇^2 denotes the operation

$$\nabla^2 = \frac{\partial^2}{\partial r^2} + \frac{1}{r}\frac{\partial}{\partial r} + \frac{1}{r^2}\frac{\partial^2}{\partial\theta^2} + \frac{\partial^2}{\partial z^2} \qquad (4)$$

We identify the displacements as u, v, and w in the r, ϕ, and ψ as follows:

$$u = \frac{1 + \mu}{E}\left(-\frac{\partial^2\phi}{\partial r^2} + \frac{2}{r}\frac{\partial\psi}{\partial\theta}\right) \qquad (5a)$$

$$v = \frac{1 + \mu}{E}\left(-\frac{1}{r}\frac{\partial^2\phi}{\partial\theta\,\partial z} - 2\frac{\partial\psi}{\partial r}\right) \qquad (5b)$$

$$w = \frac{1 + \mu}{E}\left[2(1 - \mu)\nabla^2\phi - \frac{\partial^2\phi}{\partial z^2}\right] \qquad (5c)$$

where E is the modulus of elasticity.

A unique solution to these equations requires that the stress functions ϕ and ψ satisfy the equilibrium equations and the compatibility equations. It can be shown that this is accomplished if

$$\nabla^4\phi = 0 \quad \text{and} \quad \nabla^4\psi = 0 \qquad (6)$$

Stress analyses which are of practical importance to pavement systems are concerned with a solid of revolution deformed symmetrically with respect to the axis of revolution. The deformation being symmetrical with respect to the z axis, it follows that the stress components are independent of the angle θ, and all derivatives with respect to θ vanish. The components of shearing stress $\tau_{r\theta}$ and $\tau_{\theta r}$ also vanish on account of the symmetry (see Fig. 7.2). Thus Eqs. (2) reduce to

$$\frac{\partial \sigma_r}{\partial r} + \frac{\partial \tau_{rz}}{\partial z} + \frac{\sigma_r - \sigma_\theta}{r} = 0 \tag{7a}$$

$$\frac{\partial \tau_{rz}}{\partial r} + \frac{\partial \sigma_z}{\partial z} + \frac{\tau_{rz}}{r} = 0 \tag{7b}$$

The strain components for axially symmetrical deformation are, from Eqs. (1),

$$\epsilon_r = \frac{\partial u}{\partial r} \quad \epsilon_\theta = \frac{u}{r} \quad \epsilon_z = \frac{\partial w}{\partial z} \quad \gamma_{rz} = \frac{\partial u}{\partial z} + \frac{\partial w}{\partial r} \tag{8}$$

In the general three-dimensional case in rectangular coordinates, each of the six components of strain (three normal and three shear) may be expressed in terms of the three components of displacement.

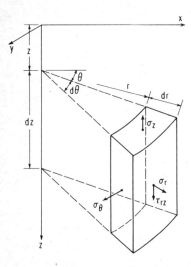

Fig. 7.2 Stress state of element in cylindrical coordinate system under a symmetrical load.

Thus these equations are not independent, and three relationships exist between the six components of strain. These are called *compatibility conditions,* and they can be expressed in terms of components of stress by using Hooke's law. In cylindrical polar coordinates with axial symmetry, the compatibility conditions become

$$\nabla^2\sigma_r - \frac{2}{r^2}(\sigma_r - \sigma_\theta) + \frac{1}{1 + \mu}\frac{\partial^2}{\partial r^2}(\sigma_r + \sigma_\theta + \sigma_z) = 0 \qquad (9a)$$

$$\nabla^2\sigma_\theta - \frac{2}{r^2}(\sigma_r - \sigma_\theta) + \frac{1}{1 + \mu}\frac{1}{r}\frac{\partial}{\partial r}(\sigma_r + \sigma_\theta + \sigma_z) = 0 \qquad (9b)$$

The components of stress and displacement may be expressed in terms of a stress function ϕ in such a way that Eqs. (1a) and (1b) are identically satisfied. These expressions are

Stress:

$$\sigma_z = \frac{\partial}{\partial z}\left[(2 - \mu)\nabla^2\phi - \frac{\partial^2\phi}{\partial z^2}\right] \qquad (10a)$$

$$\sigma_r = \frac{\partial}{\partial z}\left(\mu\nabla^2\phi - \frac{\partial^2\phi}{\partial r^2}\right) \qquad (10b)$$

$$\sigma_\theta = \frac{\partial}{\partial z}\left(\mu\nabla^2\phi - \frac{1}{r}\frac{\partial\phi}{\partial r}\right) \qquad (10c)$$

$$\tau_{rz} = \frac{\partial}{\partial r}\left[(1 - \mu)\nabla^2\phi - \frac{\partial^2\phi}{\partial z^2}\right] \qquad (10d)$$

Displacement:

$$\text{Vertical} \quad w = \frac{1 + \mu}{E}\left[(1 - 2\mu)\nabla^2\phi + \frac{\partial^2\phi}{\partial r^2} + \frac{1}{r}\frac{\partial\phi}{\partial r}\right] \qquad (10e)$$

$$\text{Horizontal} \quad u \;=\; -\frac{1 + \mu}{E}\frac{\partial^2\phi}{\partial r\,\partial z} \qquad (10f)$$

Again, a unique solution can be obtained if and only if the stress function satisfies the equilibrium equations and compatibility equations. This can be accomplished by providing that the stress function ϕ satisfies the equation

$$\left(\frac{\partial^2}{\partial r^2} + \frac{1}{r}\frac{\partial}{\partial r} + \frac{\partial^2}{\partial z^2}\right)\left(\frac{\partial^2\phi}{\partial r^2} + \frac{1}{r}\frac{\partial\phi}{\partial r} + \frac{\partial^2\phi}{\partial z^2}\right) = \nabla^2\nabla^2\phi = 0 \qquad (11)$$

The symbol ∇^2 denotes again the operation

$$\nabla^2 \;=\; \frac{\partial^2}{\partial r^2} + \frac{1}{r}\frac{\partial}{\partial r} + \frac{\partial^2}{\partial z^2} \qquad (12)$$

The problem is thus reduced to the solution of this partial differential equation subject to the boundary conditions at the surfaces, at the interfaces, and at infinite depth.

7.2. FORCE ON BOUNDARY OF A SEMI-INFINITE BODY

The solution of Eq. (11) can be obtained by assuming that the stress function ϕ is a series of polynomials. For the case of a concentrated load, the stress function can be in the form

$$\phi \;=\; B(r^2 + z^2)^{1/2} \qquad (13)$$

where B is a constant to be adjusted later to satisfy the boundary condition of the stress field. Substituting in Eqs. (10), the corresponding stress components are

$$\sigma_r \;=\; B\left[(1 - 2\mu)z(r^2 + z^2)^{-3/2} - 3r^2 z(r^2 + z^2)^{-5/2}\right] \qquad (14a)$$

$$\sigma_\theta \;=\; B(1 - 2\mu)z(r^2 + z^2)^{-3/2} \qquad (14b)$$

$$\sigma_z = -B\left[(1 - 2\mu)z(r^2 + z^2)^{-3/2} + 3z^3(r^2 + z^2)^{-5/2}\right] \quad (14c)$$

$$\tau_{rz} = -B\left[(1 - 2\mu)r(r^2 + z^2)^{-3/2} + 3rz^2(r^2 + z^2)^{-5/2}\right] \quad (14d)$$

All these stresses approach infinity when the stress point approaches the origin of coordinates, where the concentrated force is applied. To avoid the necessity of considering infinite stresses, the body is assumed to be a spherical container and the forces over the surface of the sphere are assumed to be as calculated from the above equations.

In polar coordinates, the stress distribution of an element within the sphere is (see Fig. 7.3)

$$\sigma_R = \frac{A}{R^3} \quad (15a)$$

$$\sigma_t = \frac{\partial \sigma_R}{\partial R}\frac{R}{2} + \sigma_R = -\frac{1}{2}\frac{A}{R^3} \quad (15b)$$

where A is a constant and $R = (r^2 + z^2)^{1/2}$. (See Timoshenko [97], pp. 362-365.) Converting this stress distribution in cylindrical coordinates and integrating with respect to the z axis from $z = 0$ to $z = \infty$, the stress components in an infinitely extended spherical container are

$$\sigma_r = \frac{A}{2}\left[\frac{1}{r^2} - \frac{z}{r^2}(r^2 + z^2)^{-1/2} - z(r^2 + z^2)^{-3/2}\right] \quad (16a)$$

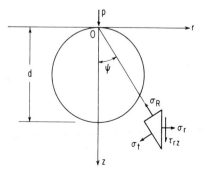

Fig. 7.3 Stress state in a container—Boussinesq solution.

$$\sigma_\theta = -\frac{1}{2} A \left[\frac{1}{r^2} - \frac{z}{r^2} (r^2 + z^2)^{-1/2} \right] \tag{16b}$$

$$\sigma_z = \frac{1}{2} Az(r^2 + z^2)^{-3/2} \tag{16c}$$

$$\tau_{rz} = \frac{1}{2} Ar(r^2 + z^2)^{-3/2} \tag{16d}$$

The summation of Eqs. (1) and (16) satisfies the boundary condition that at the surface of the semi-infinite body, for $z = 0$ stress components σ_θ and τ_{rz} will be equal to zero. Therefore, the following relation exists for the constants A and B:

$$A = 2B(1 - 2\mu) \tag{17}$$

Substituting in Eqs. (16) and adding together the stresses Eqs. (14), the total stress components are

$$\sigma_r = B \left\{ (1 - 2\mu) \left[\frac{1}{r^2} - \frac{z}{r^2} (r^2 + z^2)^{-1/2} \right] - 3r^2 z (r^2 + z^2)^{-5/2} \right\} \tag{18a}$$

$$\sigma_\theta = B(1 - 2\mu) \left[-\frac{1}{r^2} + \frac{z}{r^2} (r^2 + z^2)^{-1/2} + z(r^2 + z^2)^{-3/2} \right] \tag{18b}$$

$$\sigma_z = -3Bz^3(r^2 + z^2)^{-5/2} \tag{18c}$$

$$\tau_{rz} = -3Brz^2(r^2 + z^2)^{-5/2} \tag{18d}$$

This stress distribution satisfies the boundary conditions since $\sigma_z = \tau_{rz} = 0$ for $z = 0$. It remains now to determine the constant B so that the forces distributed over a hemispherical surface with center at the origin are statically equivalent to the force P acting along the z axis. The component in the z direction of forces on the hemispherical surface is

$$\bar{Z} = -(\tau_{rz} \sin\psi + \sigma_z \cos\psi) = 3Bz^2(r^2 + z^2)^{-2} \qquad (19)$$

The summation of $\bar{Z}$ is equal to the external force P:

$$P = 2\pi \int_0^{\pi/2} \bar{Z}r(r^2 + z^2)^{1/2}\, d\psi = 6\pi B \int_0^{\pi/2} \cos^2\psi \, \sin\psi \, d\psi \qquad (20)$$

from which

$$B = \frac{P}{2\pi} \qquad (21)$$

Finally, substituting in Eqs. (18), the following expressions are obtained for the stress components due to an external force P acting on the boundary of a semi-infinite solid:

$$\sigma_r = \frac{P}{2\pi}\left\{(1 - 2\mu)\left[\frac{1}{r^2} - \frac{z}{r^2}(r^2 + z^2)^{-1/2}\right] - 3r^2z(r^2 + z^2)^{-5/2}\right\} \qquad (22a)$$

$$\sigma_\theta = \frac{P}{2\pi}(1 - 2\mu)\left[-\frac{1}{r^2} + \frac{z}{r^2}(r^2 + z^2)^{-1/2} + z(r^2 + z^2)^{-3/2}\right] \qquad (22b)$$

$$\sigma_z = -\frac{3P}{2\pi}z^3(r^2 + z^2)^{-5/2} \qquad (22c)$$

$$\tau_{rz} = -\frac{3P}{2\pi}rz^2(r^2 + z^2)^{-5/2} \qquad (22d)$$

Consider now the displacement produced in the semi-infinite solid by the load P. From Eqs. (1) or (8) for strain components,

$$u = \epsilon_\theta r = \frac{r}{E}[\sigma_\theta - \mu(\sigma_r + \sigma_z)]$$

Substituting the values for the stress components from Eqs. (22),

$$u = \frac{(1 - 2\mu)(1 + \mu)P}{2\pi Er} \left[z(r^2 + z^2)^{-1/2} \right.$$

$$\left. - 1 + \frac{1}{1 - 2\mu} r^2 z(r^2 + z^2)^{-3/2} \right] \quad (22e)$$

For determining vertical displacements w, we have, from Eqs. (1) or (8),

$$\frac{\partial w}{\partial z} = \epsilon_z = \frac{1}{E} [\sigma_z - \mu(\sigma_r + \sigma_\theta)]$$

or

$$\frac{\partial w}{\partial r} + \gamma_{rz} - \frac{\partial u}{\partial z} = \frac{2(1 + \mu)}{E} \tau_{rz} - \frac{\partial u}{\partial z}$$

Substituting for the stress components and for the displacement u the values found above, we obtain

$$\frac{\partial w}{\partial z} = \frac{P}{2\pi E} \left\{ 3(1 + \mu)r^2 z(r^2 + z^2)^{-5/2} \right.$$

$$\left. - [3 + \mu(1 - 2\mu)] z(r^2 + z^2)^{-3/2} \right\}$$

$$\frac{\partial w}{\partial r} = - \frac{P(1 + \mu)}{2\pi E} [2(1 - \mu)r(r^2 + z^2)^{-3/2} + 3rz^2(r^2 + z^2)^{-5/2}]$$

from which, by integration,

$$w = \frac{P}{2\pi E} [(1 + \mu)z^2(r^2 + z^2)^{-3/2} + 2(1 - \mu^2)(r^2 + z^2)^{-1/2}] \quad (22f)$$

Since the solution was first given by J. Boussinesq in 1885, Eqs. (22) are known as Boussinesq theories.

7.3. EXTENSION OF BOUSSINESQ EQUATIONS

Having the solution for a concentrated force as given in Eqs. (22), the displacements and stresses produced by a distributed load can be determined by the use of superposition. Taking a small element of the loaded area, shown shaded in Fig. 7.4, bounded by two radii including the angle $\partial\theta$ and two arcs of circle with the radii r and $r + \partial r$, the load on this element is $pr\,\partial r\,\partial\theta$. Substituting this expression in Eq. (22f) for the P force and double-integrating with respect to $d\theta$ and dr, the vertical strain at the point (r, θ, z) is

$$\frac{\partial w}{\partial z} = \epsilon_z = \frac{1}{E}[\sigma_z - \mu(\sigma_r + \sigma_\theta)] \tag{23}$$

The displacement at the point (r, θ, z) is

$$W(r, \theta, z) = \frac{P}{2\pi E}\iiint\left[3(1 + \mu)\,\frac{r^3 z}{(r^2 + z^2)^{5/2}} - (3 + \mu - 2\mu^2)\,\frac{rz}{(r^2 + z^2)^{3/2}}\right] dr\,d\theta\,dz \tag{24}$$

The classic solution of the above equations is the construction of an influence chart in the form to satisfy

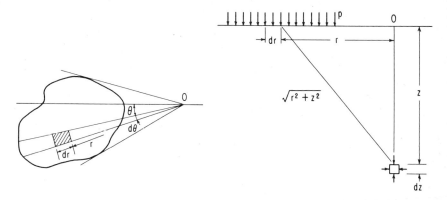

Fig. 7.4 Load distribution over a part of boundary of a semi-infinite solid.

$$\epsilon_z\left(\frac{r}{z},\theta\right) = \frac{P}{2\pi E} \int\int \left\{ 3(1 + \mu) \frac{(r/z)^3}{[1 + (r/z)^2]^{5/2}} \right.$$
$$\left. - (3 + \mu - 2\mu^2) \frac{r/z}{[1 + (r/z)^2]^{3/2}} \right\} d\left(\frac{r}{z}\right) d\theta \tag{25}$$

where r/z is a dimensionless parameter and the influence area is

$$\Delta A = \frac{1}{2\pi} \left\{ 3(1 + \mu) \frac{(r/z)^2}{[1 + (r/z)^2]^{5/2}} \right.$$
$$\left. - (3 + \mu - 2\mu^2) \frac{1}{[1 + (r/z)^2]^{3/2}} \right\} \Delta \frac{r}{z} \left(\frac{r}{z} \Delta\theta\right) \tag{26}$$

The strain at point (r, θ, z) becomes

$$\epsilon_z\left(\frac{r}{z}, \theta\right) = \frac{p}{E} \Sigma\Delta A \tag{27}$$

The displacement in a semi-infinite mass at point (r, θ, z) is

$$W(r, \theta, z) = \frac{p}{E} \int \Sigma\Delta A(z) \, dz \tag{28}$$

7.4. DISPLACEMENT ALONG AXIS OF A CIRCULAR LOAD

The solution of Eqs. (23) to (28) involves tedious and time-consuming operations. However, there are two simple conditions which are very important to the pavement analysis. The first case involves the stress and strain components along the load axis in a semi-infinite mass, subject to a circular load at the surface, as shown in Fig. 7.5. The integration of Eqs. (22) can be

$$\sigma_z = \int_0^a \frac{3}{2\pi} \frac{z^3}{(r^2 + z^2)^{5/2}} p 2\pi r \, dr = p(1 - \sin^3 \alpha) \tag{29a}$$

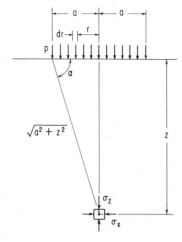

Fig. 7.5 Stresses within a semi-infinite solid.

$$\sigma_x = \int_0^a \tfrac{1}{2}(\sigma_r + \sigma_\theta)p2\pi r \; dr$$

$$= \frac{p}{2}\left[1 + 2\mu - 2(1 + \mu)\sin\alpha + \sin^3\alpha\right] \tag{29b}$$

The elastic, vertical displacement of the element dz at depth z, which is subject to the triaxial stress condition, σ_z and σ_r, has the value

$$dw = \frac{1}{E}(\sigma_z - 2\mu\sigma_x)\,dz \tag{30}$$

By substituting Eqs. (29) into Eq. (30) and integrating between $z = z$ and $z = \infty$, the total vertical displacement becomes

$$W_z = \int_z^\infty \frac{1}{E}(\sigma_z - 2\mu\sigma_x)\,dz \tag{31}$$

The most conventional approach in solving the equation is to assume that the elastic modulus E is a constant in the entire mass. The vertical displacement at depth z is

$$W_z = \frac{p}{E}(a^2 + z^2)^{1/2}[2(1 - \mu^2) - (1 - \mu - 2\mu^2)\sin\alpha$$
$$- (1 + \mu)\sin^2\alpha] \tag{32}$$

and at the surface of elastic mass, where $z = 0$

$$W_0 = \frac{2pa}{E}(1 - \mu^2) \tag{33}$$

With Eq. (33), it is possible to calculate the displacement at the surface of a semi-infinite solid at the center of the load.

If, for the pavement layers above the subgrade, it is assumed that (1) no deformation has taken place in the pavement layers and (2) the rigidity of the layers has no effect on the stress distribution in the subgrade, the displacement at the interface of pavement layer and subgrade, as expressed by the W_z value in Eq. (32), would be a useful mathematical model in evaluating the deflection of a pavement structure (see Fig. 7.6). Although assumptions seem to be hypothetical and unrealistic, many reports on pavement evaluation have substantiated that a good correlation exists between the actual measurements and the computed deflections by the Boussinesq equation (32). The advantage of the Boussinesq theory lies in the simplicity of the mathematical models which can be handled by practicing engineers in their researching for preliminary decisions. On the other hand, if the boundary conditions of the pavement structure satisfy the assumptions involved in the derivation of the Boussinesq theory, the computed result is theoretically sound and physically reliable.

In his analysis of the AASHO flexible pavement tests, Vesic reported that the modulus of deformation of the subgrade would have to vary approximately as the one-third power of the vertical stress; that is

$$E \approx \sigma_z^{1/3} \tag{34}$$

By using the relationship as given by Eq. (29a), the elastic modulus would vary proportionally with the value

$$\sigma_z^{1/3} = \left[1 - \frac{z^3}{3(a^2 + z^2)^{3/2}}\right]p^{1/3} \tag{35}$$

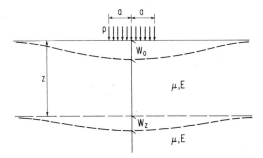

(a) Displacement in an isotropic elastic mass

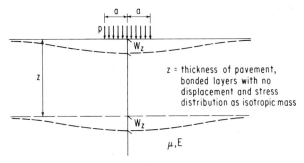

(b) Displacement in a hypothetical elastic mass

Fig. 7.6   Displacement in elastic mass.

For normal soil conditions, the overburden pressure would be encountered over a wide area, $a \gg z$. The z value becomes insignificant in the term $(a^2 + z^2)^{1/2}$, and the vertical stress is equal to the overburden pressure. In the evaluation of subgrade reaction for the pile foundation, the modulus of deformation has been found to increase with the depth z in the form

$$E = n(z + d) \tag{36}$$

where n is known as the coefficient of subgrade reaction and nd is the initial subgrade reaction at the depth $z = 0$. By substituting Eq. (36) into Eq. (31), the total vertical deformation along the load axis becomes

$$W_z = \frac{p}{n}\left\{(1 - \mu - 2\mu^2) \ln \frac{z + (a^2 + z^2)^{1/2}}{2(z + d)}\right.$$

$$+ (1 + \mu) \frac{a^2}{a^2 + d^2} \left[1 - \frac{z - d}{(a^2 + z^2)^{1/2}}\right] + (1 + \mu) \frac{d}{(a^2 + d^2)^{1/2}}$$

$$\tag{37}$$

$$\times \left(2\mu - \frac{d^2}{a^2 + d^2}\right)$$

$$\times \ln \frac{a^2(z + d)}{[d + (a^2 + d^2)^{1/2}][(a^2 + z^2)^{1/2}(a^2 + d^2)^{1/2} + a^2 - zd]}\right\}$$

For normal aircraft loading, the radius of contact area, a is much smaller than the d value and $a^2 + d^2$ can be assumed to be equal to d^2. The above equation can be rewritten as

$$W_z = \frac{p}{n}\left\{(1 + \mu) \frac{a^2}{d^2} \left[1 - \frac{z - d}{(a^2 + z^2)^{1/2}}\right] + (1 - \mu - 2\mu^2)\right.$$

$$\tag{38}$$

$$\left. \times \ln \frac{d[d + z + (a^2 + z^2)^{1/2}]}{(z + d)^2}\right\}$$

At the surface of the elastic mass, the deflection is

$$W_0 = \frac{pa}{nd}\left[(1 + \mu) \frac{a}{d} + 2(1 - \mu^2)\right] \tag{39}$$

The magnitude of deflection, as given by the above equation, will be slightly different from that in Eq. (33).

In Fig. 7.7, the displacement of elastic mass by Eq. (32) is plotted for various depths below the surface. It can be seen that the thickness of pavement structure has a significant effect on the distribution of effective stress in the subgrade, which in turn will affect the overall displacement of pavement caused by the deformation of the subgrade.

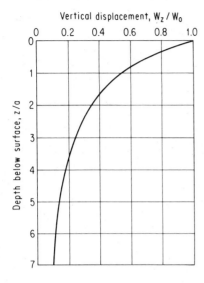

Fig. 7.7 Vertical displacement along axis of load on a semi-infinite elastic mass.

7.5. DISPLACEMENT ON BOUNDARY SURFACE

The displacement due to a circular load at the boundary surface can be obtained by integrating

$$W(r,0) = \frac{p}{\pi E}(1 - \mu^2) \iint dr\, d\theta \tag{40}$$

For a circular load, the chord length intersected by $d\theta$ is equal to $2\sqrt{a^2 - r^2 \sin^2\theta}$ (see Fig. 7.4). The above equation can be rewritten as

$$W(r,0) = \frac{2p(1 - \mu^2)}{\pi E} \int_{-\theta_1}^{\theta_1} \sqrt{a^2 - r^2 \sin^2\theta}\; d\theta \tag{41}$$

The relation between a, r, and θ must satisfy the following:

$$a \sin\psi = r \sin\theta \tag{42}$$

and Eq. (41) becomes

$$\bar{W}(r,0) = \frac{4(1-\mu^2)pr}{\pi E} \int_0^{\pi/2} \left(1 - \frac{a^2}{r^2}\sin^2\psi\right)^{1/2} d\psi$$

$$- \left(1 - \frac{a^2}{r^2}\right)\int_0^{\pi/2} \left(1 - \frac{a^2}{r^2}\sin^2\psi\right)^{-1/2} d\psi \qquad (43)$$

The integrals in this equation are known as *elliptic* integrals, and they can be expressed by means of power series:

$$W(r,0) = \frac{2(1-\mu^2)pa}{E} \frac{m}{2}\left(1 + \frac{1}{8}m^2 + \frac{3}{64}m^4 + \cdots\right) \qquad (44)$$

where $m = a/r$ and $m < 1$. At the boundary of the loaded circle, the displacement can be obtained by substituting $a/r = 1$ in Eq. (43):

$$W(a,0) = \frac{4(1-\mu^2)pa}{\pi E} \qquad (45)$$

The power series in Eq. (44)

$$\frac{m}{2}\left(1 + \frac{1}{8}m^2 + \frac{3}{64}m^4 + \cdots\right) < \frac{2}{\pi} \qquad (46)$$

will converge for all values of $m < 1$.

For points within the loaded area, the total displacement is again governed by Eq. (40) but the chord length is $2a\cos\theta$. The displacement can be computed by

$$W(r,0) = \frac{4(1-\mu^2)pa}{E} \int_0^{\pi/2} \left(1 - \frac{r^2}{a^2}\sin^2\psi\right)^{1/2} d\psi$$

$$= \frac{2(1-\mu^2)pa}{E}\left[1 - \frac{1}{4}\left(\frac{1}{m}\right)^2 - \frac{3}{64}\left(\frac{1}{m}\right)^4 - \cdots\right] \qquad (47)$$

where $1/m = r/a$ and $r/a < 1$. For $a = r$, at the boundary of the loaded circle, the power series converge:

$$\left[1 - \frac{1}{4}\left(\frac{1}{m}\right)^2 - \frac{3}{64}\left(\frac{1}{m}\right)^4 - \cdots\right]_{\frac{1}{m} \to 1} = \frac{2}{\pi} \qquad (48)$$

The maximum displacement is encountered at the center of the loaded circle:

$$W_0 = \frac{2(1 - \mu^2)pa}{E} \qquad (49)$$

Comparing W_0 with the deflection at the boundary of the circle, the maximum deflection is $\pi/2$ times the deflection at the boundary. It is also important to note that for a given intensity of the load, p, the displacement is not constant but increases in the same ratio as the radius of the loaded circle.

7.6. LAYERED SYSTEMS

Since pavements normally consist of several layers of material, it is natural to consider the theory of layered systems. The Boussinesq equations are theoretically sound for one-layer systems—a semi-infinite mass having a distributed load on the boundary surface. Although actual measurements have demonstrated that the deflection of a pavement system is in good agreement with the deflection W_z computed by the Boussinesq equation (32), the hypothesis remains theoretically unthinkable (see Fig. 7.6). In recent years, considerable effort has been expended on the analysis of stresses and displacements in multiple-layered systems such as the system shown in Fig. 7.8. Most of the analyses include certain basic assumptions, which can be summarized as follows: (1) each layer is composed of materials which are isotropic, homogeneous, and weightless; (2) the system acts as a composite system, that is, there is a continuity of stresses and/or displacements across the interfaces, depending upon the assumptions made regarding the interface conditions; and (3) most solutions assume materials which are linearly elastic.

The first solution for a generalized multiple-layered elastic system was presented by Burmister [13,14,15]. In this series of papers, Burmister formulated the problem for N-layered elastic systems and developed solutions for specific two- and three-layered systems.

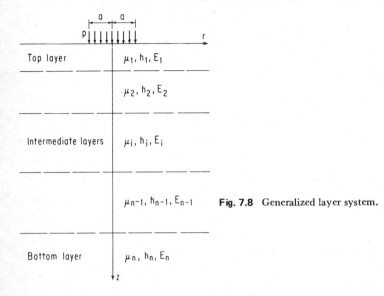

Fig. 7.8 Generalized layer system.

Burmister's work was limited to uniform, normal loads applied over a circular area. Schiffman [48] later extended Burmister's work for more generalized asymmetric loading conditions, including shear stresses at the surface.

However, the amount of tedious computation and the complexity of the mathematical models have prevented many practicing engineers from using this powerful method for solving their pavement problems. In the academic community, many investigators have advanced the solutions to specific two- and three-layered system with normal, symmetrical loads. Tables of coefficients have been presented to permit the evaluation of these systems. Perhaps the most extensive tabulation of coefficients is that given by Jones [28]. The coefficients developed by Jones were presented in graphical form on three-dimensional plots by Peattie [45]. These plots permit visual interpolation for analysis of systems not specifically solved.

In an axially symmetric cylindrical coordinate system, such as that shown in Fig. 7.2, the equations of equilibrium can be expressed in term of a stress function ϕ, as shown in Eqs. (10). A unique solution can be obtained if and only if the stress function ϕ satisfies the equilibrium equations (10) and compatibility equations (9). This can be accomplished by providing that ϕ is a solution of the biharmonic equation

$$\nabla^2\nabla^2\phi = 0 \tag{50}$$

where ∇^2 is again the Laplacian operator.

The problem is thus reduced to the solution of this partial differential equation subject to the boundary conditions at the surface, at the interfaces, and at infinite depth. The boundary conditions are expressed next; the numerical subscripts refer to the different layers.

For a multiple-layered system, each of Eqs. (10) must be solved for each layer of the system. Considering now only the symmetric case, the load normal to the 0th interface is given by the expression $p(m)J_0(mr)$, where $p(m)$ is an arbitrary function of the parameter m and $J_0(mr)$ denotes the Bessel function of the first kind of order zero. A stress function of the form

$$\phi_i = J_0(mr)(A_i + B_iz)e^{mz} + (C_i + D_iz)e^{-mz} \tag{51}$$

has been found to satisfy the necessary conditions, where the subscript i refers to the ith layer of the system and A_i, B_i, C_i, and D_i are constants of integration which must be determined from the boundary and interface conditions.

For each layer, the stress and displacement components have to be satisfied at the ith and $(i - 1)$st interface. The interface is assumed to be "rough." The components of stress and displacement are continuous at the interface; that is, $\sigma^i = \sigma^{i+1}$. At the free surface, the normal stress is equal to the load and the shear stress has to be zero. At the lower boundary, when z approaches infinity at the nth interface, all stress and displacement components are equal to zero.

The solution (see Appendix 1) of stress and displacement components at the interface are obtained by substituting Eq. (51) in Eqs. (10).

$$\sigma_z^i = m^2J_0(mr)\left\{(1 - 2\mu_i)(B_ie^{mz} + D_ie^{-mz}) - m[(A_i + B_iz)e^{mz} - (C_i + D_iz)e^{-mz}]\right\} \tag{52a}$$

$$\tau_{rz}^i = m^2J_1(mr)\left\{2\mu_i(B_ie^{mz} - D_ie^{-mz}) + m[(A_i + B_iz)e^{mz} + (C_i + D_iz)e^{-mz}]\right\} \tag{52b}$$

$$u^i = \frac{1 + \mu_i}{E_i} m J_1(mr) \Big\{ B_i e^{mz} + D_i e^{-mz} + m[(A_i + B_i z)e^{mz}$$
$$- (C_i + D_i z)e^{-mz}] \Big\}$$

(52c)

$$w^i = \frac{1 + \mu_i}{E_i} m J_0(mr) \Big\{ 2(1 - 2\mu_i)(B_i e^{mz} - D_i e^{-mz})$$
$$- m[(A_i + B_i z)e^{mz} + (C_i + D_i z)e^{-mz}] \Big\}$$

(52d)

$$\sigma_r^i = m^2 J_0(mr) \Big\{ (1 + 2\mu_i)(B_i e^{mz} + D_i e^{-mz}) + m[(A_i + B_i z)e^{mz}$$
$$- (C_i + D_i z)e^{-mz}] \Big\} - m^2 \frac{J_1(mr)}{mr} \Big\{ B_i e^{mz} + D_i e^{-mz}$$
$$+ m[(A_i + B_i z)e^{mz} - (C_i + D_i z)e^{-mz}] \Big\}$$

(52e)

$$\sigma_\theta^i = 2\mu_i m^2 J_0(mr)(B_i e^{mz} + D_i e^{-mz}) + m^2 \frac{J_1(mr)}{mr} \Big\{ B_i e^{mz}$$
$$+ D_i e^{-mz} + m[(A_i + B_i z)e^{mz} - (C_i + D_i z)e^{-mz}] \Big\}$$

(52f)

The matrix form of the above equations is

$$
\begin{bmatrix} \sigma_z^i \\ \tau_{rz}^i \\ u^i \\ w^i \end{bmatrix} = K(\mu_i, E_i) M(z, \mu_i) D(z) \begin{bmatrix} A_i \\ B_i \\ C_i \\ D_i \end{bmatrix}
$$

(52g)

where

$$
K(\mu, E) = \begin{bmatrix} -m^2 J_0(mr) & \cdots & \cdots & \cdots \\ \cdots & m^2 J_1(mr) & \cdots & \cdots \\ \cdots & \cdots & \dfrac{1+\mu}{E} m J_1(mr) & \cdots \\ \cdots & \cdots & \cdots & -\dfrac{1+\mu}{E} m J_0(mr) \end{bmatrix}
$$

$$M(z,\mu) = \begin{bmatrix} 1 & mz + 2\mu - 1 & -1 & -mz + 2\mu - 1 \\ 1 & mz + 2\mu & 1 & mz - 2\mu \\ 1 & mz + 1 & -1 & -mz + 1 \\ 1 & mz + 4\mu - 2 & 1 & mz - 4\mu + 2 \end{bmatrix}$$

and

$$D(z) = \begin{bmatrix} me^{mz} & \cdots & \cdots & \cdots \\ \cdots & e^{mz} & \cdots & \cdots \\ \cdots & \cdots & me^{-mz} & \cdots \\ \cdots & \cdots & \cdots & e^{-mz} \end{bmatrix}$$

Continuity conditions at the ith interface can be expressed by

$$K(\mu_i, E_i)M(h_i, \mu_i)D(h_i)\begin{bmatrix} A_i \\ B_i \\ C_i \\ D_i \end{bmatrix} = K(\mu_{i+1}, E_{i+1})M(h_i, \mu_{i+1})D(h_i)\begin{bmatrix} A_{i+1} \\ B_{i+1} \\ C_{i+1} \\ D_{i+1} \end{bmatrix} \quad (53a)$$

or reduced to

$$\begin{bmatrix} A_i \\ B_i \\ C_i \\ D_i \end{bmatrix} = \frac{D^{-1}(h_i)X_iD(h_i)}{4(\mu_i - 1)}\begin{bmatrix} A_{i+1} \\ B_{i+1} \\ C_{i+1} \\ D_{i+1} \end{bmatrix} \quad (53b)$$

where

$$X_i = 4M^{-1}(h_i, \mu_i)K^{-1}(\mu_i, E_i)K(\mu_{i+1}, E_{i+1})M(h_i, \mu_{i+1})(\mu_i - 1)$$

We have $4n$ constants to determine. Equations (53) give us $4(n - 1)$ equations. At the free surface, where $z = 0$, we have two more equations:

$$mA_1 + 2\mu_1B_1 - mC_1 - 2\mu_1D_1 = 0 \quad (54a)$$

and

$$mA_1 + (2\mu_1 - 1)B_1 - mC_1 + (2\mu_1 - 1)D_1 = \frac{p(m)}{m^2} \qquad (54b)$$

At the nth interface, where $z = \infty$, the boundary conditions are

$$\begin{bmatrix} A_{n-1} \\ B_{n-1} \\ C_{n-1} \\ D_{n-1} \end{bmatrix} = \frac{D^{-1}(h_{n-1})X_{n-1}D(h_{n-1})}{4(\mu_{n-1} - 1)} \begin{bmatrix} 0 \\ 0 \\ C_n \\ D_n \end{bmatrix} \qquad (54c)$$

This gives us all the required equations to determine the constants A_i, B_i, C_i, and D_i for all interfaces $i = 1, 2, \ldots, n$.

Because Eqs. (10) are linear functions in ϕ, the stress function

$$\phi(r, z) = \int_0^\infty \phi(r, z, m) \, dm$$

will solve the problem of Eqs. (10) when the system is subject to a load

$$p(m)J_0(mr) \, dm \qquad (55a)$$

The inversion formula for the Hankel transform of order zero is

$$\frac{1}{2}[f(r - 0) + f(r + 0)] = \int_0^\infty m[rf(r)J_0(mr) \, dr]J_0(rm) \, dm \qquad (55b)$$

Therefore, the load parameter is

$$p(m) = m \int_0^\infty rf(r)J_0(mr) \, dr \qquad (55c)$$

For a particular form $r = a$, the function $f(r)$ is equal to $\frac{1}{2}$. The integration of the above equation is

$$p(m) = aJ_1(ma) \tag{55d}$$

The stress and displacement components at any depth in the ith layer can be obtained by the following equations:

$$\bar{\sigma}_z^i = a \int_0^\infty J_1(ma)\sigma_z^i \, dm \tag{56a}$$

$$\bar{\sigma}_r^i = a \int_0^\infty J_1(ma)\sigma_r^i \, dm \tag{56b}$$

$$\bar{\sigma}_\theta^i = a \int_0^\infty J_1(ma)\sigma_\theta^i \, dm \tag{56c}$$

$$\bar{\tau}_{rz}^i = a \int_0^\infty J_1(ma)\tau_{rz}^i \, dm \tag{56d}$$

$$\bar{u}^i = a \int_0^\infty J_1(ma)u^i \, dm \tag{56e}$$

$$\bar{w}^i = a \int_0^\infty J_1(ma)w^i \, dm \tag{56f}$$

σ_z^i, σ_r^i, σ_θ^i, τ_{rz}^i, u^i, and w^i are stress and displacement components at ith interface, as given by Eqs. (52a) to (52f).

7.7. SYSTEMS WITH VISCOELASTIC MATERIALS**

The discussion on equilibrium equations of elastic bodies was limited to systems with linearly elastic materials. Huang [24] and

Moavenzadeh and Ashton [3,37] have extended the theory for analysis of layered systems to the case of systems having materials with viscoelastic characteristics. Moavenzadeh and Elliott [42] have extended the solution of a layered system with viscoelastic materials to a case with moving loads.

Huang [24] in his analysis started with the basic equations developed for the elastic layered system and redefined the elastic constants E and μ in terms of the more fundamental properties of shear modulus G and bulk modulus K according to the following relationships:

$$E = \frac{9KG}{3K + G} \quad \text{and} \quad \mu = \frac{3K - 2G}{2(3K + G)} \tag{57}$$

By dividing the stress components into hydrostatic and deviatoric tensors, the bulk and shear moduli can be applied as follows:

$$s_{ij} = 2Ge_{ij} \quad \text{and} \quad \sigma_{ij} = 3KE_{ij} \tag{58}$$

where s_{ij} = deviatoric stress = $\sigma_{ij} - \delta_{ij}\sigma_{kk}/3$
e_{ij} = deviatoric strain = $\epsilon_{ij} - \delta_{ij}\epsilon_{kk}/3$
σ_{ij} = generalized stress tensor
ϵ_{ij} = generalized strain tensor
δ_{ij} = Kronecker delta, which takes on the values:

$$\delta_{ij} = \begin{cases} 1 & \text{when } i = j \\ 0 & \text{when } i \neq j \end{cases}$$

Equations (57) and (58) are valid for viscoelastic materials as well as elastic materials. For viscoelastic materials, however, G and K are not constants but are linear differential operators given by

$$G = \frac{Q}{R} \quad \text{and} \quad K = \frac{Q'}{R'} \tag{59}$$

where Q, R, Q', and R' are linear operators of the form

$$\sum_{k=0}^{n} a_k D^k \tag{60}$$

and D^k is the kth derivative of D with respect to time. The coefficients a and k are different for each operator and must be specified.

Using the notation from Eq. (59), Eq. (58) can be rewritten as

$$Rs_{ij} = 2Qe_{ij} \quad \text{and} \quad R'\sigma_{ij} = 3Q'\epsilon_{ij} \tag{61}$$

These operators can now be used to describe the stress-strain relationship for either elastic or viscoelastic materials and can be applied to the layered system previously described. To eliminate the variable of time, all stress and displacement components are transformed by means of the Laplace transform and all operations are performed on the transformed components. After all operations are complete, an inversion process is used to get the components back into real time.

To apply the Laplace transform, all load and response variables must be defined as an initial-value problem; that is, all loads and responses must be expressed in terms of $t = 0$. This is not a severe limitation, however, and can be handled without great difficulty for simple functions.

Moavenzadeh and Ashton [37] and Moavenzadeh and Elliott [42] also developed equations for a layered system with viscoelastic materials, by starting from the equations for the elastic layered system with symmetric loads.

Moavenzadeh and coworkers [37,42] express the properties of the viscoelastic materials in terms of the creep compliance of the material. Creep compliance $k(t)$ is the ratio of strain to stress for a viscoelastic material subjected to a creep strain test. The strain response $\epsilon(t)$ of a material with a general stress history $\sigma(\tau)$ can be expressed as

$$\epsilon(t) = \int_{-\infty}^{t} k(t - \tau) \frac{\partial \sigma(\tau)}{\partial \tau} \, d\tau \tag{62}$$

For an elastic material, the creep compliance is just the reciprocal of the modulus of elasticity; Moavenzadeh and coworkers contend that use of the compliance is superior to use of the viscoelastic moduli because there are more data available on creep than on relaxation behavior for viscoelastic materials and the nature of the integrals is less subject to error with the compliance than with the viscoelastic moduli.

In their solution, Moavenzadeh and coworkers took the time factor into account by evaluating the stress and displacement components by means of convolution integrals. Convolution integrals are well suited for evaluation by numerical methods with high-speed computers. Computer programs were developed by Moavenzadeh et al. to solve three-layer viscoelastic systems under moving loads.

7.8. MODIFICATIONS TO LAYERED SYSTEMS**

One weakness of all the above analyses for layered systems is the application of stress functions in the formulation and solution of the problem. Because of the stress-functions approach used, the systems must be infinite in extent in the horizontal direction. Thus solutions cannot be obtained for systems limited in the horizontal direction, and hence solutions cannot be obtained for edge or corner loading conditions.

Lattes, Lions, and Bonitzer [31] formulated the layered-system problem somewhat differently. They developed the equations of equilibrium and compatibility for the layered system in cartesian coordinates and solved these equations by applying the method of Galerkin. By deviatoric components it is possible to include the viscoelastic as well as the elastic properties of the materials in the individual layers.

In discussing their approach to the problem, Lattes, Lions, and Bonitzer list the following advantages to their approach: (1) the Fourier and Bessel functions used with the Hankel transform in the previous solutions are not ideally suited for numerical computation by computers; (2) because of the symmetry of the cylindrical coordinate system used by others, it is not possible to evaluate horizontally limited structures and the effect of loads operating near the edges of the pavement; and (3) their approach is able to handle both elastic and viscoelastic properties of materials in the various layers at will. It should be noted, however, that the problem of how to handle the large number of parameters inherent in this solution has not been resolved.

7.9. COMPUTER SOLUTIONS FOR LAYERED SYSTEMS

The effort required to solve the layered-system problem for any loading is nearly overwhelming for academic researchers and creates a practically unsurmountable difficulty for many professional engineers. As displacement is a reality to the solution of any structural system, the stress outputs from the layered-system analysis remain a serious challenge to practicing engineers in translating the theoretical or fictitious stress outputs into real pavement performance. Although the layered theory was first introduced by Burmister about a quarter of a century ago, the promotion of the theory has been primarily confined to the research community, and no significant progress has been observed in the engineering profession.

Since the publication of the extensive tabulation of coefficients by Jones [28] in 1962, more appreciation has been generated on the power of layered theory in solving pavement problems. Jones' work represented the first attempt to use modern high-capacity computers in solving the differential equations of the layered system. Subsequently, significant improvement was made on the computational technique for extending the computer program to solve N-layered elastic systems for both symmetric and asymmetric loading conditions. The load function is expressed in terms of an nth-order Hankel transform, which is defined as

$$F_n(f(r,z)) = \int_0^\infty rf(r,z)J_n(mr)\ dr \qquad (63)$$

After the stress and deflection components in the system are solved in terms of the Hankel transform, the components can be inverted into real space by means of the Hankel inversion. The Hankel inversion of transform F_n is given by the equation

$$f(r,z) = \int_0^\infty mFn(f(r,z))J_n(mr)\ dm \qquad (64)$$

As a computational technique, it may be convenient to first solve for the stress and displacement components in terms of an incremental load and then integrate over the appropriate range. For example, consider the case of a symmetric load with an intensity p that is uniformly distributed over a circular area of radius a. The Hankel inversion of the stress component of the zero-order transform of σ_z at the pavement surface is

$$\sigma_z^* = -pa \int_0^\infty J_0(mr)J_1(ma) \, dm \qquad (65)$$

To solve the problem, only the following increment is used:

$$\sigma_z^{*\prime} = -J_0(mr)J_1(ma) \qquad (66)$$

After solving for all the components, the components are multiplied by pa and integrated with respect to m over a range of from zero to infinity. This procedure greatly simplifies the computations and yields the same result as that obtained by using the full-load increment throughout the calculations.

Currently (1971), there are two major computer programs available: one developed by the California Research Corporation and the other by the Shell group in Amsterdam. The latter program has a much wider utilization in Europe than in the United States, where its circulation is confined to research organizations. There are several programs now in use at universities, research centers, and public work departments. General acceptance in the engineering profession is still in the developing stage.

Prior to the adoption of a computer program, it is necessary to review the basic structure of the program. Modifications are often required to reflect the actual condition of the pavement system. The listing of a layered-system program is given in Appendix 1.

7.10. FINITE-ELEMENT TECHNIQUES FOR LAYERED SYSTEMS**

As early as in 1949, Newmark [43] stated that "the use of the [finite-element] model offers certain advantages: there is no ambiguity concerning the boundary condition; statical checks on the results have physical meaning and can be made more adequately; variations in dimensions and physical properties can be more easily treated." With the improved use of computer techniques, many investigators have turned to finite-element methods for solid continua. Although the current technology in high-speed computers remains a major limitation in accurately solving the model equations, the tremendous potential of finite-element methods should not be lightly treated.

Generalized methods for applying finite-element techniques to the analysis of layered systems have been given by Wilson [56] and Barksdale [10]. A condensed version of Barksdale's presentation is given below.

A cylindrical coordinate system represented by r, θ, and z is used throughout this development. The positive displacements u, v, and w are in positive r, θ, and z directions, respectively. Tensile stresses and strains are assumed to be positive, and compressive stresses and strains to be negative. The continuum is characterized by an assemblage of discrete circular solid ring elements having rectangular cross sections in the zr plane, as shown in Fig. 7.9. The ring-shaped elements and any internal and external loads that may be applied to the continuum are placed so as to maintain axial symmetry about the z axis. For each element, the nodal points i, j, k, and l are numbered in the clockwise direction, with node i having the smallest r and z coordinates. The general notation used to represent the finite-element idealization in cylindrical coordinates for an

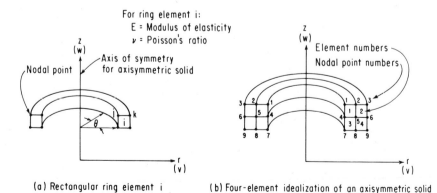

(a) Rectangular ring element i   (b) Four-element idealization of an axisymmetric solid

Fig. 7.9 Notation used in the finite-element idealization of an axisymmetric solid.

axisymmetric solid is shown in Fig. 7.9. The notation used to represent stresses and strains is illustrated in Fig. 7.1.

The displacement of nodal point i is defined as

$$\{\delta_i\} = \left\{ \begin{array}{c} v_i \\ w_i \end{array} \right\} \tag{67}$$

where v_i is the displacement of node i in the radial direction and w_i is the displacement in the z direction. With this notation, the node-displacement vector for each element may then be written as

$$\{\delta_i\} = \left\{ \begin{array}{c} \delta_i \\ \delta_j \\ \delta_k \\ \delta_l \end{array} \right\} \tag{68}$$

Similarly, the displacement vector for all n nodes in the system can be written as a displacement matrix:

$$\{\delta\} = \left\{ \begin{array}{c} \delta_1 \\ \delta_2 \\ \vdots \\ \delta_n \end{array} \right\} \tag{69}$$

In the same manner as above, the vector of external forces acting on node point i can be defined as

$$\{F_i\} = \left\{ \begin{array}{c} F_r \\ F_z \end{array} \right\} \tag{70}$$

where F_r is the force acting on node i in the r direction and F_z is the force in the z direction. With this notation, the load matrix for the system can be written as

$$\{F\} = \begin{Bmatrix} F_1 \\ F_2 \\ \vdots \\ F_n \end{Bmatrix} \tag{71}$$

From a free-body analysis of the forces acting on each node, the force-displacement relationship of the system can be written as

$$\{F\} = [K]\{\delta\} \tag{72}$$

where $\{\delta\}$ and $\{F\}$ are as defined in Eqs. (69) and (71) and $[K]$ is the stiffness matrix of the system. For isotropic materials, the relationship between stress and strain can be derived from classic elasticity theory and expressed in matrix notation as

$$\begin{Bmatrix} \sigma_z \\ \sigma_r \\ \sigma_\theta \\ \tau_{rz} \end{Bmatrix} = \frac{E(1-\mu)}{(1+\mu)(1-2\mu)} \begin{bmatrix} 1 & \dfrac{\mu}{1-\mu} & \dfrac{\mu}{1-\mu} & 0 \\ 1 & 1 & \dfrac{\mu}{1-\mu} & 0 \\ \cdots & \cdots & 1 & 0 \\ \cdots & \cdots & \cdots & \dfrac{1-2\mu}{2(1-\mu)} \end{bmatrix} \begin{Bmatrix} \epsilon_z \\ \epsilon_r \\ \epsilon_\theta \\ \gamma_{rz} \end{Bmatrix} \tag{73}$$

which can be simplified to the form

$$\{\sigma\} = [D]\{\epsilon\} \tag{74}$$

where $\{\sigma\}$ = stress vector
$\{\epsilon\}$ = strain vector
$[D]$ = elasticity matrix

The displacement components v, w in a solid continuum actually vary as complicated functions of position. However, to make the solution of elasticity problems possible by the finite-element method, simplifying assumptions are made on the manner in which displacement and stress components vary within each element of the model. Several different approaches have been proposed to accomplish this, including power series and Fourier series expansions

[10,55]. The effect of the assumptions with regard to these functions on the accuracy of the answer decreases as the size of the element decreases.

Strain at any point within an element can be determined from the displacement function by the relationship

$$
\left\{ \begin{array}{c} \epsilon_z \\ \epsilon_r \\ \epsilon_\theta \\ \gamma_{rz} \end{array} \right\} = \left\{ \begin{array}{c} \dfrac{\partial w}{\partial z} \\ \dfrac{\partial v}{\partial r} \\ \dfrac{v}{r} \\ \dfrac{\partial w}{\partial z} + \dfrac{\partial w}{\partial r} \end{array} \right\} \tag{75}
$$

which, after appropriate substitution for a displacement function and manipulation, can be written in a simplified form as

$$
\{\epsilon\} = [B]\{\delta_n\} \tag{76}
$$

where $\{\epsilon\}$ = strain vector
$\{\delta_n\}$ = node-displacement vector given in Eq. (68)
$[B]$ = coefficient matrix relating total strain to deflection

For axisymmetric systems, Wilson [56] defined the stiffness for each element in the system as

$$
[K] = \int_{vol} [B]^T [D][B] \, dv \tag{77}
$$

where $[B]$ is defined as above and $[D]$ is the elasticity matrix relating stress and strain through the elastic modulus and Poisson's ratio.

The integral in Eq. (77) must be integrated over the entire volume of the ring element. Formal integration, however, is very tedious because of the complexity of the equation. Thus, Barksdale [10] evaluates the integral numerically using a gaussian quadrature scheme.

The stiffness matrix for the entire system is determined by combining the stiffness of all elements in the solid. Barksdale [10] accomplishes this by setting up a stiffness matrix $[k_{ij}]$ and an element matrix and manipulating the two matrices to produce an equivalent stiffness matrix $[K]$ for the system.

In the formulation of the load vector, all concentrated loads applied at the node points may be placed in the load vector $\{F\}$. Distributed surface tractions, however, must be converted in some manner into equivalent concentrated loads and applied at the nodes. One method is to lump the distributed surface tractions into concentrated node loads, which perform the same amount of work as the nodal points undergo displacements.

Consider the uniformly distributed surface loading p applied normally to the r-θ surface of element n, as shown in Fig. 7.9. The equivalent concentrated force F_z^k in the z direction at node k is then equal in magnitude to the work done by the distributed loads when $w_k = 1$ and all other components of the displacements of node j and k are zero:

$$F_j^k = 2\pi p \int_{r_j}^{r_k} \frac{r - r_j}{r_k - r_j} r \, dr \tag{78}$$

which, upon integration, simplifies for the uniform normal loading to

$$F_z^k = \frac{2\pi p}{r_k - r_j} \left(\frac{1}{3} r_k^3 - \frac{1}{2} r_j r_k^2 + \frac{1}{6} r_j^3 \right) \tag{79}$$

Following the same approach, the expression for the concentrated nodal force at j can be shown to be

$$F_k^j = \frac{2\pi p}{r_k - r_j} \left(\frac{1}{6} r_k^3 - \frac{1}{2} r_k r_j^2 + \frac{1}{3} r_j^3 \right) \tag{80}$$

With the system stiffness matrix $[K]$ and the load-force vector $\{F\}$, the system can be solved for the node-displacement vector $\{\delta_n\}$. This idea was expressed earlier as Eq. (72) in the form

$$[K]\{\delta\} = \{F\} \tag{81}$$

Equation (81) is identical in form to the $[K]\{W\} = \{F\}$ developed by Hudson for the finite-element model for slabs. Equation (81) is somewhat more general, however, which accounts for the replacement of the deflection matrix $\{W\}$ with a more general deflection tensor $\{\delta\}$. Solution of equations of the form given in Eq. (81) essentially reduces to the solution of a large number of simultaneous linear equations. Many schemes are available in the literature for solutions of this type of equation, and hence these schemes will not be discussed here. It is noted, however, that selection of appropriate methods for the solution of these equations is one of the important steps when applying finite-element techniques.

The finite-element techniques developed by Barksdale [10] were based on the linear elasticity concepts. Barksdale also demonstrates, however, how the matrices can be handled on an incremental basis to solve problems in nonlinear elastic theory. The technique demonstrated by Barksdale consists essentially of breaking the stress-strain curves into a series of tangent sections and assigning a modulus for the material equal to the slope of the tangent segment for each particular stress level. Thus, with this technique it is necessary to solve the problem a number of times to develop the final answer.

The finite-element model developed by Barksdale [10] is a significant step forward in the modeling of pavement systems. The model has some advantages over the finite-element model for slabs in that it can handle nonlinear as well as linearly elastic materials and it can handle multiple-layered systems by assigning the appropriate stiffness vectors to the node points representing various layers.

The model presented also has some serious deficiencies as a generalized model for pavement systems. One of the more serious of these deficiencies is the requirement that the system be axisymmetric with respect to the physical system and the force system. This restriction limits the use of the model to static loads on systems with no realistic bounds on the horizontal plane; that is, the finite-element model developed by Barksdale cannot handle systems with edge and corner loads. Furthermore, it would appear that in order for the procedure to be adapted to a more realistic physical system, it would be necessary to start the development from the same starting point as Barksdale's.

B. Stress Analysis of Pavement Composition

With the development of equilibrium theories, stress and displacement can be evaluated for a pavement system. However, engineers still cannot design the composition of a pavement based simply on two parameters: modulus of elasticity E and Poisson's ratio μ. Westergaard and Hogg introduced the mathematical solution of bending moment and bending stress for a pavement layer supported by an idealized subgrade, that is, one in which the equilibrium condition is neglected. The influence charts subsequently introduced by Pickett and Ray have become a boon to modern pavement engineering. Many good concrete pavements have been meaningfully designed for strength and thickness.

In recent years, field observations and experiments have indicated that many pavement distresses can be related to overemphasis on material strength and inadequate equilibrium in the subgrade. The use of plate theory has been stretched beyond its limitation. Among all its shortcomings, the most serious one is the determination of the K value, the subgrade reaction. It is a fictitious parameter used for convenience in solving the differential equation of an elastic plate. The plate theory in its present form, which will be discussed in the following, is the most popular pavement equation in the engineering profession, but it can be the most misleading one in the engineering concept.

7.11. GENERAL PLATE THEORY**

Several theories have been developed for evaluating the behavior of pavement systems. Of these, the theories of continuously supported elastic slabs are the most popular. The first of the slab theories was developed by Hertz [65] in 1884, when he proposed a mathematical method for analyzing ice bridges by assuming the ice to be an elastic slab supported by a liquid. The theory was later expanded by Westergaard [100–106] for pavement analysis to include the so-called *Winkler foundation* [107]. All slab theories applied to pavement analysis are based upon the classic differential equations of equilibrium for elastic plates.

In classic plate theories, plates are divided into four main groups for analysis: membranes, medium-thick plates, plates with inplane forces, and thick plates. For most analyses, the medium-thick-plate theory is used since it is relatively simple to apply and has been shown to be valid provided the slab deflection is less than one-half the slab thickness. This review is limited primarily to the theory of medium-thick plates for pavement systems.

Development of the classic differential equation for medium-thick plates is presented in many standard references, including [97], and therefore will not be presented here. The differential equation of equilibrium is developed from free-body analyses of the typical slab elements shown in Figs. 7.10 and 7.11. Stress and forces acting on an element of the slab are shown in these figures. All forces shown in these figures are shown as acting in the positive direction.

In the development of the classic differential equation for medium-thick plates, the following assumptions are made:

1. Stresses acting normal to the upper and lower faces of the slab are small compared with other stresses in the plate and can be ignored. Thus, on an elemental slice taken parallel to the center plane of the slab, a two-dimensional state of stress exists.

2. The slab is of uniform stiffness and thickness.

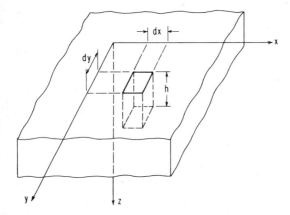

Fig. 7.10 Section of a slab showing coordinate system and differential element.

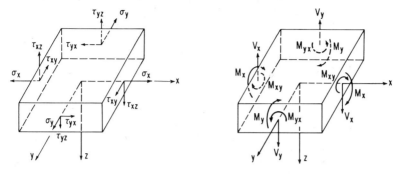

Fig. 7.11 Differential element removed from the slab with force acting.

3. The slab is supported in such a manner that axial or shear forces do not develop within the plane of the plate. (If the deflection is small compared with the thickness, this assumption is not necessary.)

4. The material in the slab is isotropic and homogeneous and has a linear stress-strain relationship.

5. Sections normal to the upper and lower faces of the slab before bending remain normal during bending. This assumption requires that the shear deformation in the x, z, and y, z directions be zero. Since shear stresses must exist in the slab, it is necessary to assume that the shear modulus of the material approaches infinity. It is obviously impossible for this physical condition to exist, but it can be shown that this assumption does not cause significant error in the results, provided the bending deformation of the slab is large when compared with the shear deformation. This same basic assumption is made in the development of the elementary equation for the bending stresses in a beam ($\sigma = MC/I$).

With these assumptions, the differential equation for an elastic slab in shorthand notation is

$$\nabla^2 \nabla^2 w = \frac{p}{D} \tag{82}$$

where ∇^2 is the Lagrange operator, given as

$$\nabla^2 = \frac{\partial^2}{\partial x^2} + \frac{\partial^2}{\partial y^2}$$

and

$$D \ = \ \frac{Eh^3}{12(1 \ - \ \mu^2)}$$

Equation (82) is the basic differential equation of equilibrium for medium-thick slabs regardless of support or loading conditions. If the plate is continuously supported and if the stress transmitted to the subgrade is assumed to act normal to the bottom surface of the plate (this requires that there be no shear stress at the interface of the slab and the support), then the reactive pressure q can be subtracted from the applied load p, so that Eq. (82) can be written as

$$D\nabla^2\nabla^2 w \ = \ p - q \tag{83}$$

With the few exceptions noted later, Eq. (83) is the basis for all theoretical treatments of pavements as elastic slabs. A number of investigators have studied pavement behavior using Eq. (82) as the starting point for their studies. In most cases, the only basic difference in such studies is the manner in which the subgrade reactive pressure q is related to pavement deflection. If it is assumed that the slab and the soil are contiguous throughout the extent of the slab, then the deformation of the subgrade surface must conform to that of the centerline of the slab. The deformation of the subgrade can be assumed to be the result of pressure applied to the subgrade by the slab.

Westergaard [100–106], following the lead of Hertz [65], assumed that deflection at any point on the subgrade surface is directly proportional to the vertical stress applied at that point. With this assumption, the slab support conditions are identical to the conditions assumed by Winkler [107]. This foundation can be represented by the individual spring system shown in Fig. 7.12a. Stiffness of the springs in the model is represented by a stiffness

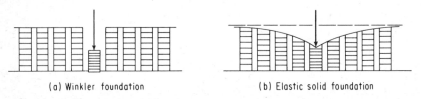

(a) Winkler foundation   (b) Elastic solid foundation

Fig. 7.12 Deflection patterns for assumed subgrade properties under a point load.

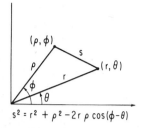

$$s^2 = r^2 + \rho^2 - 2r\rho\cos(\phi-\theta)$$

Fig. 7.13 Influence function $k(r,\theta,\rho,\phi)$ expressing the deflection at point r,θ due to a unit load at ρ,ϕ.

factor k, which is referred to as the *modulus of subgrade reaction*. Values for k can be determined from results of plate-load tests performed on the subgrade.

In 1938, Hogg and Holl [68–70] considered the problem of an elastic slab continuously supported by an elastic solid. The significant difference between an elastic solid subgrade and the Winkler foundation assumed by Hertz and Westergaard is shown in Fig. 7.12a. The most significant difference in the two support systems is that with the elastic solid, the deflection at any arbitrary point is not proportional to the stress transmitted to the subgrade at that point.

Equation (83) can be solved in general form for either the elastic solid or the Winkler support condition by using an influence function to represent the relationship between subgrade stress and pavement deflection. To utilize the influence function, however, requires that the equations for the elastic plate be converted to a system of polar coordinates (r, θ) and that only symmetric cases be considered.

The influence function $k(s) = k(r, \theta; \rho, \phi)$ shown in Fig. 7.13 illustrates the deflection at a point designated by (ρ, ϕ) due to a unit load applied to the subgrade surface at a point (r, θ). The letter s denotes the distance between the points designated (ρ, ϕ) and (r, θ). If the soil is assumed to be isotropic, the influence function is symmetrical with respect to the point of load, and hence is a function only of the distance s, which can be considered as known. For the two types of subgrade support considered in this discussion, the influence functions take the following form:

1. For the Winkler foundation,

$$k(s) = \begin{cases} 0 & \text{for } s \neq 0 \\ \lim_{\epsilon \to 0} \dfrac{1}{\pi\epsilon^2 k} & \text{for } s = 0 \end{cases} \tag{84}$$

where ϵ is an arbitrarily small increment and k is the modulus of subgrade reaction defined earlier. This expression follows directly from the definition for k given in Fig. 7.12.

2. For the elastic solid subgrade,

$$k(s) = \frac{1}{\pi C}\frac{1}{s} \tag{85}$$

where C is a function of the elastic properties of the subgrade such that $C = E/(1 - \mu^2)$. The influence function for the elastic solid is derived from Boussinesq's classic solution for the deflection of an elastic solid due to a point load. With the influence functions given in Eqs. (84) and (85), it is possible to express the subgrade reaction q at any point in terms of the slab deflection w. Transforming the influence function and the load function by means of the Hankel transform, the deflection of the pavement can be written in the form

$$w = \int_0^\infty \frac{mP(m)K(m)}{1 + Dm^4 K(m)} J_0(mr)\, dm \tag{86}$$

where m = an arbitrary dimensionless parameter ranging from zero to infinity
   $P(m)$ = loading function
   $K(m)$ = influence function for the subgrade reaction
   D = slab stiffness as defined for Eq. (82)
   J_0 = a Bessel function of the first kind of zero order
By substituting in the appropriate functions for the load and subgrade conditions, Eq. (86) can be solved for the deflection at any point in the pavement.

7.12. SPECIFIC SOLUTIONS OF GENERAL PLATE THEORY**

For a concentrated load, the transformed load function is

$$P(m) = \int_0^\infty \lim_{\epsilon \to 0} \frac{P}{\pi \epsilon^2} t J_0(mt)\, dt = \frac{P}{2\pi} \tag{87}$$

Also, for a Winkler foundation, the transformed influence function for subgrade stress is

$$K(m) = \lim_{\epsilon \to 0} \int_0^\epsilon \frac{2}{k} \frac{J_0(mt)}{\pi \epsilon^2} t \, dt = \frac{1}{k} \qquad (88)$$

Substituting from Eqs. (87) and (88) into Eq. (86) gives

$$w = \frac{P}{2\pi} \int_0^\infty \frac{m}{k + Dm^4} J_0(mr) \, dm \qquad (89)$$

which, by integration, produces

$$w = \frac{Pl^2}{2\pi D} kei(r) \qquad (90)$$

where l is the radius of relative stiffness for the slab given by the expression

$$l = \sqrt[4]{\frac{Eh^3}{12(1 - \mu^2)k}} \qquad (91)$$

kei is a tabulated function known as a *Kelvin function* and is the imaginary part of the Bessel function of the second kind of zero order, and the remaining terms are as defined earlier. It can be shown that the integral given in Eq. (90) is identical with the solution obtained by Hertz [65] when he proceeded directly from Eq. (83).

The maximum slab deflection w_0 will occur directly under the load. Noting that the value of $kei(r)$ for $r = 0$ is equal to $\pi/4$, the maximum deflection for a slab on a Winkler foundation under a concentrated load becomes

$$w_0 = \frac{Pl^2}{8D} \qquad (92)$$

This is the equation for the maximum slab deflection given by Westergaard [100], Timoshenko [97], and others. Moments in the slab due to the concentrated load P are obtained by applying basic relationships between moment and curvature:

$$M = \frac{D}{2}\left[(1 + \mu)\left(\frac{d^2w}{dr^2} + \frac{1}{r}\frac{dw}{dr}\right) + (1 - \mu)\left(\frac{d^2w}{dr^2} - \frac{1}{r}\frac{dw}{dr}\right)\cos 2\theta\right] \quad (93)$$

The stresses in the slab due to bending are calculated from the moment given Eq. (93) and the relationship $\sigma = 6M/h^2$ which follows directly from beam theory for a slab of unit width.

All the above equations were developed using only a concentrated load. For the more realistic cases of distributed loads, the equations for deflections and moments in the slab can be obtained by assuming a concentrated load acting on a finite area of the slab and integrating over the loaded area. Obviously, some of these integrations will be extremely difficult. Thus formal solutions for conditions other than symmetrical cases and for the maximum stresses are difficult to obtain.

In his original work, Westergaard [100], working directly with Eq. (83), presented solutions for the symmetrical case of an interior load and also solutions for edge and corner loading conditions. In some of his later works, Westergaard [102-106] suggested some changes in the form of the solutions to take into account stress concentrations directly under the loaded area. He also developed equations for elliptically loaded areas, which are more representative of the type of loading normally assumed on pavements, especially airfield pavements.

The fundamental equations by Westergaard and Hogg are given below for the deflection of elastic plate.

Interior—Liquid Subgrade:

$$w = \frac{Pl^2}{4D}\operatorname{Re}H_0'\ \sqrt{i}\ \frac{r}{l} \quad (94)$$

Interior—Solid Subgrade:

$$
w = \frac{Pl^2}{8D} \sum_{m=0,1}^{\infty} (-1)^m \left\{ \frac{\dfrac{8}{3\sqrt{3}} \left(\dfrac{r}{2l}\right)^{6m}}{[(3m)!]^2} \right.
$$

$$
+ \frac{\dfrac{4}{\pi} \left[\log \dfrac{\lambda r}{2l} - \left(1 + \dfrac{1}{2} + \cdots + \dfrac{1}{3m+1} \right) \right] \left(\dfrac{r}{2l}\right)^{6m+2}}{[(3m+1)!]^2} \tag{95}
$$

$$
\left. - \frac{\dfrac{8}{3\sqrt{3}} \left(\dfrac{r}{2l}\right)^{6m+4}}{[(3m+2)!]^2} + \frac{2\left(\dfrac{r}{2l}\right)^{6m+5}}{[\Gamma(3m+3.5)]^2} \right\}
$$

Edge—Liquid Subgrade:

$$
w = \frac{2P}{\pi k l^2} \int_0^\infty \frac{\gamma \cos \dfrac{\alpha x}{l} \left(\cos \dfrac{\beta y}{l} + (1-\mu)\alpha^2 \sin \dfrac{\beta y}{l} \right) e^{-\gamma y/l}}{1 + 4(1-\mu)\alpha^2\gamma^2 - (1-\mu)^2\alpha^4} \, d\alpha \tag{96}
$$

Near an Edge—Liquid Subgrade:

$$
w_{\mathrm{II}} = \frac{2P}{\pi k l^2} \int_0^\infty \frac{\gamma \cos \dfrac{\alpha x}{l} \left(A \cos \dfrac{\beta y}{l} + B \sin \dfrac{\beta y}{l} \right) e^{-\gamma y/l}}{1 + 4(1-\mu)\alpha^2\gamma^2 - (1-\mu)^2\alpha^4} \, d\alpha \tag{97}
$$

$$
w_{\mathrm{I}} = w_{\mathrm{II}} + \frac{P}{\pi k l^2} \int_0^\infty \frac{\cos \dfrac{\alpha x}{l}}{\beta^2 + \gamma^2} \left(\beta \cos \dfrac{\beta y}{l} \sinh \dfrac{\gamma y}{l} \right.
$$

$$
\left. - \gamma \sin \dfrac{\beta y}{l} \cosh \dfrac{\gamma y}{l} \right) d\alpha \tag{98}
$$

where   Re = "the real part of"

Im = "the imaginary part of"

H_0^1 = the Hankel function of zero order that vanishes for an infinite positive imaginary argument

r, θ = polar coordinates

x, y = rectangular coordinates

λ = 1.781072 = antilogarithm of Euler's constant

$\Gamma(\ )$ = a gamma function

$$\beta = \sqrt{\frac{\sqrt{1 + \alpha^4} - \alpha^2}{2}}$$

$$\gamma = \sqrt{\frac{\sqrt{1 + \alpha^4} + \alpha^2}{2}}$$

$$A = \frac{1}{2}\left\{1 + 2\gamma^2 g - (1 - \mu)^2\alpha^4\right.$$

$$\left. + \left[1 + (1 - \mu)^2\alpha^4 + g \sin\frac{2\beta c}{l} - 2\gamma^2 g \cos\frac{2\beta c}{l}\right]e^{-2\gamma c/l}\right\}$$

$$B = \frac{1}{2}\left[2(1 - \mu)\alpha^2 - g + \left(2\gamma^2 g \sin\frac{2\beta c}{l} + g \cos\frac{2\beta c}{l}\right)e^{-2\gamma c/l}\right]$$

$$g = \frac{\mu\alpha^2}{2(\gamma^2 + \beta^2)}[2\gamma^2 - (1 - \mu)\alpha^2] + \frac{3(1 - \mu)\alpha^2}{2} - \gamma^2$$

c = distance from x axis to edge of slab

The approximate solution of the fundamental equations is presented in Table 7.1.

7.13. APPLICATION TO PAVEMENT PROBLEMS

In the preceding formulas, the external load was assumed to be uniformly distributed over a circular area. Pickett and Ray [86,87] extended the work of Westergaard [100,105] and of Hogg [68] by presenting solutions of their equations in the form of influence charts, similar to those prepared by Newmark in 1942, to simplify the calculation of theoretical deflections and moments caused by distributed loads on pavement slab. In fact, with the charts one may

TABLE 7.1 Comparison of Theoretical Stresses in Pavements

	Winkler subgrade	Elastic solid subgrade
Radius of relative stiffness load-distribution parameter	$l_k = \sqrt[4]{D/k}$ $a_k = a/l_k$	$l_e = \sqrt[3]{2D/C}$ $a_e = a/l_e$
Interior loads: Deflection at $r = 0$ ..	$w_0 = \dfrac{Pl_k^2}{8D}\,[1 - a_k^2(0.217 - 0.367\log a_k)]$	$w_0 = \dfrac{Pl_e^2}{3\sqrt{3}D}\,[1 - a_e^2(0.144 - 0.238\log a_e)]$
Subgrade reaction, $r = 0$	$q_0 = \dfrac{P}{8l_k^2}\,[1 - a_k^2(0.217 - 0.367\log a_k)]$	$q_0 = \dfrac{P}{3\sqrt{3}l_e^2}\,(1 - 0.552a_e + 0.126a_e^2)$
Moment, max. at $r = 0$	$M_{max} = -P_0(1 + \mu)(0.1833\log a_k - 0.0490 - 0.0078a_k^2)$ and, using a modification suggested by Westergaard, $\sigma_{max} = 0.275(1 + \mu)\dfrac{P}{h^2}\left(4\log\dfrac{l_k}{b} + 1.069\right)$ where $b = \sqrt{1.6a^2 + h^2} - 0.675h,\ a < 1.7h$ $= a,\ a > 1.74h$	$M_{max} = -P(1 + \mu)(0.1833\log a_e - 0.049 - 0.0120a_e^2)$ $\sigma_{max} = \dfrac{6M_{max}}{h^2}$
Edge loads: Deflection.	$w_0 = \dfrac{1}{\sqrt{6}}\,(1 + 0.4\mu)\dfrac{P}{kl_k^2}$	Edge loading condition has not been solved for the elastic solid subgrade.
Stress	$\sigma_{max} = 0.497(1 + \mu)\dfrac{P}{h^2}\left(4\log\dfrac{l_k}{b} + 0.359\right)$	
Corner loads: Deflection.	$w = \dfrac{P}{kl_k^2}\,(1.1 - 0.88a_k\sqrt{2})$	Corner loading condition has not been solved for the elastic solid subgrade.
Stress	$\sigma_{max} = \dfrac{3P}{h^2}\,[1 - (\sqrt{2}\,a_k)^{0.6}]$	

readily obtain these values for any distribution of load that might be transmitted by airplane landing gears. It is only necessary to:

1. Draw the imprint of tire or tires on transparent paper to a scale that depends on the properties of the slab and its supporting subgrade.

2. Place the drawing on the appropriate chart in a position that depends on the location of the load with respect to the point for which values are desired.

3. Count the blocks of the chart covered by the diagram.

The value desired is then obtained as a product of the intensity of loading, a factor expressing properties of subgrade and slab, and the number of blocks covered by the diagram.

Pickett and Ray's contribution to pavement engineering was significant. Practicing engineers gained a powerful tool in solving the tedious mathematical equations and developed a much deeper appreciation of the stress condition in the pavement slab. The Westergaard-Pickett-Ray method has been almost *the* pavement design since the late 1940s, and often the original assumptions and limitations have been overlooked.

The fundamental equations by Westergaard and by Hogg are given in Eqs. (94) to (98). By the use of the principle of superposition and by integration, one may find the deflection at the origin due to a distributed load. For example, the deflection of an infinitely wide concrete slab on a Winkler subgrade can be expressed by Eq. (94). If the external load P is replaced by a pie-shaped load $qrdr \, d\theta$ and is integrated with respect to r from 0 to a and with respect to θ from θ_1 to θ_2, the result is

$$w = \frac{ql^2}{D} \left(\frac{\theta_z - \theta_1}{2\pi} \right) \left[1 + \frac{\pi a}{2l} \operatorname{Im} \sqrt{i} \, H_1'\left(\sqrt{i} \, \frac{a}{l} \right) \right] \qquad (99)$$

For the entire deflection field in which $\theta_2 = 2\pi$, $\theta_1 = 0$, and a/l is of a defined dimension, the uniform load area can be divided into a desirable number, say N, of equal influence blocks. Each influence block will produce $1/N$ total deflection at the origin (see Fig. 7.14). For any uniform load smaller than the dimensional scale of the influence chart, the deflection can be obtained by

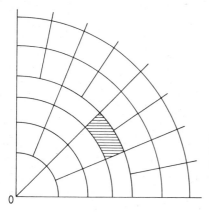

Fig. 7.14 Uniform load blocks (influence area) contributing an equal deflection at the origin.

$$w = \frac{ql^2}{D}\frac{n}{N} \qquad (100)$$

where n is the number of blocks, or equal influence areas, contributed by the designed load.

By using the same technique, Pickett and Ray prepared 48 influence charts. These charts were for both moments and deflections for four different cases. The cases are classified according to the point for which values can be determined and according to the subgrade assumption, as follows:

1. Interior, liquid subgrade
2. Interior, solid subgrade
3. Edge, liquid subgrade
4. One-half of the radius of relative stiffness from an edge, liquid subgrade

The most useful chart for practicing engineers is the influence chart for the moment in the interior of a concrete slab supported on a dense liquid subgrade (see Fig. 7.15).

7.14. COMPUTER APPLICATIONS TO PAVEMENT PROBLEMS

PCA Programs

Without exception, computer application to pavement problems has gained popular acceptance in engineering practice. The Pickett-Ray charts have been replaced by a Portland Cement Association (PCA) computer program for general use. A listing of the program is

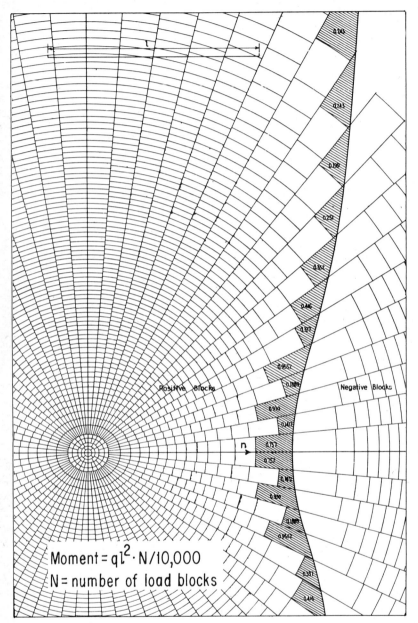

Fig. 7.15 Influence chart for moment in concrete-slab interior load on Winkler subgrade.

available from the PCA or from the printout of the IBM common library programs. The PCA program represents a refined solution of the Pickett-Ray charts. There is no major modification in theory, but the refined computer outputs provide a better understanding of the influence of multiwheel gears.

Finite-element Model**

In developing a finite-element model to simulate pavement slabs, Hudson and Matlock [72] chose to develop the equilibrium equations from the physical model rather than expressing the differential equation in finite-difference form. According to the authors, the physical-model approach was superior because:

> The model is straightforward and assists visualization of the problem. Discontinuities and freely discontinuous changes in load, bending stiffness, torsional stiffness, and other parameters are easily understood with the use of a physical model, but limitations on continuity of the differential equation make direct difference approximation suspect.

The physical model used by Hudson and Matlock to represent the slab was similar to the plate analog developed by Newmark [82]. A diagrammatic representation of this model is shown in Fig. 7.16.

The physical model is composed of rigid bars connecting elastic hinges, with torsion springs connecting adjacent parallel bars. The plate analog has the following characteristics:

1. The bars are weightless and perfectly rigid, so that no deformation occurs within the bars.

2. The mass of the plate and the external forces are concentrated at the elastic hinges.

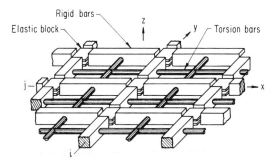

Fig. 7.16 Finite-element model of a plate or slab.

3. The result of the direct stresses are bending moments acting at the elastic hinges and at the ends of each bar.

4. The result of the vertical shearing stresses are shearing forces acting at the elastic hinges and at the ends of each bar.

5. The result of the horizontal shearing stresses are twisting moments concentrated in the torsion springs.

A typical joint taken from the slab model is shown in Fig. 7.17. In the joint model, the subgrade support for the joint is a spring acting at the joint. This spring can either support the forces exerted by the slab to the subgrade (+) or restrain the slab from lifting (−), but it cannot sustain or transmit lateral forces.

Forces acting on the members of the model at a typical joint are shown in Fig. 7.18. All forces are shown as acting in the positive direction. Equilibrium equations for the slab model can be developed from free-body analyses of the joints.

The final equation developed from the analysis of the model is given by Stelzer and Hudson [96]. Due to its length and because no particular advantage would be gained, the equation is not reproduced here.

Perhaps of greater interest than merely reproducing the equilibrium equations is the fact that the equation given by Stelzer and Hudson [96] can be presented in generalized matrix form as

$$[K]\{W\} = \{F\} \tag{101}$$

where $[K]$ is the stiffness matrix, $\{W\}$ the deflection matrix, and $\{F\}$ the load matrix. A characteristic of the finite-element approach is that the equations are ordered so that they can be expressed in the matrix

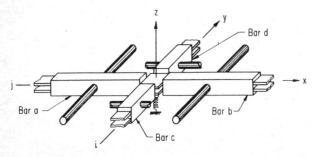

Fig. 7.17 Typical joint i, j taken from finite-element slab model.

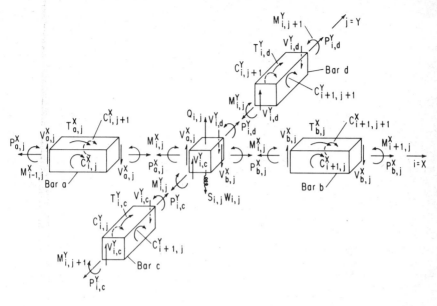

Fig. 7.18 Free body of slab mesh point.

form, which makes this approach well suited for computer application. Libraries of programs have been developed to assist the engineer in the manipulation of such matrices. Computer programs have been developed by several investigators [72,96] specifically to solve the slab problem using the finite-element model described.

The primary cause of error associated with the use of the finite-element model described above is due to approximating a continuum with a lumped-parameter model. This error can be reduced by decreasing the size of the mesh used in the analysis. Decreasing the mesh size, however, increases the computer capacity required to solve the problem and reduces the machine roundoff error. Stelzer and Hudson [96] reported a comparison of typical deflection data from analysis of the plate using the finite-element model shown and the results obtained by formal analysis of the plates. The data show that accurate results can be obtained with the model with relatively short computer times.

Use of the finite-element model shown above for evaluating pavement slabs has many advantages over the analysis of slabs as a continuum. Some advantages of the finite-element model are:

1. Freely varying loads can be applied to the slab.

2. Discontinuities in the slab can be handled in a routine manner, and even partial discontinuities can be analyzed.

3. Material properties of the slab can be varied from point to point.

4. A variety of slab support conditions can be handled with the model, including nonuniform support for continuously supported slabs.

Probably the greatest deficiency of the finite-element model described above for analysis of pavement systems is that the equilibrium condition of the pavement subgrade is not considered. Other disadvantages of the model in its present form are (1) it is not set up to handle moving traffic loads and (2) the model as presented is not developed to handle viscoelastic or inelastic materials.

In introducing the equilibrium of subgrade support, Saxena [111] treats the subgrade as an isotropic elastic half-space. The reaction under the nodal point is represented by a force instead of a spring by Hudson. The formulation of Eq. (101) becomes

$$[K]\{W\} = -\{Q\} + \{P\} \tag{102}$$

where $\{Q\}$ is a load matrix and $\{P\}$ is a reaction matrix. The equilibrium of the subgrade must satisfy the condition

$$\{W\} = [AF]\{P\} \tag{103}$$

where $[AF]$ is the flexibility matrix of the subgrade. Boussinesq equations are used by Saxena in establishing the flexibility matrix. The solution of Eq. (102) becomes

$$\left[[K][AF] - 1\right]^{-1}\{-Q\} = \{P\} \tag{104}$$

The listing of the computer program is given in Appendix 1.

7.15. REFINEMENTS OF THEORY**

Shear Stresses at Interfaces

The generalized treatment of the plate theories is based on the assumption that the forces at the upper and lower faces of the slab act normal to these faces. Klubin [75] expressed the pavement reaction by an infinite series of polynomials, which also account for two elastic constants: the modulus of elasticity and Poisson's ratio. The subgrade support is reduced to an elastic solid.

Loss of Support

The generalized treatment is also based on the assumption that the slab and subgrade are contiguous at all points. Observation of pavement in the field clearly shows that at times—either because of warping or by loss of the subgrade through pumping—the slab is not in contact with the subgrade. Leonards, Reddy, and Harr [76,91] studied the problem and introduced linear moisture and temperature gradients in a symmetrical slab partially supported by a Winkler foundation.

Moving Loads

Thompson [98], Lewis [77], and Harr [78] considered the moving loads on a warped slab supported by a viscoelastic Winkler foundation. In their investigation, the authors essentially dealt with the equation

$$D\nabla^2\nabla^2W + PH\,\frac{\partial^2 w}{\partial t^2} = P(x, y, t) - q(x, y, t) \qquad (105)$$

which is similar to Eq. (83) but with an additional term $PH(\partial^2 w/\partial t^2)$ to take into account the acceleration of the slab. When P and q are made time-dependent, the viscoelastic nature of the foundation under moving loads can also be taken into account.

Cracked Slabs

Many concrete pavements crack, but the cracks remain closed, so that partial moment and shear are transmitted across the cracks. Niu and Pickett [83,84] started with the basic differential equation for an elastic slab supported by a Winkler foundation. The pavement was divided into three regions in calculating the deflection and moments of the slab for concentrated loads at a number of locations. This condition is particularly true for continuous reinforced-concrete pavements.

7.16. INHERENT LIMITATIONS AND CONCLUSION

Extent of Slab

In the development of slab theory, the assumption is made, either implicitly or explicitly, that the slabs extend a considerable distance (approaching infinity) from the loaded area. The only exception to this is in the solutions developed by Westergaard [100] for the edge and corner load conditions. Even for the edge and corner load conditions, it was assumed that the slab was infinite in all directions away from the loaded edge or edges. Most concrete pavements, however, are relatively narrow. Conventionally reinforced pavements have transverse joints at intervals of about 15 to 40 ft, and continuously reinforced pavements usually have cracks formed at intervals of 5 to 10 ft. Thus it is important to know how far from a joint or an edge the load must be placed so that the slab will behave as a slab infinite in extent.

An analysis of circular plates by Bergstrom [61] shows that a plate approximately 9 in. thick must have the following dimensions in order to behave as a plate infinite in extent:

For maximum moment, a minimum diameter of 20 ft
For maximum deflections, a minimum diameter of 33 ft

By inference, it appears that the load must be placed at a distance of approximately 16.5 ft from the edge or joint in order for the pavement to behave as an infinite slab. With thicker pavements, the loads would have to be placed even further from the edge or joint for the desired condition to exist.

Subgrade Support

Perhaps the most important disagreement among the investigators is the assumption of subgrade support. Many investigators have criticized the Winkler-foundation assumption on the grounds that it is not representative of soil material. Others argue for its use, however, because it is possible to obtain formal solutions for edge and corner load conditions when using this support. Figure 7.19 shows a comparison of a theoretical deflection and the reaction under a concentrated load for a slab on both an elastic solid and a Winkler foundation. Note that immediately under the load, the

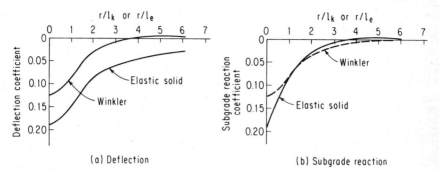

Fig. 7.19 Comparison of deflection and subgrade reaction for elastic solid and Winkler subgrade.

deflection of a slab with an elastic solid subgrade is considerably higher than that of a comparable plate on a Winkler foundation.

However, the major drawback of the Winkler foundation lies in the determination of k value. In the Bureau of Public Roads (BPR) study, reported by Kelley [73] and Teller and Sutherland [110], various diameters of test plate were used to evaluate the k value of the subgrade. It was found that the k value was influenced by the diameter of the plate, as shown in Fig. 7.20. Thus the k value is not a constant. Subsequently, Spangler [95] reported, however, that the k value under the slab decreased as the radial distance from the corner of the slabs increased. Under one slab, for example, the value for k varied from 650 psi/in. at the corner to about 50 psi/in. at a distance of approximately 40 in. from the corner. This variation in k is, of

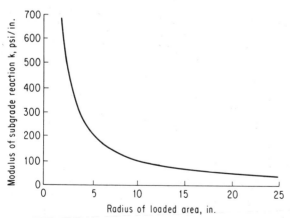

Fig. 7.20 Influence of radius of loaded plate on k value, modulus of subgrade reaction.

course, inconsistent with Westergaard's original assumption regarding the deflection and the subgrade reaction under the slab.

Inconsistent Test Results

Figure 7.21 shows a comparison of the stresses in the slabs from the AASHO road test, road test MD-1, and the BPR static load tests with Westergaard equations for the stresses under corner load conditions. The results from both the AASHO road test and the road test MD-1 are considerably lower than the results from the BPR tests. This may be partially due to the fact that the AASHO road test and the road test MD-1 results were obtained from moving loads whereas the BPR test loads were held for a duration of 5 min. When the relative effect of warping up and warping down is considered, changes in stress as high as 50 percent were noted in the results from road test MD-1 [66].

Variation of Stiffness

While most pavement slabs are reasonably uniform in thickness, the natural variability in the stiffness (E value) of the concrete probably causes considerable nonuniformity in the rigidity of the slab. This nonuniformity will cause some variation in the results from applied loads and may also result in unusual stress patterns in the slabs due to twisting of the slab. This must be kept in mind when evaluating the stress in the pavement layers.

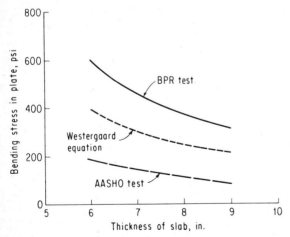

Fig. 7.21   Comparison of observed stresses with plate equations.

Conclusion

In the preceding discussions, the application of medium-thick–slab theory as a mathematical model for pavement systems has been reviewed. It is apparent that the theory is not an effective mathematical model. One of the more serious disadvantages of the medium-thick–plate theory as a mathematical model for pavements is that this treatment is valid only for systems with a stiff layer or slab supported by a softer material. This conclusion is based on the assumption that normal stresses in the slab are small compared with bending stresses and that shear deformations can be ignored. Other limitations of the slab theory as a model include the fact that the slabs in a normal pavement system are not large enough to behave as infinite slabs, as assumed in the theory, and that edge and corner load conditions can be evaluated only from semiempirical equations based on test results.

7.17. YIELD-LINE THEORY

Swedish investigators, notably Losberg [80], originated the *yield-line method* in the early 1960s for use in the design of plain and reinforced-concrete pavements. The yield-line concept involves the determination of the thickness of a slab resting on an elastic foundation such that when the slab is loaded by a static load distributed over a small circular area, the maximum radial bending moment along a critical failure line should not exceed the bending moment which would produce a continuous circumferential crack at a given radius around the loaded area. This latter bending moment is known as the *ultimate negative bending moment* of the slab section. The equilibrium of a small element at a distance r from the center of the loaded area (Fig. 7.22) would be governed by the following equations:

$$\frac{d}{dr}(rQ) = -rp \tag{106}$$

$$\frac{d}{dr}(rM_r) = M_t + rQ \tag{107}$$

where p = load intensity

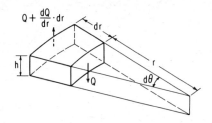

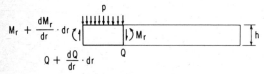

Fig. 7.22 Equilibrium of a slab element.

Q = shearing force per unit circumferential length

M_t, M_r = tangential and radial moments per unit length, respectively

Using the Rankine yield condition for the slab and ignoring membrane action, the tangential moment M_t is equal to M_o and the radial moment varies from M_o at the center of the load ($r = o$) to $-M_o$ at the circumferential yield hinge circle ($r = b$). M_o is the ultimate bending resistance of the slab and is equal to $f_o h^2/4$ in which f_o is the yield strength of the material in tension and h is the thickness of the slab.

In considering the overall equilibrium condition, Meyerhof [81] introduced that (1) the slab is supported on a Winkler foundation, $p = -kw$; (2) the slope of deflection is linear in the form

$$w = \left(1 - \frac{r}{b}\right) w_o \qquad (a)$$

in which w_o is the deflection under the load at $r = o$; and (3) the total subgrade reaction is

$$P = \frac{\pi k w_o b^2}{3} \qquad (b)$$

Consolidating Eqs. (a) and (b), the Winkler subgrade reaction becomes

$$p = - \frac{3P}{\pi b^2} \left(1 - \frac{r}{b} \right)$$ (c)

Substituting the above equation (c) in Eq. (106) and integrating with respect to r and within the boundary conditions of $rQ = o$ at $r = o$ and $r = b$, the solution of vertical shearing force for $(a < r < b)$ is

$$rQ = \frac{3Pr^2}{2\pi b^2} \left(1 - \frac{2r}{3b} \right) - \frac{P}{2\pi}$$ (108)

Substituting Eq. (108) into Eq. (107) and integrating with respect to r and within the Rankine yield conditions, the radial moment for $(a < r < b)$ is

$$M_r = M_o + \frac{Pr^2}{2\pi b^2} \left(1 - \frac{r}{2b} \right) - \frac{P}{2\pi} \left(1 - \frac{2a}{3r} \right)$$ (109)

Putting $M_r = - Mo$ at $r = b$, the solution for estimating the collapse load, P, for a central load on small slab, as given by Meyerhof [81], is

$$P = \frac{4\pi M_o}{1 - \frac{4a}{3b}} \qquad \text{for } 0.05 < \frac{a}{b} < 0.75$$ (110)

In his experiments on ice support, Meyerhof also showed that an effective radius of $b = 3.9l$ may be suggested as an approximate and safe value. Thus, the thickness of slab is

$$h = \sqrt{\frac{P}{\pi f_o} \left(1 - \frac{a}{3l} \right)} \qquad \text{for } \frac{a}{l} > 0.2$$ (111)

in which l is the radius of relative stiffness for the slab as given by Eq. (91).

C. Transient Deflection of Pavement System

In the newtonian field, the governing differential equation of motion is

$$M\ddot{w} + C\dot{w} + Kw = F(t) \qquad (112)$$

When the exciting force is removed, the viscoelastic rebound of the response system will resume:

$$M\ddot{w} + C\dot{w} + Kw = 0 \qquad (113)$$

Solutions to this equation have the form

$$w(t) = Ae^{mt} \qquad (114)$$

To satisfy the initial condition when $t = 0$, the initial deflection $w(0) = w_0$, and the initial velocity $\dot{w}(0) = 0$, the above equation becomes

$$w(t) = w_0 e^{-nt} \qquad (115)$$

where $n = C/2M$ represents the damped frequency of the response system. The parameter n is usually given by δp, where δ is the coefficient of damping and p is the natural frequency of the response system. The above equation can be rewritten as

$$w(t) = w_0 e^{-\delta pt} \qquad (116)$$

When the exciting force is a sinusoidal function of the form $1 + D \sin 2\pi\omega t$, the general equation of the response function is

$$w(t) = (1 + D \sin 2\pi\omega t)w_0 e^{-\delta pt} \cos \frac{2\pi v t}{L} \qquad (117)$$

where v = crossing velocity of exciting force

L = wavelength of response function
D = dynamic increment of exciting force
ω = frequency of exciting force

When the coefficient of damping δ and the natural frequency of the response system p are constant during the course of elastic rebound, the time factor, δpt is a linear function of t and can be expressed by the parameter $\tau = \delta pt$. Moreover, in the course of wave propagation, the products of L and p represent the theoretical velocity of the energy wave and the value δLp is the actual wave velocity, which can be expressed by C_0. By substituting the parameters τ and C_0, the above equation becomes

$$ w(\tau) = w_0 e^{-\tau} \cos \frac{2\pi v \tau}{C_0} \left(1 + D \sin 2\pi \omega \frac{L}{C_0} \tau \right) \qquad (118) $$

By the reciprocal principle, the deflection of the point A at a distance x from the load is equal to the deflection at the original point of load when the load is moved to point A (see Fig. 7.23b). The exciting force $1 + D \sin 2\pi\omega(L/C_0)\tau$ can be considered as a point load when the time interval is small, such as $[1 + D \sin 2\pi\omega(L/C_0)\tau]\,d\tau$. The effect of a moving point load can be expressed by the same reciprocal theory. As the load is moved away in a time interval τ, the deflection at the original reference point 0 is equal to the deflection at a point equivalent to the time interval τ when the load is at the original reference point (see Fig. 7.23e). For a continuous forcing function, the total dynamic deflection at the reference point, 0, will be the integration of the deflection caused by the point load:

$$ W_{\text{dyn}} = \int_0^T w(\tau)\,d\tau \qquad (119) $$

Substituting Eq. (118), the integration of the above equation is

$$ W_{\text{dyn}} = \frac{w_0}{1 + (2\pi v/C_0)^2} \left[1 - e^{-T} \left(\cos \frac{2\pi v}{C_0} T - \frac{2\pi v}{C_0} \sin \frac{2\pi v}{C_0} T \right) \right] $$
$$ - \frac{\tfrac{1}{2} D w_0}{1 + \alpha^2} \left[\sin (\alpha T + \alpha \cos \alpha T) e^{-T} - \alpha \right] $$

$$-\frac{\frac{1}{2}Dw_0}{1 + \beta^2}[(\sin\beta T + \beta\cos\beta T)e^{-T} - \beta] \tag{120}$$

where $\alpha = (2\pi/C_0)(\omega L + v)$ and $\beta = (2\pi/C_0)(\omega L - v)$. For maximum dynamic deflection in the above equation, $\cos(2\pi v/C_0)T$ will approach zero. Therefore, $(2\pi v/C_0)T = \frac{1}{2}\pi, \frac{3}{2}\pi, \frac{5}{2}\pi, \ldots$

The first peak dynamic deflection will be encountered at

$$T = \frac{C_0}{4v} \tag{121}$$

When the forcing function is a step load, the dynamic increment will

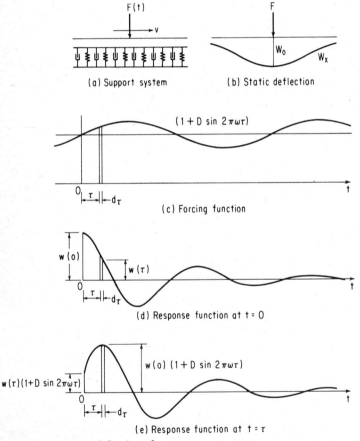

(a) Support system   (b) Static deflection

$(1 + D \sin 2\pi\omega\tau)$

(c) Forcing function

$w(0)$ $w(\tau)$

(d) Response function at t = 0

$w(\tau)(1+D \sin 2\pi\omega\tau)$ $w(0)(1+D \sin 2\pi\omega\tau)$

(e) Response function at t = τ

Fig. 7.23 Transient deflection of pavement.

be equal to zero $D = 0$, and the maximum dynamic deflection will be

$$W_{\text{dyn}} = \frac{w_0}{1 + (2\pi v/C_0)^2}\left(1 + \frac{2\pi v}{C_0}e^{-C_0/4v}\right) \qquad (122)$$

The velocity of wave propagation in the above equation can be replaced by the well-known formula

$$C_0 = \sqrt{\frac{E}{\rho}} \qquad (123)$$

where ρ is the density of the elastic mass and E is the elastic modulus. As the major portion of the pavement deflection actually takes place in the subgrade, the E and ρ values in the above equation should be designated for the subgrade soils. In the Newark test, the best modulus of deformation of the subgrade was about 10,000 psi and the adjusted mean value was 5,700 psi, with a possible low of 3,600 psi. The dynamic deflection at various vehicle velocities has been computed, as shown in Fig. 7.24. In his analysis of the AASHO

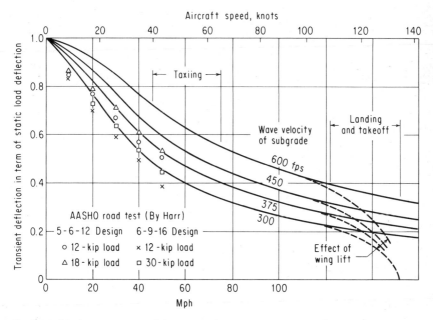

Fig. 7.24 Maximum transient deflection under a moving load.

road test, Harr reported that the dynamic deflection of highway pavements decreases with increasing crossing velocity. The result of his analysis is plotted in Fig. 7.24. It is noted that the magnitude of observed deflections is in general agreement with the theoretical curve. For airport pavements, the velocity of aircraft is excessively high and there are no test data to verify the validity of the theoretical analysis. It is hoped that in future tests, the dynamic deflection of pavements can be measured under the influence of operating aircraft.

REFERENCES

A. Equilibrium of Pavement Systems

1. W. E. A. Acum and L. Fox, Computation of Load Stresses in a Three-layer Elastic System, *Geotechnique,* vol. 2, no. 4, pp. 293–300, 1951.
2. R. G. Ahlvin and H. H. Ulrey, Tabulated Values for Determining the Complete Pattern of Stresses, Strains and Deflections beneath a Uniform Circular Load on a Homogeneous Half Space, *HRB Bull.* 342, 1962.
3. J. E. Ashton and F. Moavenzadeh, The Analysis of Stresses and Displacements in a Three-layered Viscoelastic System, *Proc. Int. Conf. Struct. Design Asphalt Pavements, 2nd, Ann Arbor, Mich.,* 1967.
4. E. S. Barber, Shear Loads on Pavements, *Proc. Int. Conf. Struct. Design Asphalt Pavements, Ann Arbor, Mich.,* 1962.
5. A. H. S. Ang, Mathematically Consistent Discrete Models for Simulating Solid Continua, in A. H. S. Ang and N. Newmark (eds.), "Computation of Underground Structural Response," DSA 1386, Final Report DA-49-146-xz-104, Urbana, Ill., 1963.
6. A. H. S. Ang and Goin N. Harper, Analysis of Contained Plastic Flow in Plane Solids, *J. Eng. Mech. Div., ASCE,* vol. 90, no. EM5, 1964.
7. A. H. S. Ang and N. M. Newmark, A Numerical Procedure for the Analysis of Continuous Plates, *Proc. Conf. Electronic Computation, Struct. Div., ASCE, 2nd,* p. 379, September, 1960.
8. D. D. Ang and M. L. Williams, Combined Stresses in an Orthotropic Plate Having a Finite Crack, *J. Appl. Mech.,* ser. E, vol. 28, no. 3, p. 372, September, 1961.
9. J. H. Argyris, Elasto-plastic Matrix Displacement Analysis of Three-dimensional Continua, *J. Roy. Aeron. Soc.,* vol. 69, pp. 633–636, 1965.
10. R. D. Barksdale, "Analysis of Layered Systems," final report of Project B-607, National Science Foundation Grant No. GK-1583, Georgia Institute of Technology, 1969.
11. A. Bastiani, The Explicit Solution of the Equations of the Elastic Deformations for a Stratified Road under Given Stresses in the Dynamic Case, *Proc. Int. Conf. Struct. Design Asphalt Pavements, Ann Arbor, Mich.,* 1962.
12. J. Boussinesq, "Application des Potentials," Paris, 1885.
13. D. M. Burmister, The General Theory of Stresses and Displacements in Layered Soil System, I, II, III, *J. Appl. Phys.,* vol. 16, no. 2, pp. 89–96, no. 3, pp. 126–127, no. 5, pp. 296–302, 1945
14. D. M. Burmister, Stress and Displacement Characteristics of a Two-layer Rigid Base Soil System: Influence Diagrams and Practical Applications, *Proc. HRB,* vol. 35, 1956.

15. D. M. Burmister, The Theory of Stresses and Displacements in Layered Systems and Application to the Design of Airport Runways, *Proc. HRB,* 1954.
16. H. Deresiewicz, "The Half-space under Pressure Distributed over an Elliptical Portion of Its Plane Boundary," American Society of Mechanical Engineers, Applied Mechanics Division, Paper 59-A-17, 1959.
17. J. M. Duncan, C. L. Monismith, and E. L. Wilson, "Finite Element Analysis of Pavements," paper presented at 47th Annual Meeting of Highway Research Board, January, 1968.
18. S. M. Fergus and W. E. Miner, Distributed Loads on Elastic Foundations: The Uniform Circular Load, *Proc. HRB,* vol. 34, pp. 582–597, 1955.
19. L. Fox, "Computation of Traffic Stresses in a Simple Road Structure," Road Research Laboratory Paper 9, London, 1948.
20. D. R. Freitag and A. J. Green, Distribution of Stresses on an Unyielding Surface beneath a Pneumatic Tire, *HRB Bull.* 342, 1962.
21. B. Gross, "Mathematical Structure of the Theories of Viscoelasticity," Ed. Herman, 1953.
22. R. J. Hank and F. H. Scrivner, Some Numerical Solutions of Stresses in Two- and Three-layered Systems, *Proc. HRB,* vol. 28, pp. 457–468, 1948.
23. L. R. Herrmann, Finite-element Bending Analysis for Plates, *J. ASCE,* vol. EM5, 1967.
24. Y. H. Huang, Stresses and Displacements in Viscoelastic Layered Systems under Circular Loaded Areas, *Proc. Int. Conf. Struct. Design Asphalt Pavements, 2nd, Ann Arbor, Mich.,* 1967.
25. W. R. Hudson and H. Matlock, Cracked Slabs with Non-uniform Support, *J. ASCE,* vol. HW1, 1967.
26. W. R. Hudson and H. Matlock, "Discontinuous Orthotropic Plates and Pavement Slabs," Research Report 56-6, Center for Highway Research, University of Texas, 1965.
27. W. R. Hudson and H. Matlock, Discrete-element Analysis for Discontinuous Plates, *J. ASCE,* vol. ST10, 1968.
28. A. Jones, Tables of Stresses in Three-layer Elastic Systems, *HRB Bull.,* 342, 1962.
29. D. C. Kraft, Analysis of a Two-layer Viscoelastic System, *Proc. ASCE, J. Eng. Mech. Div.,* vol. 91, no. EM6, part 1, December, 1965.
30. E. H. Lee and T. G. Rogers, "Solution of Viscoelastic Stress Analysis Problems Using Measured Creep or Relaxation Functions," Division of Applied Mathematics, Brown University, August, 1961.
31. R. Lattes, J. L. Lions, and J. Bonitzer, Use of Gilerksen's Method for the Study of Static and Dynamic Behavior of Road Structures, *Proc. Int. Conf. Struct. Design Asphalt Pavements, Ann Arbor, Mich.,* 1962.
32. E. H. Lee, Stress Analysis in Viscoelastic Bodies, *Quart. Appl. Math.,* vol. 13, no. 2, p. 183, 1955.
33. M. M. Lemcoe, Stresses in Layered Elastic Solids, *Proc. ASCE, J. Eng. Mech. Div.,* vol. 86, no. EM4, part 1, 1960.
34. J. R. Levey, "A Method for Determining the Effects of Random Variations in Material Properties on the Behavior of Pavement Structures," doctoral dissertation, University of Illinois, Urbana, 1968.
35. A. E. H. Love, The Stress Produced on a Semi-infinite Body by Pressure on Part of the Boundary, *Phil. Trans. Roy. Soc., London, Ser. A,* vol. 228, 1929.
36. M. R. Metha and A. S. Veletsos, Stresses and Displacements in Layered Systems, *Civil Eng. Studies, Struct. Res. Ser.* 178, University of Illinois, Urbana, 1959.
37. F. Moavenzadeh and J. E. Ashton, "Analysis of Stresses and Displacements in a Three-layer Viscoelastic System," Massachusetts Institute of Technology School of Engineering Research Report R67-31.

38. R. H. Mallett and P. V. Marcal, Finite Element Analysis of Nonlinear Structures, *J. ASCE,* vol. ST9, 1968.
39. H. Matlock and T. A. Haliburton, "A Finite-element Method of Solution for Linearly Elastic Beam-Columns," Research Report 56-1, Center for Highway Research, University of Texas, 1965.
40. H. Matlock, "Analysis of Several Beam-on-Foundation Problems," paper presented at ASCE Houston Convention, Feb. 21, 1962.
41. H. Matlock and Wayne B. Ingram, Bending and Buckling of Soil-supported Structural Elements, *Proc. Pan-Amer. Conf. Soil Mech. Found. Eng., 2nd, Brazil,* no. 32, June 1963.
42. F. Moavenzadeh and J. F. Elliott, "Moving Load on a Viscoelastic Layered System," report prepared for the U. S. Department of Transportation Federal Highway Administration, Bureau of Public Roads (Contract FH-11-6619).
43. N. M. Newmark, Influence Charts for Computation of Stresses in Elastic Foundations, *Univ. Ill. Bull.,* vol. 40, no. 12.
44. K. R. Peattie, A Fundamental Approach to the Design of Flexible Pavements, *Proc. Int. Conf. Struct. Design Asphalt Pavements, Ann Arbor, Mich.,* 1962.
45. K. R. Peattie, Stresses and Strain Factors for Three-layer Elastic Systems, *HRB Bull.* 342, 1962.
46. K. S. Pister, Viscoelastic Plates on Viscoelastic Foundations, *Proc. ASCE, J. Eng. Mech. Div.,* pp. 43–54, February, 1961.
47. K. S. Pister and R. A. Westman, Analysis of Viscoelastic Pavements, *Proc. Int. Conf. Struct. Design Asphalt Pavements, Ann Arbor, Mich.,* 1962.
48. R. L. Schiffman, General Analysis of Stresses and Displacements in Layered Elastic Systems, *Proc. Int. Conf. Struct. Design Asphalt Pavements, Ann Arbor, Mich.,* pp. 365–375, 1962.
49. R. L. Schiffman, The Numerical Solution for Stresses and Displacements in a Three-layer Soil System, *Proc. Int. Conf. Soil Mech. Found. Eng., 4th,* vol. 2, pp. 169–173, 1957.
50. G. F. Sowers and A. B. Vesic, Stress Distribution beneath Pavements of Different Rigidities, *Proc. Int. Conf. Soil Mech. Found. Eng., 5th,* vol. 2, 1961.
51. K. Terazawa, "On the Elastic Equilibrium of a Semi-infinite Elastic Solid," vol. 37, College of Science, Imperial University of Tokyo, December, 1916.
52. R. A. Westman, Layered System Subjected to Surface Shears, *Trans. ASCE,* vol. 129, 1964.
53. R. A. Westman, "Viscoelastic and Thermoelastic Analysis of Layered Systems," doctoral dissertation, Department of Civil Engineering, University of California, 1962.
54. A. C. Whiffin and N. W. Lister, The Application of Elastic Theory to Flexible Pavements, *Proc. Int. Conf. Struct. Design Asphalt Pavements, Ann Arbor, Mich.,* 1962.
55. E. L. Wilson, "Finite Element Analysis of Two-dimensional Structures," University of California, Report 68-2, Berkeley, June, 1968.
56. E. L. Wilson, Structural Analysis of Axisymmetric Solids, *J. AIAA,* vol. 3, no. 12, 1965.
57. M. L. Williams, Structural Analysis of Viscoelastic Materials, *J. AIAA,* vol. 2, no. 5, pp. 785–808, 1964.
58. O. C. Zienkiewicz and Y. K. Cheung, Finite Elements in the Solution of Field Problems, *The Engineer,* London, 1965.
59. O. C. Zienkiewicz and Y. K. Cheung, "The Finite Element Method in Structural and Continuum Mechanics," McGraw-Hill Book Company, New York, 1967.

60. O. C. Zienkiewicz and G. S. Hollister (eds.), "Stress Analysis," John Wiley & Sons, Inc., New York, 1965.

B. Elastic Plate on Idealized Subgrade

61. Sven G. Bergstrom, Ernst Fromen, and Sven Linderholm, Investigation of Wheel Load Stresses in Concrete Pavements, *Proc. Swedish Cement Concrete Res. Inst.*, no. 13, 1949.

62. M. M. Filonenko-Borodich, Some Approximate Theories of the Elastic Foundation (in Russian), *Lichenyie Zap. Mosk. Gos. Univ.-Mekhan.*, no. 46, 1940.

63. M. E. Harr and G. A. Leonards, Warping Stresses and Deflections in Concrete Pavements, *Proc. HRB,* 1959.

64. M. Hentenyi, "Beams on Elastic Foundation," The University of Michigan Press, Ann Arbor, 1946.

65. H. Hertz, Uber das Gleichgewict Schwimmender Elastischer Platten, *Ann. Physik Chem.*, vol. 22, 1884.

66. "Road Test One-MD," *HRB Special Report 4,* 1952.

67. "The AASHO Road Test, Report 5, Pavement Research," HRB Special Report 61E, 1962.

68. A. H. A. Hogg, Equilibrium of a Thin Plate Symmetrically Loaded, Resting on an Elastic Subgrade of Infinite Depth, *Phil. Mag.*, ser. 7, vol. 25, March, 1938.

69. D. L. Holl, Equilibrium of a Thin Plate Symmetrically Loaded on a Flexible Subgrade, *J. Sci.,* vol. 12, no. 4, Iowa State College, July, 1938.

70. D. L. Holl, Thin Plates on Elastic Foundations, *Proc. Int. Congr. Appl. Mech., 5th, Cambridge, Mass.,* 1938.

71. W. R. Hudson, Comparison of Concrete Pavement Load-Stresses at AASHO Road Test with Previous Work, *Highway Res. Record,* no. 42.

72. W. R. Hudson and H. Matlock, Analysis of Discontinuous Orthotropic Pavement Slabs Subjected to Combined Loads, *Highway Res. Record,* no. 131, pp. 1–48, 1966.

73. E. F. Kelley, Applications of the Results of Research to the Structural Design of Concrete Pavements, *Public Roads,* vol. 20, 1939.

74. A. D. Kerr, Viscoelastic Winkler Foundation with Shear Interactions, *Proc. ASCE, J. Eng. Mech. Div.,* June, 1961.

75. P. I. Klubin, Computations of Beams and Circular Plates on Elastic Foundations (in Russian), *Inzh. Sb.,* vol. 12, 1952.

76. G. A. Leonards and M. E. Harr, Analysis of Concrete Slabs on Ground, *Proc. ASCE, J. Soil Mech. Found. Div.,* June, 1959.

77. K. H. Lewis, "Analysis of Concrete Slabs on Ground and Subjected to Warping and Moving Loads," Joint Highway Research Project 16, Purdue University, Lafayette, Ind., June, 1967.

78. K. H. Lewis and M. E. Harr, Analysis of Concrete Slabs on Ground Subjected to Warping and Moving Loads, *Highway Res. Record* no. 291, p. 194, 1969.

79. R. K. Livesley, Some Notes on the Mathematical Theory of a Loaded Elastic Plate Resting on an Elastic Foundation, *Quart. J. Mech. Appl. Math.,* vol. 6, 1953.

80. Anders Losberg, "Structurally Reinforced Concrete Pavements," *Doktorsavhandl. Chalmers Tek. Hogskola,* GöteBorg, 1960.

81. G. G. Meyerhof, Load-carrying Capacity of Concrete Pavements, *Journal of the Soil Mechanics and Foundation Division, ASCE,* vol. 88, SM3, June 1962, pp. 89–116.

82. N. M. Newmark, Numerical Methods of Analysis of Bars, Plates, and Elastic Bodies, in L. E. Grinter (ed.), "Numerical Methods of Analysis in Engineering," The Macmillan Company, New York, 1949.

83. Hsien-Ping Niu, "Bending of a Cracked Pavement," doctoral dissertation, University of Wisconsin, Madison, 1967.
84. H. P. Niu and G. Pickett, The Effect of Degree of Continuity across a Void or Crack on Performance of Concrete Pavements, *Highway Res. Record,* no. 291, p. 186, 1969.
85. P. L. Pasternak, On a New Method of Analysis of an Elastic Foundation by Means of Two Foundation Constants (in Russian), *Gos. Izd. Lit. Stroit. Arkhitekt.,* Moscow, 1954.
86. G. Pickett, M. E. Raville, W. C. James, and F. J. McCormick, "Deflections, Moments and Reactive Pressures for Concrete Pavements," Kansas State College Engineering Experiment Station Bulletin 65, October, 1951.
87. G. Pickett and G. K. Ray, Influence Charts for Concrete Pavements, *Trans. ASCE,* vol. 116, 1951.
88. K. S. Pister and C. L. Monismith, Analysis of Viscoelastic Flexible Pavements, Flexible Pavement Design Studies 1960, *HRB Bull.* 269, pp. 1–15, 1960.
89. K. S. Pister and R. A. Westmann, Bending of Plates on an Elastic Foundation, *J. Appl. Mech.,* June, 1962.
90. K. S. Pister and M. L. Williams, Bending of Plates on a Viscoelastic Foundation, *Proc. ASCE, J. Eng. Mech. Div.,* October, 1960.
91. A. S. Reddy, G. A. Leonards, and M. E. Harr, Warping Stresses and Deflections in Concrete Pavements: Part III, *Highway Res. Record,* no. 44, 1963.
92. E. Reissner, A Note on Deflection of Plates on a Viscoelastic Foundation, *J. Appl. Mech.,* vol. 25, no. 1, March, 1958.
93. E. Reissner, A Note on Deflections of Plates on Viscoelastic Foundations, *Trans. ASME,* vol. 25, no. 1, March, 1958.
94. F. Schiel, Der Schwimmende Balken, *Z. Angew. Math. Mech.,* vol. 22, 1942.
95. M. G. Spangler, "Stresses in the Corner Region of Concrete Pavements," Iowa Engineering Experiment Station, Bulletin 157, 1942.
96. F. C. Stilzer, Jr., and R. Hudson, "A Direct Computer Solution for Plates and Pavement Slabs," Research Report 56-9, Center for Highway Research, University of Texas, Austin, 1967.
97. S. Timoshenko and S. Woinowsky-Krieger, "Theory of Plates and Shells," 2d ed., McGraw-Hill Book Company, New York, 1959.
98. W. E. Thompson, Analysis of Dynamic Behavior of Roads Subject to Longitudinally Moving Loads, *Highway Res. Record,* no. 39, 1963.
99. G. Unal, "Bending of a Semi-infinite Slab on an Elastic Half Space," doctoral dissertation, University of Wisconsin, Madison, 1965.
100. H. M. S. Westergaard, Stresses in Concrete Pavements Computed by Theoretical Analysis, *Proc. HRB,* part 1, 1925; also published in *Public Roads,* vol. 7, no. 2, April, 1926.
101. H. M. Westergaard, Analysis of Stresses in Concrete Pavements Due to Variations of Temperature, *Proc. HRB,* vol. 6, 1926.
102. H. M. S. Westergaard, Mechanics of Progressive Cracking in Concrete Pavements, *Public Roads,* vol. 10, no. 4, June, 1929.
103. H. M. Westergaard, Analytical Tools for Judging Results of Structural Tests of Concrete Pavements, *Public Roads,* vol. 14, no. 10, December, 1933.
104. H. M. Westergaard, Stresses in Concrete Runways of Airports, *Proc. HRB,* vol. 19, 1939.
105. H. M. Westergaard, New Formulas for Stresses in Concrete Pavements of Airfields, *Proc. ASCE* vol. 113, 1947.
106. H. M. Westergaard, Stress Concentrations in Plates Loaded over Small Areas, *Proc. ASCE,* vol. 68, no. 4, part 1, April, 1942, with discussion by L. W. Teller and E. C.

Sutherland; *Proc. ASCE,* vol. 68, no. 8, part 1, October, 1942. Also printed in *Trans. ASCE,* vol. 108, 1948.

107. E. Winkler, "Die Lehre von der Elastizitat und Festigkeit," Praga Dominicus, 1867.
108. J. F. Wiseman, M. E. Harr, and G. A. Leonards, Warping Stresses and Deflections in Concrete Pavements; Part II, *Proc. HRB,* vol. 39, 1960.
109. V. Z. Vlasov and N. N. Leontev, "Beams, Plates and Shells on Elastic Foundations," translated from Russian, published for the National Aeronautics and Space Administration and the National Science Foundation by the Israel Program for Scientific Translations.
110. L. W. Teller and E. C. Sutherland, The Structural Design of Concrete Pavements, *Public Roads,* vol. 16, nos. 8 to 10; vol. 17, nos. 7 and 8; vol. 23, no. 8.
111. S. K. Saxena, "Foundation Mats and Pavement Slabs Resting on an Elastic Foundation—Analyzed through a Physical Model," doctoral dissertation, Department of Civil Engineering, Duke University, Durham, N.C., 1971.

APPENDIX ONE

Computer Programs

A. Analysis of Stresses and Displacements in an N-Layered System

The original program was developed by Chevron Research Company in 1963. A revised version with free-form input has been developed by the staff of the University of Illinois for use on either the IBM 360 machine or the Burroughs 5500. The program given here is the original Chevron version, which consists of the following.

Main Routine Handles input and output (except for writing the final answer) and calls the subroutines, which do all the arithmetic operations.

PART Calculates the partition $\{x_i\}$ and the four points in each interval (x_{i-1}, x_i) at which $T(m)$ is to be evaluated for integration.

COEE and COHIGH Computes $A_i(m)$, $B_i(m)$, $C_i(m)$, and $D_i(m)$. COHIGH works in the range where A_i and B_i are effectively zero.

BESSEL Computes either $J_0(m)$ or $J_1(m)$ to within an absolute tolerance of less than 10^{-6}. This routine could be used in any program requiring evaluation of zero and first-order Bessel functions.

CALCIN Completes the evaluation of $T(m)$ and performs the integration of

$$\sum_{i=1}^{n} \int_{x_{i-1}}^{x_i} T(m)\ dm$$

It writes out the final results.

HIGHM Writes the comment "SLOW CONVERGENCE" in computing the displacements and stresses. The message gives a good idea of the quality of convergence.

Program Documents

1. J. Michelow, "Analysis of Stresses and Displacements in an N-layered Elastic System under a Load Uniformly Distributed on a Circular Area," California Research Corp., Richmond, Calif., September, 1963.
2. H. Warren, and W. L. Dieckmann, "Numerical Computation of Stresses and Strains in a Multiple Layered Asphalt Pavement System," California Research Corp., Richmond, Calif., September, 1963.

B. Finite-element Method—Analysis of Axisymmetric Solids

The program was originally developed by E. L. Wilson in 1967 based on his paper "Structural Analysis of Axisymmetric Solids," *AIAA Journal*, vol. 3, no. 12, December, 1965. There are three subroutines: (1) STIFF, forming the global stiffness matrix for quadrilaterals and triangles; (2) STRESS, computing element stresses and affective strains; and (3) BANSOL, reducing blocks of equations by the use of gaussian elimination.

The boundary conditions of the axisymmetric solids are assumed to be (1) clamped along the centerline, (2) supported on rollers at the top and bottom nodal points, and (3) under uniform compression on the fourth boundary surface. In the finite-element approximation, the continuous structure is replaced by a system of axisymmetric elements which are interconnected at circumferential joints and nodal circles. Equilibrium equations are developed at each node circle. A solution of this set of equations constitutes a solution to the system.

The program has been modified by the Naval Civil Engineering Laboratory to include a subroutine MESH, which generates element properties and material layer boundaries. It is desirable to investigate the depth, width, angle, and boundary conditions of frustums on the Boussinesq solution to centerline displacements and stresses.

C. Finite-element Method—Elastic Plate on Boussinesq Foundation

This program was originally developed by Hudson and Matlock to solve the stress and deflection of orthotropic plates and pavement slabs on a Winkler foundation by a direct finite-difference method. The solution is carried out by using a back-and-forth recursive technique. The program given here has been modified by Saxena [111] to treat the supporting subgrade as an isotropic elastic half-space. The reactive force under the node point has been accounted for from Boussinesq's solution. The equations expressed in the matrix form are

$$\{P\} = \left[[AK][AF] - 1\right]^{-1}\{-Q\} \qquad (a)$$

and

$$\{W\} = [AF]\{P\} \qquad (b)$$

where $\{Q\}$ = load matrix of slab
$\{P\}$ = reaction matrix of subgrade
$[AK]$ = stiffness matrix of slab
$\{W\}$ = deflection matrix of subgrade
$[AF]$ = flexibility matrix of subgrade

For a known reaction $\{P\}$, the deflection $\{W\}$ can be computed from Eq. (b). It is not necessary to invert the matrix $[AF]$. The inversion of a large matrix $\left[[AK][AF] - 1\right]^{-1}$ would require the use of double precision in the computer to reduce the roundoff error.

```
C          THIS PROGRAM SOLVES ORTHOTROPIC PLATES AND PAVEMENT SLABS
C          BY A DIRECT METHOD. THE DIRECT SOLUTION IS CARRIED OUT BY USING
C          A BACK AND FORTH RECURSIVE TECHNIQUE DESCRIBED BY HUDSON
C          MATLOCK
C
C          NOTATION
C          AA(),A1(),A2()        TEMPORARY A( ) TERMS
C
C          A()                   CONTINUITY OR RECURSION COEFFICIENT
C          AA4(),AA5(),AA6()     STIFFNESS MATRIX AND LOAD MATRIX
C          AAUG()                AUGMENTED MATRIX
C          AA1(),AA2(),AA3(),    TERMS WHICH MAKE THE SUBMATRICES OF THE
C          ALF                   THETA IN RADIANS
C          AN1(N)                ALPHA NUMERIC REMARK, INFORMATION ONLY
C          AN2(N)                ALPHA NUMERIC REMARK, INFORMATION ONLY
C          B()                   CONTINUITY OR RECURSION COEFFICIENT
C          BB(),BB1(),BB2()      TEMPORARY B() TERMS
C          BMX(I,J)              BENDING MOMENT IN THE X DIRECTION
C          BMY(I,J)              BENDING MOMENT IN THE Y DIRECTION
C          CX(I,J)               TORSIONAL STIFFNESS
C          CXN                   TEMPORARY INPUT VALUE OF TWISTING STIFFNESS
C          CY(I,J)               TORSIONAL STIFFNESS
C          CYN                   TEMPORARY INPUT VALUE OF TWISTING STIFFNESS
C          C()                   CONTINUITY OR RECURSION COEFFICIENT
C          CC(),CC1(),CC2()      TEMPORARY C() TERMS
C          DP(N)                 SQUARE ROOT OF PRODUCT OF BENDING STIFFNESS
C          DX(I,J)               BENDING STIFFNESS (SLAB)
C          DXN                   TEMPORARY INPUT VALUE OF BENDING STIFFNESS
C          DY(I,J)               BENDING STIFFNESS (SLAB)
C          DYN                   TEMPORARY INPUT VALUE OF BENDING STIFFNESS
C          E()                   CONTINUITY OR RECURSION COEFFICIENT
C          D()                   CONTINUITY OR RECURSION COEFFICIENT
C          HX                    INCREMENT LENGTH IN X DIRECTION
C          HXDHY3                HX DIVIDED BY HY CUBED
C          HY                    INCREMENT LENGTH IN Y DIRECTION
C          HYDHX3                HY DIVIDED BY HX CUBED
C          I                     STATION NUMBER X DIRECTION
C          II,II                 TEMPORARY VALUE OF I
C          IN1                   X COORDINATE OF THE FROM STATION
C          IN2                   X COORDINATE OF THE THRU STATION
C          ISTA                  EXTERNAL X STATION NUMBER
C          ITEST                 BLANK FIELD FOR ALPHANUMERIC ZERO
C          J                     STATION NUMBER Y DIRECTION
C          J1,J2                 TEMPORARY VALUE OF J
C          JN1                   Y COORDINATE OF THE FROM STATION
C          JN2                   Y COORDINATE OF THE THRU STATION
C          JSTA                  EXTERNAL Y STATION NUMBER
C          MX                    NUMBER OF INCREMENTS IN X DIRECTIONS
C          MXP3                  MX PLUS THREE
C          MXP5                  MX PLUS FIVE
C          MXP7                  MX PLUS SEVEN
C          MY                    NUMBER OF INCREMENTS IN Y DIRECTIONS
```

```
C          MYP5           MY PLUS FIVE
C          MYP7           MY PLUS SEVEN
C          NCT2           NUMBER OF VALUES IN TABLE 2
C          NCT3           NUMBER OF CARDS IN TABLE 3
C          NPROB          NUMBER OF PROBLEM, PROG STOPS IF ZERO
C          ODHX           ONE DIVIDED BY HX
C          ODHY           ONE DIVIDED BY HY
C          ODHXHY         ONE DIVIDED BY HX TIMES HY
C          PDHXHY         POISSON''S RATIO DIVIDED BY HX TIMES HY
C          PR             POISSON''S RATIO
C          PX(I,J)        AXIAL LOAD IN X DIRECTION
C          PXN            TEMPORARY INPUT VALUE OF AXIAL LOAD
C          Q(I,J)         TRANSVERSE FORCE PER MESH POINT
C          QBMX           HXHY * SECOND DERIV BENDING MOMENT (X)
C          QMBY           HYHX * SECOND DERIV BENDING MOMENT (Y)
C          QN             TEMPORARY INPUT VALUE OF LOAD
C          QPX            VERTICAL REACTION DUE TO VERTICAL FORCES
C          QPY            VERTICAL REACTION DUE TO AXIAL FORCES
C          QTMX           HXHY * SECOND DERIV TWIST MOMENT
C          QTMY           HYHX * SECOND DERIV TWIST MOMENT
C          REACT          NET TRANSVERSE FORCE
C          S(I,J)         SPRING SUPPORT, VALUE PER MESH POINT
C          SN             TEMPORARY INPUT VALUE OF SUPPORT SPRINGS
C          TMX            TWISTING MOMENT (YX)
C          TMY            TWISTING MOMENT (YX)
C          TX(I,J)        EXTERNAL COUPLE IN X DIRECTION
C          TXN            TEMPORARY INPUT VALUE OF EXTERNAL COUPLE
C          TY(I,J)        EXTERNAL COUPLE IN Y DIRECTION
C          TYN            TEMPORARY INPUT VALUE OF EXTERNAL COUPLE
C          W()            VERTICAL DEFLECTION
C          W1(),W2()      TEMPORARY VALUES OF W()


          COPIES OF LISTINGS
          DOUBLE PRECISION AK1,AK2,AK3,AK4,AA6,AFF
          DIMENSION AN3(14)
          DIMENSION AA1(17,17),AA2(17,17),AA3(17,17),AA4(17,17),
     2    AA5(17,17),DX(21,31),DY(21,31),CX(21,31),CY(21,31)
          DIMENSION AKP(238,238),AKF(195,375)
          COMMON DS(21,31),Q(21,31),S(21,31),W(21,31),BMX(21,31),
     2    BMY(21,31),STX(21,31),STY(21,31),DF(21,31),D1(3),F1(3),F2(3),
     3         DP(6),X(21),Y(31)
          COMMON AK1(41,41),AK2(41,195),AK3(195,41),AA6(195,1),
     2    AK4(195,195),AFF(195,195)
          EQUIVALENCE (AKP(1,1),AKF(1,1))
          REAL NUS
    1 FORMAT(20H PROGRAM MAT SAXENA)
    2 FORMAT(A5,5X,10A2,10A2,10A2,A5)
    4 FORMAT (///10H       PROB  , /5X, A5, 5X, 40A2 )
   23 FORMAT ( 4( 3X, 12 ),    6E10.3 )
   20 FORMAT(415,6E10.3)
```

```
30 FORMAT  (//30H      TABLE 1. CONTROL DATA ,              /
 1            /30H            NUM CARDS TABLE 2 ,43X, 12,    /
 2            30H            NUM CARDS TABLE 3 ,43X, 12,    /
 4            30H            NUM INCREMENTS MX ,43X, 12, /
 5            30H            NUM INCREMENTS MY ,43X, 12, /
 6            30H            INCR LENGTH HX       ,35X, E10.3,/
 7            30H            INCR LENGTH HY        ,35X,    E10.3,/
 9            30H            POISSON''S RATIO       ,35X,    E10.3,/
 8            30H            MODULUS OF PLATE     ,35X,    E10.3,/
 6            30H     POISSON''S RATIO OF SOIL    ,35X,    E10.3,/
 2            30H            THICKNESS             ,35X,    E10.3   )
 43 FORMAT  (5X, 2( 2X, 12, IX, 12)), 6E11.3)
 33 FORMAT  (//51H      TABLE 2. STIFFNESS AND LOAD DATA, FULL VALUES,
 1            35H ADDED AT ALL STAS I,J IN RECTANGLE, /
 2            / 50H         FROM   THRU   DX          DY        Q
 3            45H   CX             CY                          ,/)
 35 FORMAT  ( 25H      TABLE 4. RESULTS         ///
 1            40H             I    J          REACTION    )
 39 FORMAT  (/ 50H         I,J   DEFL        BMX      BMY
 1            45H STRESS-X    STRESS-Y                        /)
 40 FORMAT  (/ 40H         I,J    TMX         TMY
 1            45H SHEAR         STRESS-DIFFERENCE            /)
 44 FORMAT  (5X,2( 2X, 12, 1X, 12 ), 22X,4E11.3)
 45 FORMAT  (7X,12,13,9E12.3)
 24 FORMAT  ( 4(3X,12),20X,4E10.3)
 25      FORMAT(1015)
 26      FORMAT(8E10.3)
903      FORMAT(13X,215,7X,3(E10.3,2X),//)
1601     FORMAT( ' MATRIX AKP COMPUTED ' )
1602     FORMAT( ' MATRIX AKP DIVIDED ' )
1603     FORMAT( ' MATRIX AKF COMPUTED ' )
1604     FORMAT( ' MATRIX AKF NORMALISED ' )
C     PROGRAM AND PROBLEM IDENTIFICATION
1000      READ(1,2,END=1005) NPROB,(AN3(N),N=1,10)
      WRITE(3,1)
          WRITE(3,4) NPROB,(AN3(N),N=1,10)
C INPUT TABLE 1.
1001 READ(1,20)NCT2,NCT3,MX,MY,HX,HY,PR,E,NUS,TH
          READ(1,26) ES1,ES2,ES3,Z1,Z2
      WRITE(3,30)NCT2,NCT3,MX,MY,HX,HY,PR,E,NUS,TH
        WRITE(3,26) ES1,ES2,ES3,Z1,Z2
        A = MY*HY
        B = MX*HX
      MXP1 = MX + 1
      MYP1 = MY + 1
      MXP2 = MX+2
      MYP2 = MY+2
      MXY3 = MX+3
      MYP3 = MY+3

        MXP4 = MX+4
```

```
        MYP4 = MY +4
    MXP5 = MX+5
    MYP5 = MY+5
    MXP6 = MX+6
    MYP6 = MY+6
        MXP7 = MX+7
        MYP7 = MY+7
    MXP9 = MX + 9
    MYP9 = MY + 9
        MHYP1 = (MY/2)+1
        MHXP1 = (MX/2)+1
        MHYP2 = (MY/2)+2
        MHXP2 = (MX/2)+2
        MHYP3 = (MY/2)+3
        MHXP3 = (MX/2)+3
        MHXP4 = (MX/2)+4
        MHYP4 = (MY/2)+4
        MHYP5 = (MY/2)+5
        MHXP5 = (MX/2)+5
        NNU = MXP1*MYP1
        NND = MHYP2*MXP3
        NNB = MXP1*MHYP1
        NNC = NND-NNB-2
        MXP32 = MXP3*2
HYDHX3 = HY/HX**3
HXDHY3 = HX/HY**3
PDHXHY = PD/(HX*HY)
ODHXHY = 1.0/HY*HX)
ODHX = 1.0/HX
ODHY = 1.0/HY
        DO 1002 I = 1,MXP7
        X(I) = 0.0
        DO 1002 J = 1,MYP7
DX(I,J) = 0.0
DY(I,J) = 0.0
CX(I,J) = 0.0
CY(I,J) = 0.0
Y(J) = 0.0
DS(I,J) = 0.0
Q(I,J) = 0.0
W(I,J) = 0.0
        BMX(I,J) = 0.0
        BMY(I,J) = 0.0
        STX(I,J) = 0.0
        STY(I,J) = 0.0
        DF(I,J) = 0.0
        S(I,J) = 0.0
1002 CONTINUE
        DO 1122 I = 1,NND
        DO 1122 J = 1,NND
        AKP(I,J) = 0.0
```

```
1122 CONTINUE
          DO 1012 I = 1,NNB
      AA6(I,1) = 0.0
          DO 1012 J = 1,NNB
          AK4(I,J) = 0.0
          AFF(I,J) = 0.0
1012      CONTINUE
          DO 91 I = 1,MXP3
          DO 91 J = 1,MXP3
      AA1(I,J) = 0.0
      AA2(I,J) = 0.0
      AA3(I,J) = 0.0
      AA4(I,J) = 0.0
      AA5(I,J) = 0.0
   91     CONTINUE
          DO 1003 I = 1,NNC
          DO 1003 J = 1,NNC
          AK1(I,J) = 0.0
1003      CONTINUE
          DO 1022 I = 1,NNC
          DO 1022 J = 1,NNB
          AK2(I,J) = 0.0
1022      CONTINUE
          DO 1032 I = 1,NNB
          DO 1032 J = 1,NNC
          AK3(I,J) = 0.0
1032      CONTINUE
          DO 401 I = 2,MXP7
      X(I) = X(I-1)+HX
  401 CONTINUE
          DO 402 I = 2,MYP7
      Y(I) = Y(I-1)+HY
  402 CONTINUE
C INPUT TABLE 2.
          QAV = 0.0
      WRITE(3,33)
      DO 360 N = 1,NCT2
      READ(1,23)IN1,JN1,IN2,JN2,DXN,DYN,QN,CXN,CYN
      WRITE(3,43)IN1,JN1,IN2,JN2,DXN,DYN,QN,CXN,CYN
      I1 = IN1+4
      J1 = JN1+4
      I2 = IN2+4
      J2 = JN2+4
      IF(I1.GT.I2) GO TO 360
      DO 350 I = I1,I2
      IF(J1.GT.J2) GO TO 360
      DO 350 J = J1,J2
      DX(I,J)=DX(I,J)+DXN
      DY(I,J)=DY(I,J)+DYN
       Q(I,J)=  Q(I,J)+  QN
      CX(I,J)=CX(I,J)+CXN
```

```
        CY(I,J)=CY(I,J)+CYN
            QAV = QAV + Q(I,J)/(A*B)
350 CONTINUE
360 CONTINUE
C-----FORM SUB MATRICES
        IA = 1
        DO 1600 J = 3,MHYP4
        DO 100 I = 3,MXP5
    II=I-2
    AA1(II,II)=DY(I,J-1)*HXDHY3
            AA2(II,II) = -2.0 * ( PDHXHY * ( DX(I,J) + DY(I,J-1))
    1               + HXDHY3 * ( DY(I,J-1) + DY(I,J) ) )
    2               + ODHXHY * ( - CX(I,J) - CX(I+1,J)
    3               - CY(I,J) - CY(I+1,J))
        AA3 (II,II) = HYDHX3 * (DX(I-1,J)+ 4.0 * DX(I,J)
    1       + DX(I+1,J) + HXDHY3 * (DY(I,J-1) + 4.0
    2       * DY(I,J)+ DY(I,J+1))+ PDHXHY * 4.0
    3       * ( DX(I,J) + DY(I,J)) + ODHXHY
    4       * ( CX(I,J) + CX(I,J+1) + CX(I+1,J)
    5       + CX(I+1,J+1) + CY(I,J) + CY(I,+1,J)
    6       + CY(I,J+1) + CY(I+1,J+1))
    AA4(II,II)=-2.0*(HXDHY3*(DY(I,J)+DY(I,J+1))
    1+PDHXHY*(DX(I,J)+DY(I,J+1)))
    2+ODHXHY*(-CX(I,J+1)-CX(I+1,J+1)
    3 -CY(I,J+1)-CY(I+1,J+1))
        AA5(II,II) =HXDHY3 *DY(I,J+1)
            IF (J.EQ.3.OR.J.EQ.MYP5) GO TO 116
            IF (I.EQ.3.OR.I.EQ.MXP5) GO TO 116
            AA6(IA,1) = Q(I,J)
            IA = IA + 1
116         IF(II-1) 110,110,115
115             AA2(II,II-1) =DX(I-1,J)*PDHXHY +DY(I,J-1)*PDHXHY+
    1       ODHXHY *(CX(I,J)+ CY(I,J))
    AA3(II,II-1)=-2.0*(HYDHX3*(DX(I-1,J)+DX(I,J))
    1       +PDHXHY *(DX(I-1,J)+DY(I,J)))
    3       +ODHXHY * (-CX(I,J)-CX(I,J+1)
    3 -CY(I,J)-CY(I,J+1)
    AA4(II,II-1) =PDHXHY *(DX(I-1,J)+DY(I,J+1))
    1       +ODHXHY *(CX(I,J+1)+CY(I,J+1))
110     IF(II-2) 150,150,125
125     AA3(II,II-2)=DX(I-1,J)*HYDHX3
150     IF(II-MXP3) 155,160,160
155     AA2(II,II+1)=PDHXHY *(DX(I+1,J)+DY(I,J-1))
    1   + ODHXHY *(CX(I+1,J)+CY(I+1,J))
        AA3(II,II+1) = -2.0 * ( HYDHX3 * (DX(I,J) +DX(I+1,J))
    1       +PDHXHY *(DX(I+1,J) +DY(I,J)))
    2       +ODHXHY*(-CX(I+1,J)-CX(I+1,J+1)
    4       -CY(I+1,J) -CY(I+1,J+1))
    AA4(II,II+1)=PDHXHY *(DX(I+1,J)+DY(I,J+1))
    1       +ODHXHY *(CX(I+1,J+1)+CY(I+1,J+1))
160     IF(II+1-MXP3) 165,100,100
```

```
  165      AA3(II,II+2)=HYDHX3 *DX(I+1,J)
  100   CONTINUE
           IF (J.EQ.MHYP3) GO TO 166
           IF (J.EQ.MHYP4) GO TO 168
           GO TO 99
  166      WRITE(3,25)J
           DO 167 K = 1,MXP3
           DO 167 L = 1,MXP3
           AA3(K,L) = AA3(K,L)+AA5(K,L)
  167      CONTINUE
           GO TO 99
  168      WRITE(3,25)J
           DO 169 K = 1,MXP3
           DO 169 L = 1,MXP3
           AA1(K,L) = AA1(K,L)+AA5(K,L)
           AA2(K,L) = AA2(K,L)+AA4(K,L)
  169      CONTINUE
   99      CONTINUE
           N=J-3
           IF(N-2) 198,198,197
  197      LA = (N-2)*MXP3+1
           L = LA
           GO TO 1199
  198      LA = 1
           L = LA
 1199      KA1 = 1
           KA2 = 1
           KA3 = 1
           KA4 = 1
           KA5 = 1
  199      K = (MXP3)*N+1
           IF (L.EQ.K) GO TO 206
           IF (L.EQ.(K+MXP3)) GO TO 209
           IF (L.EQ.(K±MXP32))GO TO 212
           IF (L.EQ.(K-MXP3)) GO TO 203
           IF (L.EQ.(K-MXP32)) GO TO 200
  200      DO 202 I = 1,MXP3
           IF (KA1.LE.MXP3) L = LA
           DO 201 JB = 1,MXP3
           AKP(K,L) = AA1(I,JB)
           L = L+1
  201      CONTINUE
           K = K+1
           KA1 = KA1+1
  202      CONTINUE
           GO TO 199
  203      LB = L
           IF (LB.GT.NND) GO TO 1600
           DO 205 I = 1,MXP3
           IF(KA2.LE.MXP3) L = LB
           DO 204 JB = 1,MXP3
```

```
               AKP(K,L) = AA2(I,JB)
               L = L+1
204            CONTINUE
               K = K+1
               KA2 = KA2+1
205            CONTINUE
               GO TO 199
206            LC = L
               IF (LC.GT.NND) GO TO 1600
               DO 208 I = 1,MXP3
               IF(KA3.LE.MXP3) L = LC
               DO 207 JB = 1,MXP3
               AKP(K,L) = AA3(I,JB)
               L = L+1
207            CONTINUE
               K = K+1
               KA3 = KA3+1
208            CONTINUE
               GO TO 199
209            LD = L
               IF (LD.GT.NND) GO TO 1600
               DO 211 I = 1,MXP3
               IF(KA4.LE.MXP3) L = LD
               DO 210 JB = 1,MXP3)
               AKP(K,L) = AA4(I,JB)
               L = L+1
210            CONTINUE
               K = K+1
               KA4 = KA4+1
211            CONTINUE
               GO TO 199
212            LEE = L
               IF (LEE.GT.NND) GO TO 1600
               DO 213 I = 1,MXP3
               IF ( KA5.LE.MXP3)  L = LEE
               DO 214 JB = 1,MXP3
               AKP(K,L) = AA5(I,JB)
               L = L+1
214            CONTINUE
               K = K+1
               KA5 = KA5 + 1
213            CONTINUE
1600  CONTINUE
               WRITE(3,1601)
C        DIVIDE MATRIX AKP INTO FOUR MATRICES
889            I = 1
               I1 = 1
               I2 = 1
               I3 = 1
               I4 = 1
               DO 369 IA = 3,MHYP4
```

```
            DO 369 JA = 3,MXP5
            J = 1
            J1 = 1
            J2 = 1
            J3 = 1
            J4 = 1
            DO 371 IB = 3,MHYP4
            DO 371 JB = 3,MXP5
            TEMP = AKP(I,J)
            IF(IA.EQ.3.OR.IA.EQ.MYP5) GO TO 373
            IF(JA.EQ.3.OR.JA.EQ.MXP5) GO TO 373
            IF(IB.EQ.3.OR.IB.EQ.MYP5) GO TO 372
            IF(JB.EQ.3.OR.JB.EQ.MXP5) GO TO 372
            AK4(I4,J4) = TEMP
            J4 = J4+1
            GO TO 375
372         IF(IB.EQ.3.AND.JB.EQ.3) GO TO 375
            IF(IB.EQ.3.AND.JB.EQ.MXP5) GO TO 375
            AK3(I3,J3) = TEMP
            J3 = J3+1
            GO TO 375
373         IF(IA.EQ.3.AND.JA.EQ.3) GO TO 375
            IF(IA.EQ.3.AND.JA.EQ.MXP5) GO TO 375
            IF(IB.EQ.3.OR.IB.EQ.MYP5) GO TO 374
            IF(JB.EQ.3.OR.JB.EQ.MXP5) GO TO 374
            AK2(I2,J2) = TEMP
            J2 = J2+1
            GO TO 375
374         IF(IB.EQ.3.AND.JB.EQ.3) GO TO 375
            IF(IB.EQ.3.AND.JB.EQ.MXP5) GO TO 375
            AK1(I1,J1) = TEMP
            J1 = J1+1
375         J = J+1
377         CONTINUE
            I = I+1
            IF(IA.EQ.3.OR.IA.EQ.MYP5) GO TO 400
            IF(JA.EQ.3.OR.JA.EQ.MXP5) GO TO 400
            I4 = I4+1
            I3 = I3+1
            GO TO 369
400         IF(IA.EQ.3.AND.JA.EQ.3) GO TO 369
            IF(IA.EQ.3.AND.JA.EQ.MXP5) GO TO 369
            I2 = I2+1
            I1 = I1+1
369         CONTINUE
            WRITE(3,1602)
            DO 555 I = 1,NNB
            DO 555 J = 1,NNU
555         AKF(I,J) = 0.0
            AL = HY/HX
            D1(1) = 0.0
```

```
          D1(2) = Z1/HX
          D1(3) = Z2/HX
          DO 589 I = 2,3
          FI(I) = (AL/3.14)*(AL*ALOG((1.+(SQRT(AL**2+1.)))*
2         (SQRT(AL**2+D1(I)**2))/(AL*(1.+SQRT(AL**2+D1(I)**2+1.))))
3         +ALOG((AL+SQRT(1+AL**2))*SQRT(1+D1(I)**2) /
4         (AL+SQRT(AL**2+D1(I)**2+1.))))
          F2(I) = (D1(I)/(2*3.14))*(ATAN(AL/(D1(I)*SQRT(AL**2+D1(I)**2
2              +1.))))
589       CONTINUE
          F1(1) = (AL*ALOG((1.+SQRT( AL**2+1.))/AL)
2              +ALOG(AL+SQRT(AL**2+1)))
          AN1 = 1-NUS**2
          AN2 = 1-NUS-NUS**2
          CC1 = (1-NUS**2)/(3.14*ES1)
          CC2 = (1-NUS**2)/(3.14*ES2)
          CC3 = (1-NUS**2)/(3.14*ES3)
          CD1 = (1+NUS)/(2*3.14*ES1)
          CD2 = (1+NUS)/(2*3.14*ES2)
          CD3 = (1+NUS)/(2*3.14*ES3)
          I = 1
          DO 222 NN = 4,MHYP4
          DO 221 KK = 4,MXP4
          J = 1
          DO 220 N = 4,MYP4
          DO 220 K = 4,MXP4
          DS(K,N) = ABS(SQRT((X(K)-X(KK))**2+(Y(N)-Y(NN))**2))
          IF(DS(K,N)) 225,228,225
228       AKF(I,J) = (4./HY)*(- (1./ES1)*(F1(2)*AN1+F2(2)*AN2
2         -(1./ES2)*((F1(3)*AN1+F2(3)*AN2)-(F1(2)*AN1+F2(2)*AN2))
3         +(1./ES3)*(F1(3)*AN1+F2(3)*AN2)-CC3*F1(1))
          GO TO 224
225       AKF(I,J) = -CC1*(1./(SQRT(DS(K,N)**2+Z1**2))-1./DS(K,N))
2                   -CC2*(1./(SQRT(DS(K,N)**2+Z2**2))
3                        -1./(SQRT(DS(K,N)**2+Z1**2)))
4                   -CC3*(1./(SQRT(DS(K,N)**2+Z2**2)))
5         -CD1*((Z1**2)/((DS(K,N)**2+Z1**2)**1.5))
6         -CD2*((Z2**2)/((DS(K,N)**2+Z2**2)**1.5)
7             -(Z1**2)/((DS(K,N)**2+Z1**2)**1.5))
8         -CD3*((Z2**2)/((DS(K,N)**2+Z2**2)**1.5))
224       J = J+1
220       CONTINUE
          I = I+1
221       CONTINUE
222       CONTINUE
          WRITE(3,1603)
          I = 1
          L = NNU-NNB
70        J = I
          DO 71 NA = 1,MHYP1
72        K = J+2*(L-NA-1)*MXP1)
```

```
          IF(J.GT.L) GO TO 79
          AFF(I,J) = AKF(I,J) + AKF(I,K)
          GO TO 80
79        AFF(I,J) = AKF(I,J)
80        J = J+1
          NB = NA*MXP1
          IF(J.LE.NB) GO TO 72
71        CONTINUE
          I = I+1
          IF (I.LE.NNB) GO TO 70
          CONTINUE
          WRITE(3,1604)
          CALL MATINV( AK1,NNC)
          CALL MTB(NNC,NNC,AK1,NNB,AK2)
          CALL ADJUST(AK3,AK2,AK4,NNB,NNC,NNB)
          CALL MTA(NNB,NNB,AK4,NNB,NNB,AFF)
          CALL MADD(NNB,NNB,AK4)
          CALL INVRSL(AK4,NNB,NNB)
          DO 77 I =1,NNB
          G = 0.0
          DO 75 J = 1,NNB
75        G = G + AK4(I,J)*AA6(J,1)
          AK4(I,1) = G
      WRITE  (3,35)
          K = 1
      SUM = 0.0
          DO 405 J = 4,MHYP4
          DO 405 I = 4,MXP4
          S(I,J) = AK4(K,1)
          SUM = SUM + S(I,J)
          IF(J.EQ.4) GO TO 407
          IF(I.EQ.4.OR.I.EQ.MXP4) GO TO 407
          AREA = HX*HY
          GO TO 408
407       IF(J.EQ.4.AND.I.EQ.4) GO TO 406
          IF(J.EQ.4.AND.I.EQ.MXP4) GO TO 406
          AREA = 0.5*(HX*HY)
          GO TO 408
406       AREA = 0.25*(HX*HY)
408       SM = S(I,J)/(QAV*AREA)
          K = K+1
          ISA = I -4
          JSA = J -4
          WRITE(3,003) ISA,JSA,S(I,J),SM,SUM
405       CONTINUE
890       DO 76 I = 1,NNB
76        AA6(I,1) = AK4(I,1)
          CALL MTH(NNB,NNB,AFF,AA6)
          K = 1
          DO 305 J = 4,MHYP4
          DO 305 I = 4,MXP4
```

```
          DO 305 J = 4,MHYP4
          DO 305 I = 4,MXP4
          W(I,J) = AA6(K,1)
          K = K+1
305       CONTINUE
          DO 7 I = 1,NNC
          G = 0.0
          DO 5 J = 1,NNB
  5       G = G±AK2(I,J)*AA6(J,1)
  7       AK2(I,1) = G
          K = 1
          DO 505 J = 3,MHYP4
          DO 505 I = 3,MXP5
          CONTINUE
          IF(J.EQ.3) GO TO 506
          IF(I.EQ.3.OR.I.EQ.MXP5) GO TO 506
          GO TO 505
506       IF (J.EQ.3.AND.I.EQ.3) GO TO 505
          IF(J.EQ.3.AND.I.EQ.MXP5) GO TO 505
          W(I,J) = -AK2(K,1)
          K = K+1
505       CONTINUE
          DO 706 J = MHYP5,MYP5
          DO 706 I = 3,MXP5
          K = J-2*(J-MHYP4)
          W(I,J) = W(I,K)
706       CONTINUE
          WRITE(3,39)
          DO 1800 J = 3,MYP5
          DO 1700 I = 3,MXP5
     ISTA = I-4
     JSTA = J-4
     DO 1650 N = 1,3
       K = I+N-2
     DP(N+3) = SQRT(DX(K,J)*DY(K,J))
     BMX(K,J) = DX(K,J) * (W(K-1,J)-2.*W(K,J)+W(K+1,J))/(HX*HX) +
  1       DP(N+3)*PR*(W(K,J-1)-2.*W(K,J)+W(K,J+1))/(HY*HY)
     STX(K,J) = (BMX(K,J)*6)/(TH*TH)
     L = J+N-2
     DP(N) = SQRT(DX(I,L)*DY(I,L))
     BMY(I,L) = DY(I,L)*(W(I,L-1)-2.*W(I,L)+W(I,L+1))/(HY*HY) +
  1       PR*DP(N)*(W(I-1,L)-2.*W(I,L)+W(I+1,L))/(HX*HX)
     STY(I,L) = (BMY(I,L)*6)/(TH*TH)
1650 CONTINUE
          WRITE(3,45) ISTA,JSTA,W(I,J),BMX(I,J),BMY(I,J),STX(I,J),
  1                    STY(I,J)
1700 CONTINUE
1800 CONTINUE
     WRITE (3,40)
          DO 1960 J = 4,MYP4
          DO 1950 I = 4,MXP4
```

```
        ISTA = I-4
        JSTA = J-4
        TMX = (CX(I,J)+CX(I,J+1)+CX(I+1,J)+CX(I+1,J+1))
     1       *0.250*(W(I-1,J-1)-W(I-1,J+1)-W(I+1,J-1)
     2       +W(I+1,J+I))/(4.*HX*HY)
        TMY = (CY(I,J)+CY(I,J+1)+CY(I+1,J)+CY(I+1,J+1))
     1       *(-0.250)*(W(I-1,J-1)-W(I-1,J+1)
     2       -W(I+1,J-1)+W(I+1,J+1))/(4.*HX*HY)
        SHEAR = (TMX*6)/(TH*TH)
        DF(I,J) = SQRT((STX(I,J)-STY(I,J))**2+4*SHEAR*SHEAR)
        WRITE(3,45) ISTA,JSTA,TMX,TMY,SHEAR,DF(I,J)
1950 CONTINUE
1960 CONTINUE
        WRITE(3,11)
  11 FORMAT( ' END OF PROBLEM                    ////')
        GO TO 1000
1005 RETURN
        END
            SUBROUTINE MADD(N,M,AK4)
            DOUBLE PRECISION AK4,TEMP
            DIMENSION AK4(195,195)
            DO 100 I = 1,N
            DO 100 J = 1,M
            TEMP = AK4(I,J)
            AK4(I,J) = 0.0
            IF(I.EQ.J) GO TO 101
            AK4(I,J) = TEMP
            GO TO 100
101         AK4(I,J) = -1.0+TEMP
100         CONTINUE
            RETURN
            END
            SUBROUTINE INVRSL(AK4,N,L)
            DOUBLE PRECISION AK4,G,H,D,E
            DIMENSION AK4(195,195),G(195),H(195)
            NN = N-1
            AK4(1,1) = 1.0/AK4(1,1)
            DO 110 M = 1,NN
            K = M+1
            DO 60 I = 1,M
            G(I) = 0.0
            DO 60 J = 1,M
60          G(I) = G(I) +AK4(I,J) * AK4(J,K)
            D = 0.0
            DO 70 I = 1,M
70          D = D + AK4(K,I)*G(I)
            E = AK4(K,K)-D
            AK4(K,K) = 1,0/E
            DO 80 I = 1,M
80          AK4(I,K) = -G(I)*AK4(K,K)
            DO 90 J = 1,M
```

```
                    H(J) = 0.0
                    DO 90 I = 1,M
       90           H(J) = AK4(K,I)*AK4(I,J)+H(J)
                    DO 100 J = 1,M
       100          AK4(K,J) = -H(J)*AK4(K,K)
                    DO 110 I = 1,M
                    DO 110 J = 1,M
       110          AK4(I,J) = AK4(I,J)-G(I)*AK4(K,J)
                    RETURN
                    END
                    SUBROUTINE MTA(IMAX,JMAX,AK4,LMAX,KMAX,AFF)
C        A*B,RESULT IN A WITH SIZE OF A(IMAX,JMAX), ANDB(JMAX,KMAX)
C            VALUE OF B MATRIX IS PRESERVED
                    DOUBLE PRECISION AK4,AFF,G
                    DIMENSION AK4(195,195),AFF(195,195),G(195)
                    DO 6 I = 1,IMAX
                    DO 5 K = 1,KMAX
                    G(K) = 0.0
                    DO 5 J = 1,JMAX
       5            G(K) = G(K) + AK4(I,J)*AFF(J,K)
                    DO 6 K = 1,KMAX
       6            AK4(I,K) = G(K)
                    RETURN
                    END
                    SUBROUTINE MTH(IMAX,JMAX,AFF,AA6)
                    DOUBLE PRECISION AFF,AA6
                    DIMENSION AFF(195,195),AA6(195,1)
                    DO 7 I = 1,IMAX
                    G = 0.0
                    DO 5 J = 1,JMAX
       5            G = G + AFF(I,J)*AA6(J,1)
       7            AFF(I,1) = G
                    DO 6 I = 1,IMAX
       6            AA6(I,1) = AFF(I,1)
                    RETURN
                    END
                    SUBROUTINE MTB(IMAX,JMAX,AK1,KMAX,AK2)
                    DOUBLE PRECISION AK1,AK2,G
                    DIMENSION AK1(41,41),AK2(41,195),G(41)
                    DO 6 K = 1,KMAX
                    DO 5 I = 1,IMAX
                    G(I) = 0.0
                    DO 5 J = 1,JMAX
       5            G(I) = G(I) + AK1(I,J)*AK2(J,K)
                    DO 6 I = 1,IMAX
       6            AK2(I,K) = G(I)
                    RETURN
                    END
                    SUBROUTINE MATINV(AK1,N)
                    DOUBLE PRECISION AK1,INDEX,TEMP
                    DIMENSION AK1(41,41),INDEX(41,2)
```

```
              DO 108 I = 1,N
108           INDEX(I,1) = 0
              II = 0
109           AMAX = -1.0
              DO 110 I = 1,N
              IF(INDEX(I,1)) 110,111,110
111           DO 112 J = 1,N
              IF (INDEX(J,1)) 112,113,112
113           TEMP = DABS(AK1(I,J))
              IF (TEMP-AMAX) 112,112,114
114           IROW = I
              ICOL = J
              AMAX = TEMP
112           CONTINUE
110           CONTINUE
              IF (AMAX ) 225,115,116
116           INDEX(ICOL,1) = IROW
              IF(IROW-ICOL) 119,118,119
119           DO 120 J = 1,N
              TEMP = AK1(IROW,J)
              AK1(IROW,J) = AK1(ICOL,J)
120           AK1(ICOL,J) = TEMP
              II = II+1
              INDEX(II,2) = ICOL
118           PIVOT = AK1(ICOL,ICOL)
              AK1(ICOL,ICOL) = 1.0
              PIVOT = 1.0/PIVOT
              DO 121 J = 1,N
121           AK1(ICOL,J) = AK1(ICOL,J)*PIVOT
              DO 122 I = 1,N
              IF (I-ICOL) 123,122,123
123           TEMP = AK1(I,ICOL)
              AK1(I,ICOL) = 0.0
              DO 124 J = 1,N
124           AK1(I,J) = AK1(I,J) - AK1(ICOL,J)*TEMP
122           CONTINUE
              GO TO 109
125           ICOL = INDEX(II,2)
              IROW = INDEX(ICOL,1)
              DO 126 I = 1,N
              TEMP = AK1(I,IROW)
              AK1(I,IROW) = AK1(I,ICOL)
126           AK1(I,ICOL) = TEMP
              II = II-1
225           IF(II) 125,127,125
127           GO TO 134
115           WRITE(3,133)
133           FORMAT( ' ZERO PIVOT ')
134           RETURN
              END
              SUBROUTINE ADJUST(AK3,AK2,AK4 ,L,M,N)
```

```
        DOUBLE PRECISION AK3,AK2,AK4
        DIMENSION AK3(195,41),AK2(41,195),AK4(195,195)
        DO 6 K = 1,N
        DO 5 I = 1,L
        DO 5 J = 1,M
5       AK4(I,K) = AK4(I,K) - AK3(I,J)*AK2(J,K)
6       CONTINUE
        RETURN
        END
```

CHAPTER EIGHT

Properties of Landing Gears Relating to Pavement Design

The stated purpose of a landing-gear system, in general terms, is to enable an aircraft to (1) taxi or roll up to its takeoff position and away from the end of the runway; (2) take off without the use of a launching facility and change incidence for lift off; (3) change its direction of motion at the moment of landing from a downward glide to a horizontal run along the runway; and (4) carry its own means of retarding forward motion or braking without the need for external arresting equipment. Now let us look at some of these individual operational conditions.

Static Support with Relatively Small Displacement

For a simple air-cylinder type of shock strut (see Fig. 8.1), external pressure P and stroke extension E are governed by Boyle's and Charles' laws for perfect gases in the form

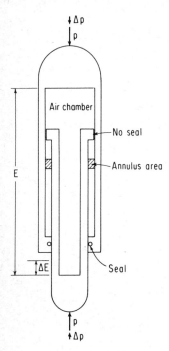

Fig. 8.1 Schematic arrangement of a simple air cylinder.

$$\frac{\Delta P}{P} = -\frac{\Delta E}{E} \qquad (1)$$

The oleo-pneumatic type of strut is somewhat more complicated, but a similar pressure-stroke relation exists.

Slow Taxi Speeds

Each landing-gear unit may be considered as two springs in series—the strut spring and the tire spring. The series combination results in the gear unit having a lower spring rate than either the strut or the tire. For a given load P, the strut deflects $\Delta_s = P/K_s$ and the tire $\Delta_t = P/K_t$. The spring rate of the combination is

$$K = \frac{P}{\Delta_s + \Delta_t} = \frac{K_s K_t}{K_s + K_t} \qquad (2)$$

This condition is satisfactory as far as the main landing gear goes, and for passengers around the center of gravity, the ride is fine, but low spring rates for nose gears (typical practice) mean large deflections over undulated roughness and rather unpleasant accelerations for forward-cabin passengers.

Moderate Taxi Speeds

At moderate taxi speeds, say up to about 40 knots, oil damping forces come more strongly into play. The schematic arrangement is shown in Fig. 8.2a. For this condition, the ground load P deflects the tire an amount P/K_t and the shock strut deflects an amount

$$\frac{P - cv_z^2}{K_s} \tag{3}$$

It is seen that most of the strut load can be resisted by damping forces.

High Taxi Speeds

At high taxi speeds, another factor comes into play to further reduce strut stroking. This is the mass damping effect of the wheel tire, brake, and piston assembly (the unsprung mass). Consequently, significant deflections will be taken by the tires with negligible strut stroking. The system works as shown in Fig. 8.2b.

Airplane Kinetic Energy

At the instant of touchdown, the aircraft has completed the flare, its weight is assumed to be supported by wing lift, and it is descending at a sinking velocity, v_z fps (see Fig. 8.3). Energy to be absorbed by the landing-gear shock strut is

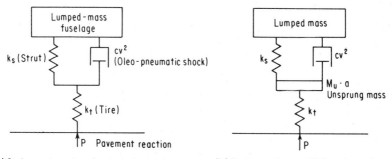

(a) Spring system at moderate taxi speeds   (b) Spring system at high taxi speeds

Fig. 8.2 Schematic arrangement of shock strut system.

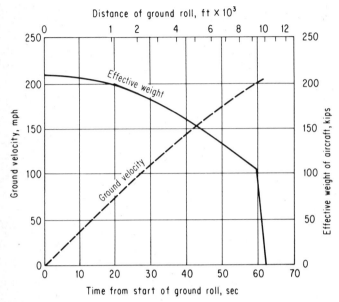

Fig. 8.3 Take-off of an aircraft.

$$KE_{gear} = \frac{Wv_z^2}{2g} \tag{4}$$

Energy to be absorbed by the landing-gear brakes is

$$KE_{brakes} = \frac{Wv_x^2}{2g} \tag{5}$$

Actually, some of the kinetic energy is absorbed by aerodynamic drag, reverse thrust, etc., but for order-of-magnitude purposes, Eq. (5) is close enough. Since kinetic energy may be equated with work performed by a device that produces a force F over a distance S, a rough estimate can be obtained of the forces involved in the shock strut and brakes.

Shock Strut

Assuming the efficiency of shock strut and tire (tire deflection plus shock-strut stroke) to be n and an effective strut stroke distance to be h, the equilibrium of strut is

$$nPh = \frac{Wv_z^2}{2g} \qquad (6)$$

Brakes

For a landing roll of L distance, the constant braking force F is

$$FL = \frac{Wv_x^2}{2g} \qquad (7)$$

or in term of g's of deceleration, $a = F/W$, or

$$a = \frac{v_x^2}{2gL} \qquad (8)$$

Landing Impact at Touchdown

Figure 8.4a shows the strut fully extended, with the "air" and the oil at the same pressure. Note that the oil is a few diameters above the orifice plate.

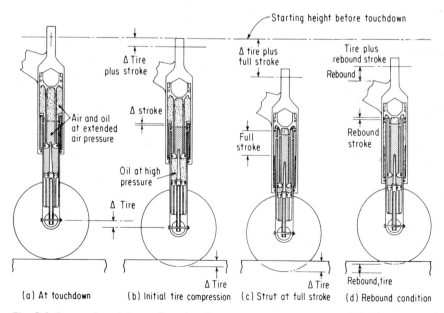

(a) At touchdown   (b) Initial tire compression   (c) Strut at full stroke   (d) Rebound condition

Fig. 8.4 Stages of touchdown of an aircraft.

Figure 8.4b shows the tire compressed to an amount just exceeding the amount required to overcome the extended airload plus friction. At this point, there can be a sudden increase in stroke velocity due to reduced friction—this is usually damped by an increased oil pressure.

Figure 8.4c shows the strut at full stroke and one-half tire deflection. At this point, the vertical downward velocity of the airplane should be zero. The air and oil pressures are again equal. But since kinetic energy is stored in two springs, the strut air spring, and the tires, the airplane starts to rebound.

At the instant of rebound, air pressure higher than static is acting on the piston area to extend the strut. This accelerating force leads to a strut extension velocity which is damped by high-pressure oil in the rebound check annulus (see Fig. 8.4d).

Natural Frequency

The complete cycle of loading and rebound represents the natural frequency of the tire–shock-absorber system. It is an important parameter in evaluating the interaction of aircraft and pavement.

Insofar as pavement engineering is concerned, the energy absorbed in the gear system will reflect the magnitude of contact pressure between gear wheels and supporting surface. As the construction of landing gear involves many mechanical components, there is no exact method for estimating accurately the performance of a gear assembly. In general practice, arbitrary rules are used by the industry, based on experience acquired over the years, and proof tests are specified by the governing agencies who determine the policy of airworthiness. As a consequence, various types of landing gear have been developed by the manufacturers for adoption by the aircraft industry. In the field of airframe design, landing-gear design has a peculiar fascination for many engineers. It combines the best in mechanical and structural engineering and requires high precision in manufacture. As for pavement design, there is practically no study of the dynamic interaction of landing gear and supporting structure. As the landing gears are the only contact with the runway structure, a better understanding of the structural as well as dynamic properties of landing gears is important in evaluating the envelope of pavement loads.

8.1. STRUCTURAL CHARACTERISTICS OF LANDING GEAR

Although the construction and physical properties of landing gears vary somewhat among aircraft manufacturers, the actual performance is similar in a general way. From the viewpoint of a structural engineer, the physical properties of a landing gear can be described as follows.

The major components of a landing gear are rubber tires, shock absorber, and retraction mechanisms. The first requirement of the landing gear is to support the load arising from taxiing, turning, swinging, braking, and the associated dynamic acceleration due to rough rolling. According to FAA regulations, the landing gear must be designed for loads in which (1) the limiting vertical load factor is 1.2 of normal landing weight, or 1.0 of the maximum takeoff weight; (2) an effective horizontal drag force is 0.8 times the vertical reaction; and (3) the side ground reaction due to turning of the aircraft is 0.5 of the vertical reaction at each wheel.

The second requirement of the landing gear is met by absorbing the vertical component of the kinetic energy of a descending aircraft. The total energy per landing gear is governed by Eq. (6), in which h can be considered as the equivalent drop height. To bring the aircraft to rest, the velocity must be reduced to zero and the total dynamic energy absorbed by the pneumatic tires, the shock absorber, and the deflection of gear mechanisms. The approximate mathematical expression can be shown as

$$\frac{W v_z^2}{2g} = n W_L (\Delta_s + \Delta_t + \Delta_g) \tag{9}$$

where Δg is the deflection of gear mechanisms and W_L is the maximum load developed during the landing. It should be noted that Eq. (9) is derived on the assumption that the descending energy is completely absorbed by the landing gear. In reality, there is always some energy absorbed by the supporting surface, particularly in landings on a runway pavement. The energy absorbed by the supporting structure will not affect the ultimate strength of a landing gear but will delay its potential rupture.

As shown in Eq. (9), the significant factors governing the stress-strain relation of a landing gear are the physical properties of the pneumatic tires and the shock absorber and the sinking velocity of an aircraft. Without describing the mechanical details, a brief review of the structural performance of each component will be given.

Tires

A pneumatic tire is able to carry load only when the inflation pressure can distribute the load evenly over the rim of a wheel. For modern air-transport tires, the normal inflation pressure ranges from 140 to 250 psi, and an increase of load capacity will result in the enlargement of contact area on the ground. The relation between external load and tire deflection is usually determined by empirical formula provided by the tire manufacturers. The effects of wall stiffness and tire thickness make exact determination impractical. An approximate flotation formula can be rewritten as

$$W = k\Delta_t p \sqrt{wd} \tag{10}$$

where W = static load on tire
k = empirical coefficient
p = inflation pressure
w, d = width and outside diameter of tire, respectively

Modern air-transport tires are normally designed to be 28 to 35 percent deflected under static load and thus to be fully squashed at about three times the static load, that is, to have a dynamic load factor of $3.0g$. Road-vehicle tires operate at much lower deflection—14 to 18 percent—and thus can carry relatively much higher dynamic loads.

Under a dynamic load, the kinetic energy of a descending aircraft is partly absorbed by the work done in tire deflection. The ratio of actual energy absorbed by the tire to the maximum kinetic energy of landing impact is called the *absorption efficiency*, n_t, of the tire and is governed by

$$n_t = \frac{Wv_z^2}{2g\Delta_t W_L} \tag{11}$$

For modern aircraft tires, the absorption efficiency is normally 0.45 to 0.47. In Fig. 8.5, the geometry and load distribution of a tire are illustrated.

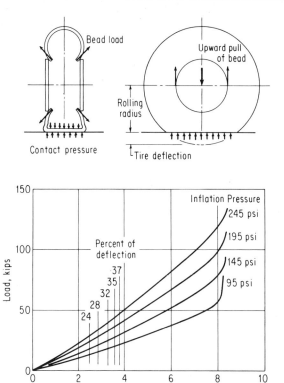

Fig. 8.5 Characteristics of pneumatic tire.

Tire Test on Pavement

For ordinary pavement construction, expansion joints vary from 1/2 to 1 in. in width. The effect of rubber tires passing over such gaps is not quantitatively established. A series of tire tests was conducted at LaGuardia Airport. The aircraft used was a DC-6, having two twin-wheel landing gears. Measurements were taken while the tires were centered on a template having a specified gap setting at each test. The test results are summarized in Table 8.1. It is noted that the rigidity of the tire wall makes a significant contribution in bridging over joint gaps without causing excessive deflection of the tire.

Shock Absorber

Several types of shock absorbers have been used by airframe industries. Oleo-pneumatic designs are more efficient in absorbing landing energy and are used on the landing gear of modern aircraft.

TABLE 8.1 Tire Imprint Area over Joints

Joint gap, in. (1)	Edge of rim to ground, in. (2)	Depth of penetration, in. (3)	Tire imprint	
			Length, in. (4)	Width, in. (5)
0	11.12	0.00	15.38	12.38
2	11.12	0.19	15.62	12.75
3	11.12	0.25	16.00	12.75
4	11.12	0.38	16.38	12.50
5	11.06	0.50	16.62	12.38
6	11.00	0.75	16.50	12.75
7	10.94	0.94	17.25	12.50

The basic construction of the shock absorber consists of a series of oil dashpots and nitrogen gas. During a compression stroke, oil inside the absorber is forced through orifices and compresses the gas in the upper chamber of the absorber. The travel of the floating piston, under dashpot action, represents the energy stored in the shock absorber. According to the manufacturer's specification, the efficiency of the oleo-pneumatic shock absorber ranges from 0.75 to 0.80. Thus the combination of tire and oleo-pneumatic strut will have an efficiency of 86 to 90 percent. Since the introduction of systems of pavement design and aircraft-pavement interaction by the writer, the airframe industry has reorganized the advantage of having soft landing gear, which means a lower natural frequency and a more efficient oleo-pneumatic system to increase the payload of an aircraft. A new generation of landing gear is in the developing stage. In the design of the new Lockheed L-500 and possibly L-1011, a double oleo system is included, as shown in Fig. 8.6. For the first time, the development of a new pavement design method will have a significant effect in improving the efficiency of landing-gear systems.

The energy stored in the compressed shock absorber and in the deflected tire is dissipated by recoil damping caused by expansion of the gas. Consequently, an extension of piston stroke results. The static deflection of a shock absorber under the maximum static gear load is usually designed to be about one-third of the fully closed position, and the ultimate compression at full closure is therefore about $3.0g$ at the maximum ramp weight. A typical static-load deflection test of the tire and gear mechanism of the Boeing 747 is shown in Fig. 8.7.

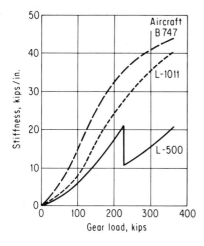

Fig. 8.6 Gear stiffness character-istics of jumbo aircraft.

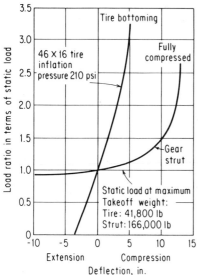

Fig. 8.7 Load deflection limit of main landing gear of B747 aircraft.

The performance of a landing-gear assembly cannot be determined accurately by any theoretical or empirical methods. It is always desirable to determine the energy-absorbing efficiency by experimental drop tests. Since an aircraft lands at a sinking velocity, a free drop of this type gives an appropriate potential energy at the moment of impact. However, during vertical deflection of the landing gear, an additional potential energy due to this travel is also produced. Thus the weight dropped should be adjusted so that

$$W_e(h + \Delta_t + \Delta_s) = W(h + \Delta_t + \Delta_s) - F_L(\Delta_t + \Delta_s) \qquad (12)$$

where W_e = effective jig weight to be used in drop test
   W = static landing weight of aircraft per gear
   h = free drop height
   F_L = wing lift to aircraft weight
If a wing lift equal to the airplane weight is simulated, the effective
mass to be used in the drop test should be

$$W_e = W \frac{h}{h + \Delta_t + \Delta_s} \qquad (13)$$

Under the FAA specification, the landing gear must withstand a
descending velocity of 10 fps at the design landing weight. The
dynamic-load factor at such landing is usually $2.0g$ at the designed
landing weight, and the factor of safety against ultimate failure
should be greater than 1.5. In addition to this limiting load factor,
the landing gear should demonstrate in the drop test that it can
withstand a sinking velocity of 12 fps at the designed landing weight.
Hence, the ultimate strength of a landing gear is about $3.0g$ at the
maximum ramp weight. The results of a typical main-landing-gear
drop test are shown in Figs. 8.8 and 8.9.

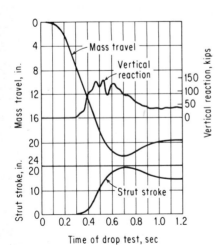

Fig. 8.8 Drop test of Boeing
707 main landing gear.

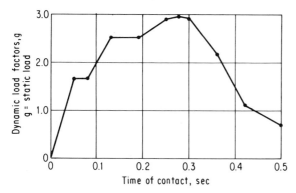

Fig. 8.9 History curve of drop test.

8.2. OPERATIONAL LOAD OF VEHICLE

In the planning of airports, flight patterns, ground aviation facilities, and major types of aircraft operations have been thoroughly studied. For pavement design, it is necessary to study only the characteristics of certain specified aircraft. Airplanes come and go. Many of them did not survive very long. The most popular airplanes, which have made significant contributions to the development of modern air transportation, are the DC-3s, 727, 707, and DC-8. The introduction of the B747 in 1969 created a sensation in the aviation industry, but the coming DC-10 and L-1011 will be truly beautiful birds for the next generation. The development of the Concorde SST and the jumbo cargo plane L-500 and their impact on the aviation industry remain to be seen. Insofar as aircraft load is concerned, within a foreseeable future, say 1980 to 1985, the predominant aircraft gear load will resemble those of the B747 and DC-10. The development of 1-million- or 2-million-lb aircraft will take some time, but there are good reasons for designing today's pavement for the future operation of 2-million-lb aircraft. The excessive investment in capital construction should be justified in the overall economic study relating to capital outlay and projected revenue. In Table 8.2, the geometric and load characteristics are given for aircraft now in common operation. The information is compiled from the point of view of pavement engineers and is not necessarily the same as that presented by the aircraft manufacturer. It is suggested that in the actual pavement design, engineers should consult with the aircraft industry to obtain the latest information on the aircraft concerned.

As mentioned earlier, the performance and design of an aircraft depend largely on the result of proof tests specified by the governing agency. Consequently, the working stress and ultimate strength of all structural components are within an identical range of stress variation. The maximum ramp weight of an aircraft is usually considered as the basic load in static analysis. The limit load anticipated in normal operation conditions, such as taxiing impact, ground turning, brake roll, takeoff, and structurally hard landing, is usually designed to be less than $2.0g$. At such a limit load, the deformation of structural elements will have no effect on the safe operation of an airplane. For the ultimate load condition, the FAA specification requires that the landing gear be demonstrated by a test simulating a sinking velocity of 12 fps at design landing weight. In this test, the landing gear must support the load for at least 3 sec before failure. It is a general design practice that the ultimate strength of a landing gear is slightly greater than $3.0g$ at the maximum ramp weight. Fatigue strength of metal is usually considered. Thus the ultimate strength of a new aircraft is about 10 to 20 percent greater than that of an aged one.

In designing the structural components of an airplane, several load criteria and test experiments have been developed by the industry. These are primarily concerned with the load conditions imposed on the aircraft itself. Consequently, there is practically no information on the interaction of landing gear and the supporting ground structure. In order to develop the load concept for pavement structures, a detailed review is given in the following of airplane loads under various operating conditions.

Vertical Ground Loads

There are three groups of vertical loadings: static, taxiing, and landing loads. The maximum ramp weight represents the airplane carrying the full payload and fuel. This is the maximum static load at taxiing and takeoff. At normal landing, an aircraft is much lighter because part of the fuel has been consumed during the flight. There is no definite relation between the maximum ramp weight and the landing weight. The landing weight decreases with increasing range (distance) of operation. Manufacturers specify only the maximum limit of landing weight. Prior to any emergency landing, excess fuel is usually jettisoned whenever possible. The maximum landing weight, as specified by the manufacturers, is shown in Table 8.2. This is the basic load for determining the dynamic impact in drop tests (see Fig. 8.9).

During the taxiing of an aircraft, the effective main-gear load will be slightly reduced because of the forward motion and wing lift of the airplane. The static load at taxiing is an academic value for a moving aircraft. At the moment of takeoff, the lift of nose wheel plus the inertia of the airplane will cause a slight increase of main-gear reaction. The most serious load condition is usually developed during ground turning. The centrifugal force at turning, according to the FAA specification, is 0.5 of the maximum takeoff weight. This is equivalent to a turning radius of 100 ft at a taxiing speed of 30 knots. The elevation of the center of gravity of the aircraft and the spacing of the landing gear determine the increase of outer-gear reaction.

The landing impact of the main gear is commonly determined by drop tests. Measurements of deceleration and travel provide the information for determining the energy absorption of the gear assembly. Since an aircraft lands at a certain sinking velocity, a free drop of this type gives the correct potential energy at the moment of impact, but there is an additional potential energy due to the travel of the mass and the compression of the shock absorber and rubber tires. Thus the weight dropped should be adjusted by Eq. (13) to reflect the static landing weight of an aircraft at a specified ratio of wing lift. In Fig. 8.10, the results of a series of drop tests are computed to indicate the load factor between the measured dynamic reaction and the adjusted static weight of an assumed wing lift. It can be seen that the landing gear tested will withstand the impact at a sinking velocity of 12 fps and that the maximum dynamic stress in the landing gear is about $2.7g$, which is about 10 percent below its ultimate strength of $3.0g$. At a crash landing, the gear is assumed to fail. There are three ways in which a gear can fail: collapse under vertical impact, breaking due to horizontal drag, and crippling due to side load. The magnitude of ultimate design capacity, as given by the manufacturers, is about $3.0g$.

On occasions, an aircraft may land without gear. As the fuselage is always less rigid than the landing gear, there will be more damage to the airplane than to the supporting structure. In another type of emergency landing, the airplane returns immediately after takeoff.

TABLE 8.2A General Dimensions of Aircraft (in feet-inches)

Dimension	Aircraft type									
	DC-8-63	DC-10-10	DC-10-30	B727-200	B707-320C	B747	B747-B	L-500	L-1011-385-1	Concorde
Overall length, A	187–5	181–4½	181–11	153–2	152–11	229–2	229–2	247–10	176–4	193–0
Wingspan, B	148–5	155–4	161–4	108–0	145–9	195–8	195–8	222–8	155–4	83–10
Nose-to-nose gear, C	15–11	27–10½	27–11	15–1	17–5	25–5	25–5	34–10	29–9	61–6
Height of fuselage, D	19–8	28–1	28–2	17–6	18–1	34–1	34–1	33–0	26–7	21–5
Body height, E	13–7	19–9	19–9	13–2	14–3	23–0	23–0	26–7	19–7	11–4
Body width, F	12–3	19–9	19–9	12–4	12–4	21–4	21–4	23–8	19–7	10–0
Inboard engine, G	2–7	2–11	3–0	11–0	2–9	4–7	4–7	13–4	2–11	6–6
Outboard engine, H,	4–2	---	---	---	4–7	6–11	6–11	10–9	---	6–6
Inboard engine, J	26–0	27–7	27–7	10–0	32–0	39–2	39–2	39–8	34–10	16–0
Outboard engine, K	45–0	---	---	---	50–0	69–6	69–6	61–11	---	20–0
Tail height, L	41–10	58–0	58–0	34–0	41–10	64–3	64–3	66–3	55–4	38–0
Tail engine, M	---	27–6	27–6	18–0	---	---	---	---	29–2	---

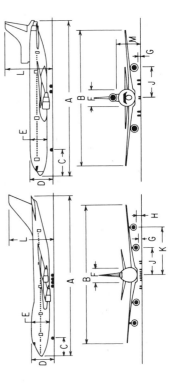

TABLE 8.2B General Dimensions of Landing Gear (in feet-inches)

Dimension	Aircraft type									
	DC-8-63	DC-10-10	DC-10-30	B727-200	B707-320C	B747	B747-B	L-500	L-1011-385-1	Concorde
a	1-6½	2-0	2-1	2-0	1-10	3-0	3-0	2-6¾	2-0	5-1
b	77-6	72-4½	72-4½	66-3	59-0	78-11½	78-11½	63-9	70-0	59-8
c	4-7	5-4	5-4	0	4-8	4-10	4-10	6-0	5-10	5-5½
d	2-8	4-6	4-6	2-10	2-10½	3-8	3-8	4-5	4-4	2-2¼
e	20-10	35-0	35-0	18-9	22-1	36-2	36-2	25-10½	36-0	25-4
f	---	---	---	---	---	12-6	12-6	---	---	---
g	---	---	2-6	---	---	10-1	10-1	18-4	---	---
h	---	---	3-1½	---	---	---	---	---	---	---

L-500

B747

DC-8, DC-10, B727,
B707, L-1011, and
Concorde

TABLE 8.2C General Weights of Aircraft and Landing Gear

Measurement	Aircraft type									
	DC-8-63	DC-10-10	DC-10-30	B727-200	B707-320C	B747	B747-B	L-500	L-1011-385-1	Concorde
Max. ramp weight, 1,000 lb	358	433	558	173	336	713	778	862	428	379
Max. takeoff weight, 1,000 lb	355	430	555	172	334	710	775	859	426	376
Max. landing weight, 1,000 lb	258	364	403	150	247	564	564	700	358	240
Operating empty weight, 1,000 lb	159	235	264	97	136	353	366	329	234	166
Nose gear max. static, 1,000 lb	30.5	42.0	69.2	16.5	34.6	74.0	89.2	93.0	40.2	29.3
Nose gear braking, 1,000 lb	64.5	71.6	107.5	23.7	51.8	109.8	129.9	171	68.1	46
Main gear max. static, 1,000 lb	172.1	203.5	210.5	79.9	157	166.5	181.8	202	203	180
Main gear braking (H), 1,000 lb	162	188.7	194.8	26.9	52.2	50	60	186	66.4	42.75
Main gear max. traction (H), 1,000 lb	121.9	138.4	143.1	63.9	125.5	133.2	145.4	162	139	144
C.A.R. takeoff length, ft	11,800	8,500	11,100	8,100	10,500	10,500	12,250	10,600	7,800	9,600
Min. landing length, ft	7,200	5,600	6,000	5,200	7,200	7,750	7,750	7,700	5,800	8,000
Min. turn radius, ft	87.8	78.8	79.8	66	70.5	85	85	83	77.9	71
Max thrust per engine, 1,000 lb	19.0	40.0	49.0	15.0	18.0	43.5	43.5	45.5	42.0	38.4
Main landing gear, size of tires, in.	45×17	50×20	52×21	49×17	46×16	46×16	49×17	56×16	50×20	47×16
MLG static load per tire, 1,000 lb	43.0	50.9	52.6	40.0	39.3	41.6	45.5	51.0	50.8	45.0
MLG tire pressure, psi	196	173	158	168	180	204	185	210	180	184
Nose wheel, size of tires, in.	34×11	37×15	40×16	32×12	39×13	46×16	49×17	49×17	36×11	31×11
Nose wheel static load per tire, 1,000 lb	15.2	21.0	34.6	8.3	17.3	37.0	44.6	23.3	20.1	14.6
Nose wheel tire pressure, psi	143	163	185	100	115	165	180	130	185	174

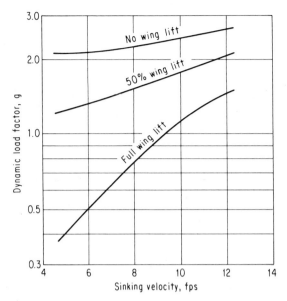

Fig. 8.10   Result of drop test of main landing gear.

There is then not sufficient time to jettison the excess fuel, and the landing impact will be far greater than with the normal landing load. As the ultimate strength of the landing gear remains the same under all landing conditions, the crash landing will not cause impact greater than the failure of the landing gear.

Horizontal Drag Force

There are several types of horizontal drag forces encountered by a moving aircraft: rolling friction, acceleration or deceleration, aerodynamic drag or engine thrust, and gear braking force. The coefficient of rolling friction varies considerably on different types of ground surface. From the discussion of Chap. 6, the coefficient of rolling friction is about 0.02 to 0.04 for good concrete runway surfaces. During low-speed taxiing, the maximum coefficient of drag friction of a treaded tire can be 0.6 to 0.8 on dry concrete surface. At the normal touchdown velocity, the coefficient of drag friction is reduced to 0.3 to 0.4. For the purpose of designing landing gear under maximum drag conditions, a tire friction of 0.8 is usually assumed. Slight variation is usually made by manufacturers for various types of aircraft.

Aircraft are fitted with brakes for the purpose of (1) stopping the aircraft at landing, (2) steering the aircraft on ground, (3) restricting forward speed at taxiing, and (4) holding against engine thrust when the engines are run up for testing purposes. According to the experience of aircraft operation, a good brake can give a steady deceleration of 9 to 11 ft/sec^2, which is equivalent to $0.3g$. Such a braking capacity is in proportion to the full engine thrust, which is about 30 percent of the aircraft takeoff weight. As usual, the ultimate capacity of a brake is about three times the normal brake capacity.

Under normal landing conditions, the aircraft is subject to a retarding force due to the aerodynamic drag which is in addition to the rolling and braking resistances. Such trim control is known to be less than 1 mph/sec, equivalent to a deceleration of $0.05g$. The maximum drag force anticipated at landing is $0.37g$, which consists of the braking force of $0.3g$, the aerodynamic drag of $0.05g$, and the rolling friction of $0.02g$. At a crash landing, if the gears are fully braked, the ultimate strength of a brake will be the governing load factor. The maximum vertical reaction which can be developed in the landing gear is equal to the ultimate braking strength of the gear divided by the coefficient of drag friction. The landing gear, in such a case, will collapse at the point of contact.

Aircraft Load Tests

The above load descriptions are made available by the airframe manufacturers. They are the loads the manufacturers use in their aircraft design. There is much information that is not available from the aircraft manufacturer but is of vital importance to pavement engineers. The most important of these data are the actual vibratory forces transmitted through the aircraft tires to the supporting structure. No such information has been developed by the aircraft industry. It therefore became necessary to initiate several tests to ascertain the nature and magnitude of vibratory forces emitting through the rubber tire.

Since the landing gears of most aircraft all have the same general type of construction and use a shock strut to transmit the aircraft forces to the tires, it was considered indicative to test only the three basic types of airplane: piston, turboprop, and jet. For availability of the various aircraft, the test was conducted at the John F. Kennedy

International Airport. A special test platform was constructed to support one of the main landing gears of the aircraft and to measure the vibration in the aircraft's tires as well as in the supporting structure.

The structure provided for the tests consisted of a swing-type platform, which supported the weight of the aircraft, and a horizontal anchor plate, which was laid on top of the platform to resist the thrust forces in the landing gear (see Fig. 8.11). The swing platform isolated the vertical forces from the horizontal forces and directed them to the platform supports which were instrumented to measure the vertical vibrations. The thrust forces were absorbed by the anchor plate, and the horizontal vibrations were measured by instrumenting the anchor plate. To allow the aircraft easy access to the platform, it was set in a pit with its top flush with grade. Additional instrumentation was planned for the landing gear of the aircraft to measure the vibrations in the aircraft before they entered the strut of the landing gear. The difference between the vibrations measured on the aircraft and those measured on the test platform would indicate the interaction of landing gear and supporting structure.

Vibrations are forces which rhythmically change their magnitude. The instruments used in these tests to measure the vibrations included strain gages and an oscillograph. The testing procedure required the aircraft to be towed to a position which would center its right landing gear on the test platform. The brakes were then locked,

Fig. 8.11 Aircraft vibration test. View of right landing gear on test platform. Strain gages are used for measuring vibration.

and the engines were started and run at various power settings and in different combinations to simulate an aircraft transversing the runway structure.

The test data indicate that the piston-type DC-7 produces the largest-amplitude vibration and the jet aircraft DC-8 produces the smallest amplitude. For the DC-8, no vibratory amplitude greater than ±1.6 kips (±0.024g) was found and the major amplitude signals were below 2.5 cps. Secondary dynamic amplitudes occurred at both lower and higher frequencies. For frequencies between 3.0 and 5.0 cps, no amplitude greater than 0.60 kip (0.009g) was recorded, and for frequencies above 5.0 cps, the highest amplitude that was noted was 0.13 kip (0.002g) at 8.0 cycles. For frequencies above 8 cycles, practically no force was transmitted.

The fact that high-frequency oscillatory forces cannot be transmitted through the landing gear and tires is predictable. For the lower frequencies, of course, resonance with supporting structures might be of importance. However, pavement structures generally have natural frequencies of the order of 6 to 15 cps or higher.

It should be noted that, even at resonance, the effect of an oscillatory force is determined by the relative damping of the supporting structure. The dynamic magnification of an oscillatory force at resonance is given by the relation

$$\text{Dynamic magnification} = \frac{1}{2\beta} \tag{14}$$

where β is the coefficient of critical damping. For ordinary construction materials at moderate stress levels, β is of the order of 2 percent, and at high stress levels, it is 5 to 10 percent. Thus, even for moderate stress levels, the magnification is less than 25. Hence, the maximum effective force due to the resonant vibration of an aircraft at a natural frequency of greater than 5.0 cps would be less than $25 \times 0.13 = 3.2$ kips (0.05g).

The horizontal dynamic forces would be substantially less than this, although the quasi-static horizontal-force components are not negligible and must be considered. The thrust magnitudes are predictable from engine characteristics and the system geometry. In the case of the entire aircraft resting on a single slab rather than one landing-gear section, as in this test, the values should be precisely computable. Vibratory thrust amplitudes are negligibly small (the worst case tested, the DC-7, was less than 1 kip, or 0.03g) and are of the same frequency specturm as the measured load frequencies. They are of little, if any, importance in structural design.

Sinking Velocity

At normal aircraft landings, the descending slope varies from 1 on 20 to 1 on 50 and the approaching velocity at touchdown is about 130 to 150 knots, equivalent to 200 to 240 fps. The sinking velocity is the vertical component of a descending aircraft. It can be seen that the sinking velocity can have a divergent value, depending on the pilot and the environment of landing. At the moment of landing contact, the pilot usually brings the aircraft to a flare position. Average commercial landings are at a sinking velocity of 2 to 3 fps, and commercial landings are seldom at a sinking velocity of 6 fps. An actual observation of the distribution of landing velocity is shown in Fig. 8.12. It should be kept in mind that each airport has its own environment, which will affect the landing condition of an aircraft.

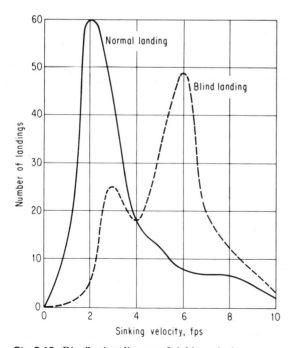

Fig. 8.12   Distribution diagram of sinking velocity.

The distribution of sinking velocity shown in Fig. 8.12 is not valid for every airport. Nevertheless, the statistical approach to sinking velocity demonstrates that the occurrence of hard landings at a sinking velocity greater than 10 fps is very remote and that the maximum sinking velocity at normal landings is likely to be less than 6 fps.

Speed of Aircraft

The foregoing analysis is for a static wheel load; that is, the aircraft does not move. As the aircraft design is considerably different from the ground vehicle, the airlift under the wings will reduce the effective wheel load at different aircraft speeds. The airlift is a function of aerodynamic pressure and engine performance. A detailed discussion will be found in Sec. 8.4 and Chap. 9.

8.3. DISTRIBUTION AND FREQUENCY OF LOADINGS

The performance of a pavement depends on the number of load repetitions over the pavement structure. The load repetition not only reduces the work stress of material but also results in progressive deformation of the pavement structure. Under actual service conditions, the load repetition at a given point of the pavement is a function of four variables: (1) traffic intensity on the pavement, (2) type of aircraft operation, (3) traffic distribution across the transverse direction, and (4) longitudinal distribution of aircraft load on runways and taxiways. The longitudinal distribution of aircraft load can be divided into landing impact and ground roll. A detailed description of these variables follows.

Pattern of Traffic Movement

The pattern of traffic movement largely depends on the geometric configuration of the airport. It is difficult to make a generalized statement about traffic patterns. Such information should be developed in the planning of the airport operation. However, in the design of airport ground facilities, it is necessary to study the traffic distribution on taxiways, runways, and apron areas. In general, the growth of an airport is indicated by the movement of aircraft in and out of the airport. The total movement of aircraft should be distributed among different portions of the airport pavement. For instance, more than 50 percent of the takeoff movement at Kennedy Airport is on runway 31L. The traffic distribution should be worked out by the density of traffic movement, such as shown in Fig. 8.13. The density of aircraft movement is represented by the percentage of total movement of the airport. For instance, in the area marked "30 to 60 percent movement," the actual aircraft movement is equal to 30 to 60 percent of the total movement at the airport.

Type of Vehicle

The effect of stress repetition on the pavement is partly governed by the type of operational vehicle. A heavy vehicle causes more damage than a lighter vehicle. Pavement damage due to one operation of a heavy vehicle may be equivalent to that due to several operations of a light vehicle. The relation between the number of repetitions and the vehicle weight should be established for the pavement design. In airport or highway operation, actual traffic is always a combination of various types of vehicles. For engineering design, it is customary to use one type of vehicle as the design standard and evaluate its

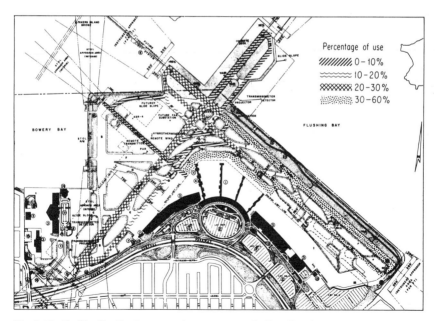

Fig. 8.13   Traffic distribution on taxiway system at LaGuardia Airport.

effect due to a given number of load repetitions. Therefore, it is necessary to convert the number of operations of other types of vehicle into an equivalent number of operations of the design vehicle. In the conversion, the theory of fatigue strength, as given by Eq. (4) in Chap. 5, can be rewritten as

$$\sigma_N = \sigma_y - (\sigma_y - \sigma_e)c \, \log N \qquad (15)$$

If the endurance limit is considered to occur at N_e number of repetitions, the relationship between the fatigue stress σ_N and the corresponding number of the load cycles N is governed by

$$\log N = \frac{\sigma_y - \sigma_N}{\sigma_y - \sigma_e} \log N_e \qquad (16)$$

For ordinary construction materials, the endurance limit is about one-half of yield strength. Assuming that stress in a pavement structure is a linear function of external vehicle load W, the equivalent number of repetitions based on the weight of a designed vehicle is equal to

$$\log N_{des} = \frac{\log N_{opt}}{1 + (W_{des} - W_{opt})/W_{des}} \qquad (17)$$

where N is the number of load repetitions, W is the weight of vehicle, and the subscripts des and opt represent design and actual operation, respectively. In reality, the stress and wheel load are not a linear function. There is no definite relation between the gross weight of an aircraft and the actual wheel load of the landing gear. For more precise analysis, the W value in the above equation should be the equivalent single wheel load of an aircraft and the stress in the pavement should be a function of wheel load, tire pressure, and contact area, as well as pavement structure. Even when the analysis is very complicated, the result still may not be reliable. For a wide variation of aircraft operation, Eq. (17) will yield a reasonable computation. An example of LaGuardia Airport operation is shown in Table 8.3. In a given day, the total operation comprises 900 aircraft movements. The type of aircraft, number of

TABLE 8.3 Movement of Fixed-wing Aircraft in a Given Day

Type of aircraft	Range of gross weight, 1,000 lb	Number of actual movements	Equivalent movement for designed aircraft
B727, etc.	153–172	370	370
DC-9, etc.	85–108	225	52
Turboprop	<64	100	17
General aviation. . .	>12.5	55	9
General aviation . .	<12.5	150	22
Total movement .		900	470

operations, and gross weight of aircraft are given. The equivalent number of movements of the designed aircraft, say B727, is reduced to 470 movements in accordance with Eq. (17).

Transverse Distribution of Wheel Load

The movement of a vehicle on a road, taxiway, or runway assumes a random distribution across the transverse direction. The load repetition at a given point is governed by the width of tire and the width of wheel path, or so-called *bandwidth*. For instance, the probability of load repetition is higher in a narrow bandwidth than in a wide band. Using a normal-distribution curve, the probability of load repetitions of B747 gears is as shown in Fig. 8.14. For various bandwidths, the probability of load repetitions is shown in Fig. 8.15 for the current generation of aircraft.

The bandwidths of an operating pavement system depend on the physical condition of the airport as well as the characteristics of the aircraft. For taxiways with centerline lights, the bandwidth has been observed to be 6 to 12 ft for 98 percent of operation; for the taxiway without centerline lights, it is 12 to 20 ft. For runways with centerline lights and instrumented navigation aids, the bandwidths are in the range of 15 to 25 ft. However, for normal landings, the bandwidths can be as much as 35 to 45 ft. With the conditions shown in Fig. 8.15, it is possible to relate the probability of load repetition with the width of the wheel path. For instance, the possibility of load repetition is 0.42 per movement of B747 aircraft in a 20-ft-wide band of wheel path. This probability is denoted as $P(y)$.

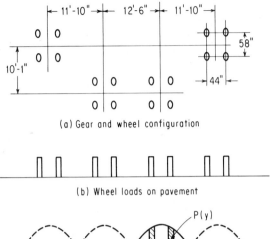

(a) Gear and wheel configuration

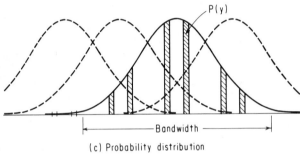

(b) Wheel loads on pavement

(c) Probability distribution

Fig. 8.14 Transverse probability distribution of wheel load—
B747 aircraft.

Longitudinal Distribution of Landing Impact

The landing impact of an aircraft depends largely on the climatic and
geometric environment of the airport as well as the navigation aid of
aircraft and ground facilities. For a typical landing condition, the
aircraft comes down at a glide slope of between 2 and 3°. Over the
threshold, the aircraft is about 50 ft above the landing surface, and
the pilot brings the aircraft into landing position and the aircraft
flares to horizontal position within a distance of about 1,200 to
1,300 ft from the threshold. At the moment of touchdown, the
horizontal velocity of the aircraft is about 125 to 145 knots and the
vertical velocity, known as the *sinking velocity,* is about 2 to 6 fps.
The landing impact zone is clearly marked on the runway surface, as
shown in Fig. 8.16. Therefore, the center of the landing impact can
be considered to be located at a distance of about 1,200 to 1,300 ft

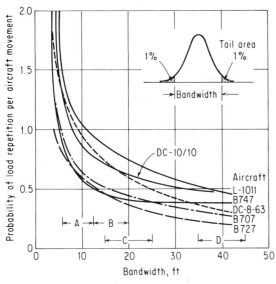

Band A: Taxiway with centerline lights
Band B: Normal taxiing
Band C: Runway with centerline lights
Band D: Normal landing

Fig. 8.15 Transverse probability distribution of wheel load against width of wheel path.

from the threshold. There are some variations in actual operation. For large aircraft, if the runway is long enough, the pilot usually overshoots the marked touchdown zone and lands at 1,500 ft or more from the end of the runway. On the other hand, if the runway is short, the pilot normally brings down the aircraft between the threshold and the marked touchdown zone. A similar landing condition is applied to general-aviation aircraft. Smaller aircraft usually land far beyond the marked touchdown zone. In general, the landing impact assumes a random distribution within the marked landing length of the runway. From observations at many airports, more than 90 percent of landings take place within a distance of about 750 ft on both sides of the designed touchdown zone. In Fig. 8.17, the probability of longitudinal landing impact $P(x)$ is given for various types of aircraft operation. The chart is constructed on the basis of the following assumptions: (1) the significant length of the touchdown zone is 1,500 ft, (2) the probability distribution of landing impact is normal (gaussian), and (3) the probability of tail

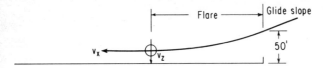

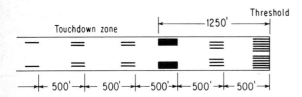

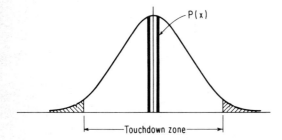

Fig. 8.16 Longitudinal distribution of wheel load.

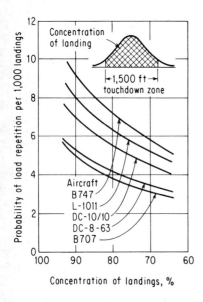

Fig. 8.17 Longitudinal probability distribution at touchdown.

areas is a variable. By combining the transverse probability of the load distribution $P(y)$ and the longitudinal distribution $P(x)$, the overall probability of landing impact is equal to $P(x)P(y)$. For B747 aircraft, the probability of normal landing when the aircraft load repeats at the same location is $0.42 \times 0.0092 = 0.0038$ load repetition per landing. This means that, one load repetition is equivalent to 265 landings.

Longitudinal Distribution of Moving Load

The longitudinal distribution of a moving load is influenced by traffic volume as well as the speed of the vehicle. For ordinary highway and taxiway design, the effect of speed is not as significant as for the runways. Therefore, the longitudinal distribution of load repetition along a taxiway is governed primarily by the transverse distribution of wheel load, as shown in Fig. 8.15. On the other hand, the effective weight of landing or takeoff aircraft is governed by such factors as speed of aircraft, point of touchdown, and body rotation. For normal takeoff, the point of body rotation is about 5,000 to 7,000 ft from the end of the runway. On runways, the first 3,000 ft from the ends are subject to heavy load intensity and large repetitions. Beyond this length, the effective aircraft load decreases due to wing lift, and the resulting stress condition in the pavement is also reduced. In general, based on the frequency and distribution of aircraft operation, the center 40-ft strip of a taxiway is more critical than the sides. For runways, the width of critical strip is about 80 ft. In the design of runway and taxiway pavements, it is necessary to consider the probability of load repetition and to design the pavement appropriately in order to achieve an economical utilization of pavement material as well as the safe operation of aircraft.

8.4. DYNAMIC IMPACT OF VEHICLE LOAD

In dealing with the dynamic interaction of vehicle and pavement, there are three distinct dynamic forces which should be clearly defined: (1) the response of the supporting structure due to rolling of the vehicle, (2) the response of the same supporting structure due to forced displacement of the vehicle, such as landing or collision condition of vehicle, and (3) the response of the vehicle due to rolling on a rough surface. The first two kinds of dynamic forces are commonly known as *dynamic impact* and will be discussed here in detail. The third kind of dynamic force has been introduced by the writer [5] and will be discussed in detail in Chap. 9.

Rolling Impact

This is the dynamic response of a structure due to the rolling of a vehicle. The vehicle is the forcing function, and the pavement structure is the responding one. For a single degree of freedom, the dynamic deflection configuration of the supporting structure is assumed to be a half sine wave. The effects of shearing deformation and torsional deflection are neglected. The computation can be further simplified by the use of the Maxwell's theorem of reciprocal deflections.

The governing equation for the dynamic response of a supporting structure to a smoothly rolling wheel load is

$$- z_d = \left(1 + \frac{F}{M_b g} \sin p_v t \right) \frac{\ddot{z}}{p_b{}^2} \tag{18}$$

The derivation of this equation is given in Chap. 9. When the mass of the vehicle remains unchanged during the motion, the dynamic deflection will be a function of the natural frequency of the supporting system. The dynamic increment of deflection is expressed in the form of z_d/z_{st}, where z_{st}, is the static deflection due to the weight of the vehicle. The magnitude of dynamic increment is commonly known as the *impact factor* of a moving load.

When the rolling vehicle acquires oscillation due to rough ridings, the initial vibration of the vehicle becomes a part of the forcing function. By assuming a complete transfer of kinetic energy, the following relation exists:

$$z_d k_b = z_v k_v \tag{19}$$

where z_v is the maximum initial amplitude of vehicle oscillation. The initial force of vehicle oscillation ΔF can be expressed by

$$\Delta F = z_v k_v \tag{20}$$

As the spring constant of the supporting structure is

$$k_b = \frac{F}{z_{st}} \qquad (21)$$

the dynamic increment due to initial vertical oscillation is

$$\frac{z_d}{z_{st}} = \frac{\Delta F}{F} \qquad (22)$$

The pulsating force of the vehicle can be expressed by

$$F(t) = F[1 + \overline{DI} \sin(p_v t - \theta)] \qquad (23)$$

where $\overline{DI}$ represents the dynamic increment due to smooth rolling and initial oscillation $\Delta F/F$ acquired during rough rolling. In computing the dynamic increment for deflection of a supporting structure, time function $F(t)$ should be used to replace the static force F in Eq. (18). An example of such a computation is shown in Fig. 8.18, which reflects the dynamic impact of a B727 aircraft

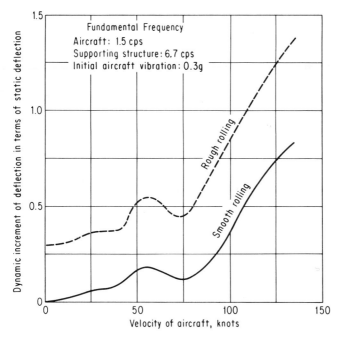

Fig. 8.18 Dynamic increment of supporting structure.

rolling on a runway structure at LaGuardia Airport. As the dynamic increment for deflection increases with the increasing static load of the exciting system, the change of effective aircraft load due to wing lift will certainly cause a corresponding change of dynamic deflection of the supporting structure. In Fig. 8.19, the solid curve represents the more realistic dynamic-load condition when the wing-lift factor is considered in the evaluation of the effective wheel load at various aircraft speeds. It is noted that the governing aircraft speed for creating the maximum dynamic response ranges from 50 to 60 knots, which is the normal range of high-speed taxiing. The body rotation of aircraft and wind flap at takeoffs would have a significant effect on the evaluation of ground loads. From the studies of aircraft operation, about 80 percent of the time of a single takeoff or landing is spent in taxiing. Insofar as rolling impact is concerned, the taxiway is actually more critical in developing dynamic forces than the runway.

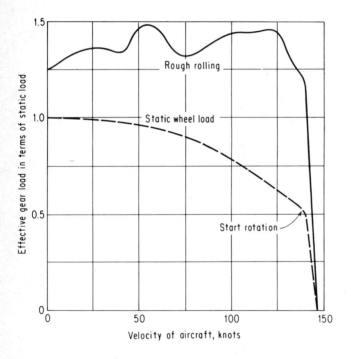

Fig. 8.19 Effect of wing lift on rolling impact.

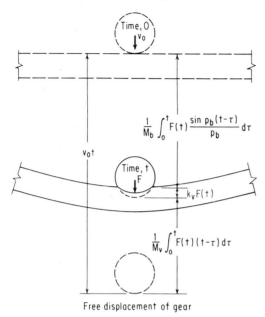

Fig. 8.20 Displacement of elastic system during collision.

Landing Impact

For the collision condition of aircraft, the displacement of mass and structure is as shown in Fig. 8.20. The general equation of displacement is

$$K_v F(t) = v_0 t - \frac{1}{M_v} \int_0^t F(t)(t - \tau)\, d\tau - \frac{1}{M_b} \int_0^t F(t) \frac{\sin p_b(t - \tau)}{p_b}\, d\tau \quad (24)$$

The first term on the right, $v_0 t$, represents the free displacement of a striking mass traveling at a velocity v_0 during the time interval t. The second term on the right represents the restraining of free displacement during the development of contact compression. The third term represents the dynamic deflection of the beam during its first mode of vibration. The term on the left side of the equation, $K_v F(t)$, represents the contact force developed in the aircraft landing gears. For $F(t) = F \sin p_v t$, the solution of the above equation gives the contact force

$$F = \frac{M_b\left(p_b^2 - p_v^2\right)}{M_b\left(p_b^2 - p_v^2\right) - M_v p_v^2} M_v p_v v_0 \tag{25}$$

For an aircraft in collision with rigid ground, the natural frequency of the ground vibration p_b is extremely high and Eq. (25) becomes

$$F = M_v p_v v_0 \tag{26}$$

or

$$F = \frac{W_v}{g} p_v v_0 \tag{27}$$

where W_v is the weight of the colliding vehicle. For a landing gear having a fundamental frequency of 1.1 cps or 6.9 radius/sec, the impact force at a sinking velocity of 12 fps is 2.58 W_v. As the ultimate strength of the landing gear is three times the maximum ramp load (about four times the landing weight, depending on the type of aircraft), the gear will not collapse at a critical sinking velocity of 12 fps if the aircraft is landed on both gears. On the other hand, if the aircraft comes in on one landing gear, the maximum collision impact will be the ultimate strength of the landing gear. The aircraft gear will collapse before developing the full dynamic impact as given in Eq. (26).

Steady State of Vibration

Another important vibration problem in runway design involves the steady state of vibration when the aircraft is completely braked and all engines are at full power, immediately before takeoff. The runway structure is subject to a continuous input of exciting energy. Assuming that there is no initial structure oscillation, that is, that $z = 0$ and $\dot{z} = 0$ when $t = 0$, the dynamic deflection of structure is

$$z = \frac{F}{M_b p_b^2} \frac{\left(p_b^2 - p_v^2\right) \sin p_v t - 2\beta p_b p_v \cos p_v t}{\left(p_b^2 - p_v^2\right)^2 + (2\beta p_b p_v)^2} p_b^2 \tag{28}$$

For the particular case in which $p_b = p_v$ at the resonant vibration, the maximum deflection is

$$z_{max} = -\frac{1}{2\beta} \frac{F}{M_b p_b^2} \tag{29}$$

where $F/M_b p_b^2$ represents the static deflection of the structure and $1/2\beta$ is the dynamic magnification factor at resonance for an infinitely long period of vibration.

Impact Factors

For the purpose of practical design analysis, the governing impact factor should be developed for the bending moment and shear forces. The relation between deflection and moment can be expressed by

$$M = -EI \frac{d^2}{dx^2} z_d \tag{30}$$

where $x = vt$ is the horizontal coordinate of the structure. As $p_v = \pi v/L$, the $p_v t$ value in Eq. (18) becomes $\pi x/L$. The bending moment at any section of the structure is obtained from Eq. (18) by differentiation with respect to x. The dynamic moment in excess of the static moment is

$$M_d = -EI \frac{\pi^2}{L^2} z_d \tag{31}$$

By the definition that the dynamic increment for deflection $DI_d = z_d/z_{st}$ and for moment $DI_m = M_d/M_{st}$, the following relation is obtained:

$$DI_m = \frac{z_{st}}{M_{st}} EI \frac{\pi^2}{L^2} DI_d \tag{32}$$

For a simple supported beam, the ratio $z_{st}/M_{st} = L^2/(12EI)$ and, therefore,

$$DI_m = \frac{\pi^2}{12} DI_d \tag{33}$$

The dynamic increment for shear and reaction can be developed in a similar manner by equating

$$V_d = EI \frac{d^3}{dx^3} z_d \qquad (34)$$

The corresponding dynamic increment for shear is

$$DI_v = \frac{\pi^3}{48} DI_d \qquad (35)$$

It is noted that there is a linear relationship between the dynamic increments for deflection, moment, and shear. Equations (33) and (35) are useful because they allow a single response quantity of deflection to be related to bending moment and shear.

8.5. DYNAMIC RESISTANCE OF MATERIAL

The dynamic strength of material is governed by the physical property of plastic flow. An exact analysis requires that the stress-strain-time-temperature relation be known for the material. In deriving theoretical relations for the stress-strain-time curves, the temperature of the material is assumed to be constant. A family of constant time-creep curves, known as an *isochronous stress-strain diagram,* is shown in Fig. 8.21. The stress-strain-time relations in tension or compression can be defined by

$$\epsilon(\sigma, t) = \frac{\sigma}{E} = K\sigma^n(1 - e^{-at}) \qquad (36)$$

where ϵ is the total strain under the stress σ during the time period t and the parameters E, K, n, and a are experimental constants of the material.

For constant stress, the time-creep relation is given by

$$\epsilon(t) = \epsilon_0 + A(1 - e^{-at}) \qquad (37)$$

By introducing the relation of strain ϵ_0 for an instantaneous load $(t = 0)$ and ϵ_e for a sustained load $(t = \infty)$, the change of strain within

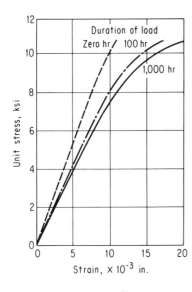

Unit stress, ksi

Duration of load
Zero hr / 100 hr

1,000 hr

Strain, X 10^{-3} in.

Fig. 8.21 Isochronous stress-strain diagrams.

the time range 0 to ∞ is an exponential function:

$$\epsilon_t = \epsilon_e + (\epsilon_0 - \epsilon_e)e^{-at} \tag{38}$$

The particular solution of the above equation for expressing time-stress relation can be written as

$$\sigma_{dy} = \sigma_y(1 + D \cdot 10^{-t}) \tag{39}$$

where σ_{dy} = yield strength under dynamic loadings
σ_y = yield strength for static loadings
t = loading period
D = an experimental coefficient of the material
The term $D \cdot 10^{-t}$ can be called the *dynamic increment of yield strength*. Laboratory tests on structural steel and concrete confirm that the dynamic increment is an exponential function of loading time to reach the yield point. At the lower range of time variation, that is, when the actual loading time is close to the standard ASTM rate of loading, the dynamic increment changes proportionally to the value of $\log t$. For common steel and concrete, the D value is in the range of 0.10 to 0.12. The increase of dynamic yield strength is about 10 percent when the maximum loading time to reach yield point is one-tenth of the standard ASTM rate of loading.

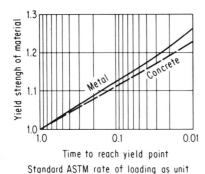

Fig. 8.22 Dynamic strength of construction material.

Time to reach yield point
Standard ASTM rate of loading as unit

For the design of runway pavements, the rate of dynamic loading ranges from 0.1 to 0.3 sec to reach peak response (see Fig. 8.9). Such dynamic loading is only two to five times faster than the rate of standard ASTM testing. The corresponding increase of dynamic yield strength amounts to only 3 to 8 percent above the static yield strength. For practical design purposes, it is a negligible amount of strength increment. Hence the dynamic strength of material employed is not a significant factor in the design of runway structures.

On the other hand, if the rate of static loading is slower than the rate prescribed in the ASTM specification, the yield stress will be somewhat under the ASTM testings. Even though the difference is only a negligible fraction, the deficiency should be taken care of in the safety factor.

REFERENCES

1. J. M. Briggs, H. S. Sver, and J. M. Louw, Vibration of Simple Span Highway Bridges, *Trans. ASCE,* vol. 124, pp. 293–294, 1959.
2. G. H. Conway, "Landing Gear Design," Chapman & Hall, Ltd., London, 1958.
3. Federal Aviation Administration, "Airplane Airworthiness, Transport Categories," Civil Aeronautic Manual 4b, Government Printing Office, 1960.
4. O. M. Sidebottom, G. A. Costello, and S. Dharmarajan, "Theoretical and Experimental Analysis of Members Made of Materials that Creep," Engineering Experiment Station Bulletin 460, University of Illinois, Urbana, 1961.
5. N. C. Yang, Interaction of Aircraft and Ground Structure, *Proc. ASCE,* vol. 95, no. ST6, June, 1970.

CHAPTER NINE
Vehicle-Pavement Interaction

In discussing the functional requirements of pavement, the question of speed and vibration of an aircraft moving over a pavement surface has been brought up. The response of an aircraft may vary on the same pavement surface, depending on the velocity and characteristics of the aircraft. There have been studies to determine the vibration problems related to the automobile and highway pavement. No definite conclusion can be derived from these studies because the nature of aircraft operations was unknown at the time and the variables involved were so numerous. The normal speed of an automobile does not exceed 80 mph, and the construction of the automobile is rather simple as compared with today's aircraft, which operates at speeds greater than 140 knots on the ground. Therefore, it becomes imperative to define the nature of aircraft operations on a specific pavement surface geometry. Without this knowledge, the engineer may not be in a position to properly design the functional surface of the pavement required for the operation of the aircraft.

9.1. VEHICLE VIBRATION

An aircraft is an assembly of structural components. Each component, from the viewpoint of dynamic analysis, can be represented by its spring-dashpot action and the effective mass participating in the vibration. For instance, the basic airplane consists of fuselage and wings of framed construction and the more flexible struts and tires for absorbing impacts. These major components have their own distinctive natural frequencies and damping characteristics.

There have been many discussions about vibration in the pilot's cabin and the discomfort of passengers, but very little consideration has been given to the contribution of the aircraft structure to the response of the pilot and passengers. Therefore, if the criteria of pilot and passengers are used to judge the random vibration of the aircraft, they will automatically include the vibration caused by rough pavement, transmitted through wheels and struts, and magnified by wing frame and fuselage. It can be seen that the pilot and passenger criteria cannot be used to describe the interaction of airplane and pavement. However, the pavement should be designed for the comfort and safety of the passengers and airline crew. This aim can be achieved only by dividing the responsibility into two groups. For the vibration and the magnification of dynamic response developed above the landing gear, the responsibility should be assigned to the aircraft manufacturer. The vibration encountered at the interface of airplane and pavement is the joint responsibility of the pavement engineer and the aircraft manufacturer. The civil engineer is responsible for supplying an operational pavement surface to accommodate the type of aircraft involved and to limit its dynamic response at the interface of landing gear and pavement. With this clarification, the interaction of pavement and aircraft is hereby limited to the study of dynamic response at the interface.

The aircraft on a pavement surface can be represented by a simple mechanical model, such as that shown in Fig. 9.1. The m represents the effective mass of the airplane above the interface. k is the spring constant of the aircraft at the interface, which is primarily governed by the spring constant of the aircraft tires. β is the dashpot action of the airplane above the interface, which is possibly governed by the damping of the pneumatic tires and the oleo-pneumatic shock absorbers. While the aircraft moves at a speed v, there is a vertical

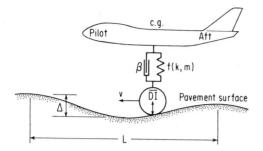

Fig. 9.1 Simplest model of aircraft vibration.

displacement of the mass z and an acceleration $\ddot{z}$. This is the simplest model of the vibration of an aircraft. Consequently, its application should be limited to conceptual studies. In describing this model of vibration, the basic property of an aircraft can be represented by the natural or fundamental frequency f, in cycles per second, which may be written

$$f = \frac{(k/m)^{1/2}}{2\pi} \tag{1}$$

or

$$f = \frac{(g/z)^{1/2}}{2\pi} \tag{2}$$

In this formula, the dashpot action or damping is assumed to be equal to zero. The damping coefficient of the dashpot is a little more difficult to define than the natural frequency, but they are closely related to the energy-absorbing efficiency of the tires and shock absorbers. For modern aircraft design, the landing-strut design is dependent primarily upon the prototype tests. The FAA, in its airworthiness requirements, specifies that landing struts meet the impact of the drop tests. Therefore, the energy-absorbing efficiency of that gear can be defined by the equation

$$\delta = \frac{\Delta}{v^2/2g} \tag{3}$$

where δ = coefficient of energy-absorbing efficiency
Δ = deflection of landing gear at drop test
v = sinking velocity, fps
$v^2/2g$ = height of drop test, in.

For modern aircraft tires, the shock-absorbing efficiency ranges from 0.45 to 0.47, and the efficiency of oleo-pneumatic shock absorbers varies from 0.75 to 0.80. With this simple dynamic model of the aircraft, it becomes easier to study the interaction of pavement and aircraft in motion.

In studying the dynamics of an elastic system, Newton's second law of motion describes the change of momentum of the mass due to the influence of the external force, in the form

$$\frac{d}{dt}(mv) = F \tag{4}$$

If the mass of the vehicle is constant during the transition and the velocity is to be represented by the displacement, the following equation is obtained:

$$m\frac{dv}{dt} = m\frac{d^2z}{dt^2} = F \tag{5}$$

In considering the equilibrium of an elastic body, a general equation of motion can be set up:

$$m\ddot{z} + c\dot{z} + kz = F(t) \tag{6}$$

where z = displacement.
$\dot{z}$ = velocity, or first derivative of displacement with respect to time t.
$\ddot{z}$ = acceleration of mass, or second derivative.
$F(t)$ = a forcing function, variable with time.

9.2. TRANSIENT VIBRATION

With this basic equation of motion in mind, the theory of transient vibration can be developed. The transient vibration is defined as the vibration of an elastic body caused by an external force, which can be a periodic function, such as

$$F(t) = F\sin 2\pi\omega t \tag{7}$$

where ω is the frequency of the forcing function in cycles per second and F is the amplitude of the forcing function. With this forcing function moving on the elastic surface, the supporting surface is the response function, which can be expressed by its mass, the spring constant, and its damping characteristics. The mass and the spring constant can be consolidated by the parameter of the fundamental frequency f, in cycles per second. The relationship between the displacement of the response function and the forcing function is (Fig. 9.2a)

$$-k_b z = \frac{F}{g} \ddot{z} \, \sin 2\pi \omega t \tag{8}$$

In words, the mass of the forcing function times the acceleration $\ddot{z}$ of the forcing function is proportional to the spring constant times the displacement of the response function. The above equation can be rewritten as

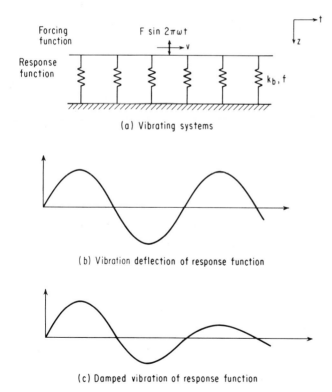

(a) Vibrating systems

(b) Vibration deflection of response function

(c) Damped vibration of response function

Fig. 9.2 Systems of transient vibration.

$$-z = \frac{M_v}{M_b} \frac{1}{(2\pi f)^2} \ddot{z} \sin 2\pi \omega t \tag{9}$$

where $M_v = F/g$ = mass of vehicle

$\qquad M_b$ = mass of response function

$\qquad f$ = natural frequency of response system, cps

For elastic systems, without the damping effect, the deflection of the response system assumes the shape shown in Fig. 9.2b. For damped vibration, the deflection of the response system will decrease with time, as shown in Fig. 9.2c. The above analysis does not include the vibration of the response system. For a more precise analysis, the motion of the response system—assuming that the surfaces of the two systems remain in contact—should be included:

$$-z = \frac{M_v}{M_b} \frac{1}{(2\pi f)^2} \ddot{z} \sin 2\pi \omega t + \frac{\ddot{z}}{(2\pi f)^2} \tag{9a}$$

9.3 STEADY STATE OF VIBRATION

Under this condition, the characteristic of the response system is closely related to its fundamental frequency f (which includes k_b and mass M_b) and its damping characteristic β (see Fig. 9.3a). Thus there is a condition of both high- and low-frequency vibration, and the steady state of vibration is derived from these basic characteristics of the vibration. If either the forcing frequency or the response frequency is much greater than the other, the resultant steady-state dynamic response is as shown in Fig. 9.3b. However, if the forcing and response frequencies are in close range, the forcing function assumes the characteristic figure shown in Fig. 9.4a and the response system assumes the characteristic shown in Fig. 9.4b. The resulting interaction of these two systems is shown in Fig. 9.4c. At a definite interval, a peak will appear, such as FH, where H is the magnification factor of the forced vibration. The magnification of the vibration is commonly known as the *resonance* of the steady-state vibration. The mathematical relationship for the magnification H can be expressed as

$$H^2 = \frac{1}{[1 - (\omega/f)^2]^2 + (2\beta\omega/f)^2} \tag{10}$$

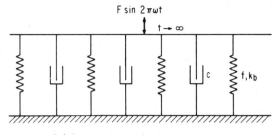

(a) Characteristics of vibration system

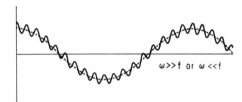

(b) Deflection of response system

Fig. 9.3 Systems of steady state of vibration.

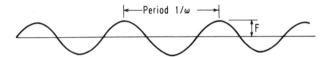

(a) Characteristics of forcing functions

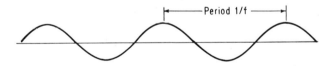

(b) Characteristics of response functions

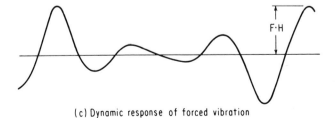

(c) Dynamic response of forced vibration

Fig. 9.4 Steady state of forced vibration.

where ω and f are in cycles per second. When ω equals f and β equals zero, H approaches infinity. This particular case is known as *mathematical resonance;* any vibration in this range will approach an infinite magnification, such as the asymptotes of the curve $\beta = 0$ shown in Fig. 9.5. In reality, every vibrating system has damping characteristics, and therefore there is always a definite magnification factor. The more realistic magnification is shown by curves such as for $\beta^2 = 1/8$ and $\beta^2 = 1/2$. For a wide range of aircraft operations, the natural frequency of the aircraft ranges from 1.0 to 2.0 cps, the natural frequency of the pavement and subgrade ranges from 7 to 50 cycles, and the natural frequency of the framed structure is in the range of 6 to 10 cps. The magnification factor is therefore in the range of 1.003 to 1.03. The maximum response at the steady state of vibration will not exceed $0.03g$, or 3 percent of the static weight.

9.4. RANDOM VIBRATION

The above analysis of the force of vibration was based on the condition that the aircraft is a forcing function and the pavement structure a response function. These are classic vibration problems. The pavement surface has been assumed to be perfectly smooth; that is, there is no vibration caused by the roughness of the pavement

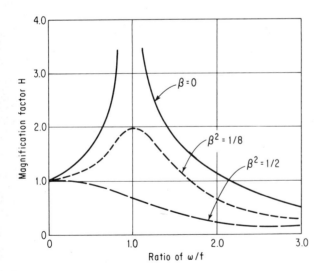

Fig. 9.5 Magnification factor of steady state of vibration.

structure. In reality, when an aircraft moves on a pavement surface which is not perfectly smooth, the roughness of the pavement becomes the cause of the vibration. This means that the roughness of the pavement is the forcing function of a vibrating system. The aircraft becomes the response function since the vibrations of the aircraft are due to rough riding. This is another level of vibration. From the viewpoint of dynamic analysis, the pavement roughness is a multifrequency random vibration, which assumes uneven wavelengths and uneven amplitudes. In treating the problem of random vibration, the phase of the forcing function has little effect on the final outcome. The output of a random event can be described statistically. In Fig. 9.6a, the abscissa represents the traveling time of the aircraft; for a certain speed, the time can be related to the distance traveled. The ordinate x represents the vertical deviation of the pavement surface from the mean. The mean square of the deviation is

$$\bar{x}^2 = \frac{1}{T} \int_0^T x^2 \, dt \tag{11}$$

For a harmonic force

$$F = F_0 \sin 2\pi\omega t \tag{12}$$

the mean square of the force $\bar{F}^2$ is equal to one-half the amplitude squared. Thus,

$$\bar{F}^2 = \frac{\omega}{n} \int_0^{n/\omega} F_0^2 \sin^2 2\pi\omega t \, dt = \frac{F_0^2}{2} \tag{13}$$

For a multifrequency function, the mean square of the force is the summation of the amplitude of all the frequencies, squared and divided by 2:

$$\bar{F}^2 = \sum^n \frac{F_n^2}{2} \tag{14}$$

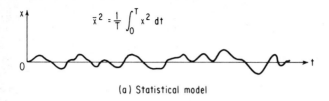

(a) Statistical model

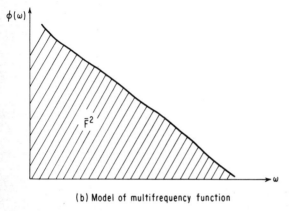

(b) Model of multifrequency function

Fig. 9.6   Characterizing a random event.

This is a very important relation in characterizing random vibration consisting of multifrequency harmonic forces. The contribution to the total mean-square value of the multifrequency function by each frequency interval can be expressed as

$$\phi(\omega) = \lim_{\Delta\omega \to 0} \frac{\Delta(\overline{F^2})}{\Delta\omega} \qquad (15)$$

where $\Delta(\overline{F^2})$ is the contribution of the mean-square value in a frequency interval of $\Delta\omega$. When the frequency interval is small and $\Delta\omega$ approaches zero, the limit of the mean-square value of the frequency segment is defined as the *power spectral density* (PSD) function of a multiple-frequency function $\phi(\omega)$. For a continuous function, the mean-square value of the random function can be expressed as

$$\overline{F^2} = \int_0^\omega \phi(\omega)\, d\omega \qquad (16)$$

In this equation, it can be seen that the total area under the ω versus $\phi(\omega)$ curve is equal to the mean-square value $\overline{F^2}$ (see Fig. 9.6b). Graphically, the definition of the PSD of a continuous function can be expressed as shown in Fig. 9.7a and the discrete function as shown in Fig. 9.7b. With this method, the random surface deviation of the pavement can be condensed into a PSD function consisting of multifrequency harmonic functions.

The response of an aircraft on rough pavement is a summation of the force vibrations at various frequency ranges. For a discrete forcing function, the response of the aircraft can be expressed by

$$\overline{X^2} = \sum_{}^{n} \frac{F_n^2}{2} H_n^2 \qquad (17)$$

where F_n is the amplitude of one particular frequency and H_n is the respective magnification factor for that frequency of vibration. For

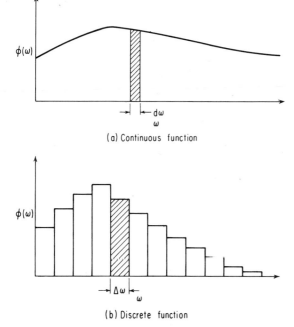

(a) Continuous function

(b) Discrete function

Fig. 9.7 Illustration of PSD function.

continuous functions, the above equation can be rewritten as

$$\overline{X}^2 = \int_0^\omega \phi(\omega) \frac{1}{[1 - (\omega/f)^2]^2 + (2\beta\omega/f)^2} \, d\omega \tag{18}$$

If $\phi(\omega)$ is in a very narrow range of variation and is considered to be a constant, the integration of the above equation results in

$$\overline{X}^2 = \phi(\omega) \frac{\pi f}{4\beta} \tag{19}$$

That is, the mean-square response of the aircraft, $\overline{X}^2$, to be expressed by $\overline{DI}^2$ in subsequent analysis, is equal to the PSD function of pavement roughness, $\phi(\omega)$, times the transfer function, $\pi f/4\beta$. In a mechanical sense, the input is the PSD of pavement roughness, the "black box" is the transfer function, and the output is the mean-square response of the aircraft. The transfer function in this study is a function of the natural frequency of the aircraft and the damping coefficient of its landing-gear assembly.

9.5. THIRD LEVEL OF FORCED VIBRATION

When an aircraft acquires a dynamic increment $\overline{DI}$ due to rough riding on the pavement surface, it acts again as a forcing function, with the pavement surface responding. In this case, the forcing function becomes

$$F(t) = (1 + \overline{DI})F \sin 2\pi\omega t \tag{20}$$

Substituting the above equation in the first level of vibration appearing in Eq. (8) results in the equation

$$-k_b z = \frac{(1 + \overline{DI})F}{g} \ddot{z} \sin 2\pi\omega t \tag{21}$$

Thus, including the effect of random vibration, the final level of the dynamic interaction can be established.

9.6. DAMPING OF VIBRATION

In the above analysis, the damping of the pavement system has not been taken into consideration, nor is the initial vibration of the aircraft and pavement considered. For a damped vibration system, the amplitude at n cycles of vibration can be assumed to be

$$z_n = z_0 e^{-n\delta} \tag{22}$$

where z_0 is the original amplitude of vibration, $\delta = 2\pi\beta$ is the coefficient of damping, and n is the number of vibration cycles (see Fig. 9.8). The residual amplitude of the vibration at n cycles of vibration is z_n. Equation (22) is known as the *logarithmic decrement* of damped vibration. For $\beta = 0.025$, $z_1 = 0.85 z_0$; that is, the amplitude of damped vibration is about 85 percent of the preceding vibration cycle.

In considering actual runway operation, it is noted that at the peak volume of air traffic, the interval of aircraft landing or takeoff is always kept to at least 30 sec. As the fundamental frequency of the runway structure is usually higher than 6 cps, the ground structure will vibrate at least 200 cycles during the intervals of aircraft movement. The residual vibration at the end of this time interval is practically zero. Consequently, there is no accumulation of dynamic force in the pavement structure. The initial oscillation of the pavement structure can be neglected when the aircraft enters the runway structure.

For modern aircraft, the shock-absorbing efficiency of pneumatic tires is normally greater than 0.45. This means that e^{δ} is less than

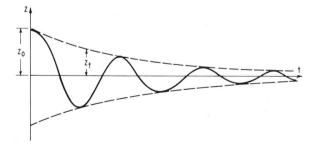

Fig. 9.8 Logarithmic decrement of damped vibration.

0.55. The corresponding damping coefficient β exceeds 0.10. At the end of 5.5 cycles of free vibration, corresponding to a time interval of 5 sec (waiting for tower clearance to enter the runway), the residual vibration of the aircraft is less than 3 percent of the initial vibration. It is appropriate that the vibration of the aircraft can be assumed to be equal to zero at the beginning of aircraft-pavement interaction.

9.7. DATA ACQUISITION AND REDUCTION

With the PSD to characterize the pavement surface, the actual analysis involves data acquisition and reduction. Without developing a perfect mechanical monitoring system, at the present state of knowledge the most reliable method of obtaining the pavement surface condition is by rod-and-level surveys. Before surveying, it is necessary to review the process of data reduction. The PSD analysis has been used extensively by mechanical and communications engineers. The continuous record of the vibration is either processed by analog computers or discretized for digital computation. The first step in digital computation is the determination of the auto-correlation function:

$$R(\tau) = \lim_{T \to \infty} \frac{1}{T} \int_0^T x(t)x(t + \tau)\, d\tau \tag{23}$$

where x is the elevation of the level reading deviated from means, t is the distance from the original point of zero, and τ is the shift of sampling distance, representing the bandwidth of frequency resolution. The PSD is then obtained by the use of Fourier transformation:

$$\phi(\omega) = 2 \int_0^\infty R(\tau) \cos 2\pi\omega\tau\, d\tau \tag{24}$$

The method for computing the PSD function by Fourier transformation of the autocorrelation function is applicable, theoretically, to an infinite length of record of random gaussian nature. For the practical purposes of data acquisition, a concise picture of the random nature is usually employed in the computation, and therefore, the statistical characteristics of the time history are of vital importance to the reliability of computation results.

In the planning of experiments and the subsequent data processing, the following steps are suggested.

Sampling Interval

The sampling interval is closely related to the selection of a frequency f such that the power above this value is negligible. The relation between the upper frequency range and the sampling interval ϵ can be expressed by

$$\epsilon = \frac{\pi}{\omega_0} = \frac{\lambda}{2V} \tag{25}$$

where V is the crossing velocity of aircraft, which is equal to the true aircraft velocity v divided by its fundamental frequency f, and λ is the length of the spatial wave. This equation evolves from a communication-sampling theorem which states that sampling with this interval will resolve frequencies up to ω_0. For the profile analysis of a pavement surface, the significant shortwave length (corresponding high spatial frequency) is in the range of 4 to 8 ft for ground-vehicle operation (ground speed varies from 3 to 5 knots, and fundamental frequency ranges from 1 to 1½ cps). The sampling interval is therefore in the range of 2 to 4 ft. For higher operational speeds, such as on runway-taxiways, the sampling interval can be increased in proportion to the operational speed.

Frequency Resolution

The frequency resolution $\Delta\omega$ is determined by

$$\Delta\omega = \frac{2\pi}{\tau_{max}} \tag{26}$$

Independent spectral estimates cannot be made closer than $2\pi/\tau_{max}$, where τ_{max} is the maximum shift of the correlation function. In digital computation, it is expressed by

$$\tau_{max} = M\epsilon \qquad (27)$$

By increasing the number of correlation-function estimates M, the bandwidth of frequency resolution $\Delta\omega$ decreases; that is, the frequency resolution is good. For profile analysis of the pavement surface, the significant long wavelength is in the range of 80 to 200 ft (operation speed of the aircraft ranges from 50 to 140 knots). The desirable number of shifts of correlation function ranges from 30 to 60.

Statistical Reliability

The statistical reliability of the power-spectrum estimates depends on the length of the records T_L and the effective bandwidth of the frequency resolution τ_{max}. In his study of the sampling theory of power-spectrum estimates, Tukey indicates that the spread of the distribution is measured by

$$\frac{\sigma_i}{\phi_i} = \left(\frac{T_L}{2\pi}\Delta\omega\right)^{-1/2} = \left(\frac{\tau_{max}}{T_L}\right)^{1/2} = \left(\frac{M}{N}\right)^{1/2} \qquad (28)$$

where $N = T_L/\epsilon$ is the total number of samples. Based on normal, or gaussian, distribution, the average value of ϕ_i may be expected to fall within the confidence band defined by $\phi_i \pm k\sigma_i$, with probabilities as follows:

k	Confidence level within band
1.0	0.68
1.65	0.90
1.96	0.95

For given values of M/N and for probability set by k, the following confidence band exists:

$$1 - k\left(\frac{M}{N}\right)^{1/2} < \frac{\phi_M}{\phi_{AV}} < 1 + k\left(\frac{M}{N}\right)^{1/2} \qquad (29)$$

With a probability of 95 percent and a sampling variance of 0.125, the average power is expected in the range of

$$0.80\phi_M < \phi_{AV} < 1.33\phi_M \tag{30}$$

The corresponding spread of distribution $k(M/N)^{1/2} = 0.25$, and the total number of samples is $62M$. Houbout suggests that the total number N be in the range of 1,500 to 4,000 samples. For the profile analysis of the pavement surface, the total length of pavement to be measured ranges from 6,000 to 10,000 ft. Rod readings should be taken at intervals ranging from 2 to 4 ft. For airport construction, intersections of taxiways and runways are always encountered within the length to be studied. The geometric requirement of the pavement surface creates local disturbances which have a probability distribution that are nongaussian in nature. The statistical meaning of the autocorrelation function exhibits a serious limitation under these practical applications.

Folding-frequency Method

In the analysis of random vibration, the total power, the sum of the power of all spectral density, is equal to the mean-square value of the response function, in the form

$$\overline{X}^2 = \sum \phi(f)\,\Delta f \tag{31}$$

For a discrete function, the mean-square value is expressed by

$$\overline{X}^2 = \frac{1}{N} \sum_{}^{n} X^2 \tag{32}$$

By using the binomial expansion, the following relation exists:

$$\left(1 - \frac{1}{2}\right)^{-1} = 1 + \frac{1}{2^1} + \frac{1}{2^2} + \frac{1}{2^3} + \cdots \tag{33}$$

and Eq. (32) can be rewritten as

$$\frac{1}{N} \sum_{}^{n} X^2 = \frac{1}{N} \sum_{}^{m} \sum_{}^{N/2^m} \frac{1}{2^m} (DS^{m-1}X)^2 \qquad m = 1, 2, 3, \ldots$$

$$= \frac{1}{N} \sum_{}^{N/2} \frac{1}{2} (DX)^2 + \frac{1}{N} \sum_{}^{N/4} \frac{1}{4} (DSX)^2 + \frac{1}{N} \sum_{}^{N/8} \frac{1}{8} (DS^2X)^2 + \cdots \tag{34}$$

$$= \frac{1}{N} \sum^{N/2} \frac{(DX)^2}{f} \frac{f}{2} + \frac{1}{N} \sum^{N/4} \frac{(DSX)^2}{f} \frac{f}{4} + \frac{1}{N} \sum^{N/8} \frac{(DS^2X)^2}{f} \frac{f}{8} + \cdots \quad (34)$$

where X = experimental reading

$f = 1/\epsilon$

ϵ = sampling interval

$DX_q = X_{q+1} - X_q$

$SX_q = X_{q+1} + X_q$

$DSX_q = SX_{q+2} - SX_q$

$S^2X_q = SX_{q+2} + SX_q$

$DS^2X_q = S^2X_{q+2} - S^2X_q$

Since the mean-square value is the sum of all PSD, the following relation exists for a binomial folding-frequency function:

$$\frac{1}{N} \sum^N X^2 = \phi\left(\frac{f}{2}\right)\frac{f}{2} + \phi\left(\frac{f}{4}\right)\frac{f}{4} + \phi\left(\frac{f}{8}\right)\frac{f}{8} + \cdots \quad (35)$$

By combining Eqs. (34) and (35), the PSD of a time function is, therefore, expressed by

$$\phi\left(\frac{f}{2}\right) = \frac{1}{Nf} \sum^{N/2} (DX)^2$$

$$\phi\left(\frac{f}{4}\right) = \frac{1}{Nf} \sum^{N/4} (DSX)^2 \qquad (36)$$

$$\phi\left(\frac{f}{2^m}\right) = \frac{1}{Nf} \sum (DS^{m-1}X)^2 \qquad N = 2^m$$

The computation of the PSD by the folding-frequency method is much simpler than by the autocorrelation function. For massive data processing, a computer program has been developed for the folding-frequency and PSD computation.

9.8. EXPERIMENTS ON VEHICLE-PAVEMENT INTERACTION

In cooperation with the FAA and the BPR, experiments on vehicle-pavement interaction were conducted on the test track at Newark Airport as well as on four active runways at Kennedy Airport. The experiment consisted of measuring the surface elevation along the wheel path and monitoring the dynamic increment of the vehicle at the wheel base, which represents the vibration at the vehicle-pavement interface.

Characterizing Surface Roughness

The Newark test track consisted of 16 sections, each 30 ft wide and 75 ft long, of different pavement compositions, varying from 8 to 34 in. in thickness (see Fig. 9.9). Two longitudinal profiles, 1 ft on each side of the centerline of the test track, were maintained as permanent reference lines during the testing period. Profiles were established by taking precise rod-level readings at intervals of 3.75 ft. Therefore, 20 rod readings were taken for each test section. Since the composition

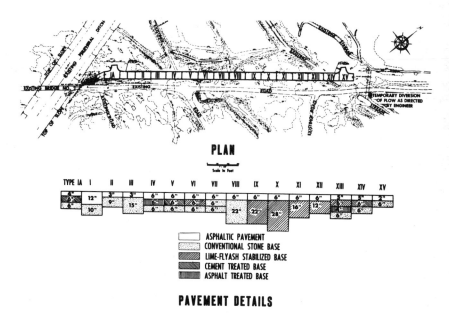

Fig. 9.9 Plan and details of test pavement—Newark Airport.

of the test sections varied, the uneven settlement of the subgrade created diversified surface characteristics of the pavement. In particular, a drastic change of pavement surface was usually observed at the transition of two adjacent sections.

As demonstrated by Eq. (19), the interaction of vehicle and pavement is not related to the physical properties of the pavement. The Newark test track was a good site at which to determine the validity of this concept. In processing the surveying record, it is noted that the pavement surface consisted of a series of local variations, such as the transitions of test sections and several built-in bumps. The roughness of the pavement surface was not of a gaussian (normal) distribution. The method that utilizes the autocorrelation function for computing the power PSD function was therefore not applicable to this analysis. Although the folding-frequency method yields only an estimate of the PSD, the computation does not involve the assumption of gaussian distribution.

Because the test sections were initially of a level grade, the surveyed elevations were used directly in the computations. (If the profile is an inclined surface, the survey record should be adjusted to the actual general slope and the deviation from the general slope should be used in the computation.) The PSD function of the longitudinal profile computed by the folding-frequency method is plotted on the left side of Fig. 9.10. Note that at the frequencies ranging from 0.017 to 0.0041 ft^{-1} , equivalent to wavelengths of 60 to 240 ft, the nonlinear change of the PSD is very significant. Evidently, it reflects the effect of the length of the test sections. For wavelengths shorter than 75 ft, the PSD can be expressed by a linear function of the crossing frequency in a log-log plot having the form

$$\phi\left(\frac{1}{L}\right) = \frac{C}{(1/L)^m} \qquad (37)$$

where L is the wavelength, in feet, and $\phi(1/L)$ is the PSD of the longitudinal profile. The observed C and m values are 1.1×10^{-5} and 2.0, respectively.

At Kennedy Airport, the longitudinal profile was established at 10-ft intervals along the centerline of the four active runways. All runway pavements consist of 12-in. portland-cement concrete having transverse joints 20 ft apart. By using the same computation

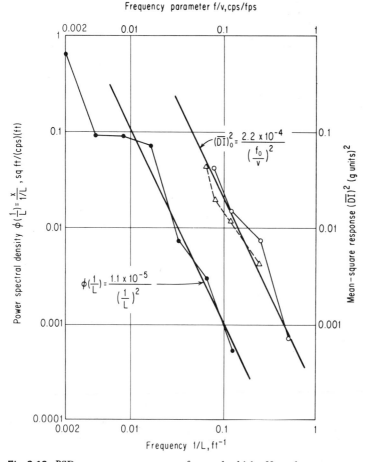

Frequency parameter f/v, cps/fps

$$\overline{(DI)}_0^2 = \frac{2.2 \times 10^{-4}}{\left(\frac{f_0}{v}\right)^2}$$

$$\phi\left(\frac{1}{L}\right) = \frac{1.1 \times 10^{-5}}{\left(\frac{1}{L}\right)^2}$$

Power spectral density $\phi\left(\frac{1}{L}\right) = \frac{x}{1/L}$, sq ft/(cps)(ft)

Mean-square response $\overline{(DI)}^2$ (g units)^2

Frequency 1/L, ft^{-1}

Fig. 9.10 PSD, mean-square response of ground vehicle—Newark test.

program, a similar PSD function was calculated. In Fig. 9.11, the longitudinal profile of runway 4L-22R is plotted. The corresponding C and m values are 8.6×10^{-5} and 1.3, respectively.

The ordinate of a frequency-spectral density function (Figs. 9.10 and 9.11) represents the power of excitation at a defined frequency domain. The total power of excitation will be the summation of power below the frequency-spectral density curve. For a sinusoidal surface configuration, the wavelength L is commonly referred to as the length of a straightedge in actual pavement construction. The maximum surface deviation from a straightedge, expressed by Δ in feet, is equal to $2F$, as given in Eqs. (13) to (15). The PSD of a

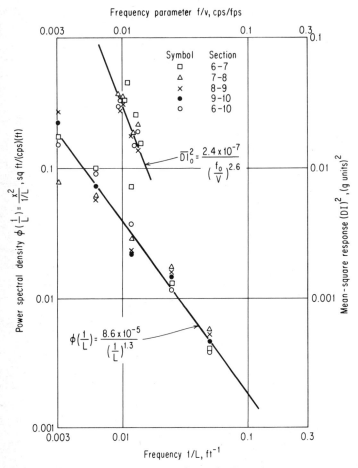

Fig. 9.11 PSD mean-square response of aircraft—Kennedy test.

truncated sinusoidal wave measured by the straightedge method is

$$\phi\left(\frac{1}{L}\right) = \frac{\frac{1}{2}(\frac{1}{2}\Delta)^2}{1/L} = \frac{1}{8}\Delta^2 L \tag{38}$$

For a pavement surface which consists of a series of sinusoidal waves of various wavelengths, the relation between the frequency and PSD is a discrete function, similar to those shown in Figs. 9.10 and 9.11. Usually, a discrete function inherits more error in computation than a continuous function. However, in the study of pavement roughness, the error due to nongaussian characteristics of the pavement surface is more critical than the error due to truncation.

Characterizing Vehicle Response

During the interaction tests conducted with the BPR test vehicle on the Newark Airport test pavement, the trailer axle was instrumented with four SR-4 strain gages to measure the transient strain response during the passage of the vehicle. As the dual wheels of the test vehicle would have introduced a more complex mode of loading than is desirable for pavement studies, the test vehicle instrumentation system was designed to provide an indication of the frequency, phase, and order of magnitude of the total force variation about the static-load level for the dual tires acting together on the pavement structure. The magnitude of the total force variation is here called the *dynamic increment,* or *DI*. It is known that instantaneous forces applied to a pavement are not uniformly distributed between the dual wheels; however, only the total force of the dual wheels is of primary significance for pavement loadings.

The load strain was, therefore, based on this requirement. The axle-housing strain oscillograms provided an accurate measure of the frequency and phase of the total force variation and an indication of the dynamic increment.

A segment of the continuous roll record of run 139-6 has been reproduced in Fig. 9.12. Each of the traces represents the output of a Wheatstone bridge circuit with four active arms. Each oscillogram is accompanied by an initial base reading representing the nulled Wheatstone bridge circuit while the test vehicle was at equilibrium prior to the start of the run. The outputs of the strain gages were

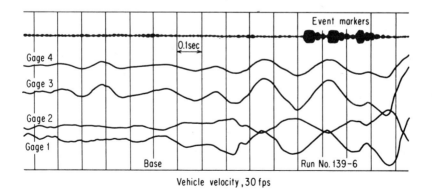

Fig. 9.12 Oscillogram record of test run 139, segment 6.

directly calibrated with the static-load increment of the test vehicle. Therefore, the ordinate of each oscillogram can be converted into the *DI* value of the vehicle.

There are three event mark locations on the upper trace of each oscillogram. The timing lines are spaced at 0.1-sec intervals. By knowing the station of event mark locations and the time interval on the oscillogram, the speed of the test vehicle can be computed. It should be noted that the occasional nonuniformity of the spacing of the time lines on the oscillograms is largely due to a sluggish paper-drive motor in the recorder. The test runs, all northbound, are identified and given in Table 9.1.

TABLE 9.1   Dynamic Response of Moving Vehicle

Run No.	Vehicle offset pavement, in.	Vehicle speed v, fps	Response frequency f_0, cps	Frequency parameter f_0/v, cycle/ft	Root mean square DI(rms), g unit	Arithmetic mean DI, g unit	Average $(DI)^2$, $(g$ unit$)^2$
090	9 rt.	6.81	3.54	0.520	0.030	0.027	0.00073
094	7 lt.	15.06	3.68	0.245	0.090	0.075	
096	7 lt.	15.59	3.75	0.241	0.106	0.097	
123	9 rt.	14.14	3.62	0.256	0.092	0.084	0.0073
100	7 lt.	31.40	3.49	0.111	0.137	0.128	
111	9 rt.	30.46	3.79	0.124	0.134	0.123	0.0157
104	7 lt.	46.50	3.61	0.078	0.197	0.192	
107	9 rt.	47.80	3.82	0.080	0.199	0.192	
109	9 rt.	45.35	3.55	0.078	0.223	0.220	0.0405
092	57 lt.	15.47	3.72	0.241	0.076	0.068	
121	37 lt.	15.02	3.67	0.244	0.073	0.064	0.0043
098	57 lt.	31.35	3.79	0.121	0.120	0.107	
113	37 lt.	32.10	3.58	0.111	0.120	0.110	
139	37 lt.	30.10	3.82	0.127	0.113	0.104	0.0114
102	57 lt.	46.23	3.49	0.076	0.149	0.142	
137	37 lt.	42.75	3.71	0.087	0.145	0.136	0.0193
133	37 lt.	56.50	3.83	0.068	0.221	0.203	
135	57 lt.	58.15	3.62	0.062	0.217	0.207	0.0420

NOTE: "g unit" denotes the stress level under the static wheel load.

According to Eq. (19), the output due to pavement roughness is reflected by the mean-square response of the test vehicle, and the transfer function between the input and output is related to the natural frequency of the vehicle and the damping coefficient of the response system. Therefore, in the data processing of the oscillo-grams, the envelope of the continuous trace is more representative than the actual phase of the response variation. Using the upper and lower envelopes of the response function also reduces the inherent error caused by the fluctuation of the initial base reading. The mean-square response of each test run, sampled at a time interval of 0.1 sec, is shown in Table 9.1. In order to correlate the PSD of the pavement profile with the mean-square response, as expressed by $\overline{DI_0^2}$, of the test vehicle, the frequency parameter f_0/v is used for plotting Fig. 9.10. The f_0 value represents the fundamental frequency of the test vehicle. The approximate relation between $\overline{DI}_0$ and v can be expressed as

$$(\overline{DI}_0)^2 = \frac{b}{(f_0/v)^n} \tag{39}$$

The observed b and n values at the Newark test are 2.2×10^{-4} and 2.0, respectively.

In the landing and takeoff tests conducted by the FAA, a Convair 440 was instrumented with accelerometers mounted on the hub housing of the nose wheel. On each main gear assembly, acceler-ometer and strain gages were mounted on the lower strut of the oleo system to measure the vertical acceleration and dynamic-load increment. Two landings and takeoffs were conducted at each end of four active runways. According to Eq. (19), the speed of the aircraft cannot be introduced in the transfer function; therefore, all tests should be conducted at a constant speed. At landings, all instruments were monitored at a constant speed of 75 or 85 knots. At takeoffs, the monitoring speed was 55 or 65 knots. A portion of the oscillogram is reproduced in Fig. 9.13. Event markers, as shown on the bottom of the oscillogram, are used in identifying the location of the test. As noted on the oscillogram, the output of the acceler-ometer inherits considerable noise and dynamic magnification. Some measurements indicate a dynamic force greater than $6g$, which is not a realistic measurement. Therefore, the accelerometer readings were

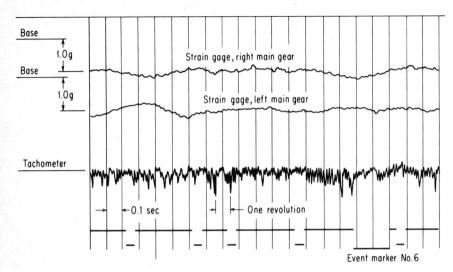

Fig. 9.13 Oscillogram record of test run 225, runway 22R, Kennedy.

not used in the data processing. The strain-gage outputs were directly calibrated with the static load of the airplane, and the oscillogram record was processed for the mean-square response by the same procedure as that shown in Table 9.1.

A similar plot of mean-square response and crossing frequency is shown on the right side of Fig. 9.11. The approximate relation as given by Eq. (39) is also applicable, and the observed b and n values for the Kennedy test are 2.4×10^{-7} and 2.6, respectively.

9.9. TRANSFER FUNCTION

As indicated by Eq. (10), the maximum dynamic response of a moving vehicle occurs when the response frequency f of the vehicle coincides with the forcing frequency ω of the pavement roughness. Using wavelength as a computation method, this statement means that the maximum dynamic response of a moving vehicle occurs when its crossing wavelength is equal to the wavelength of the pavement roughness. It can be expressed as

$$\frac{f_0}{v} = \frac{1}{L} \tag{40}$$

where f_0 is the fundamental frequency of the test vehicle, v is the crossing velocity of the vehicle, and L is the wavelength of the pavement roughness. The respective units are cycles per second, feet per second, and feet per cycle.

The transfer function, defined as the ratio of the output mean-square dynamic response to the input PSD, can be expressed as

$$T^2 = \frac{\overline{DI_0^2}}{\phi(1/L)} \tag{41}$$

For a moving vehicle having a fundamental frequency f other than f_0 of the test vehicle, the mean-square response $\overline{DI^2}$ is equal to $\overline{DI_0^2}$ times f/f_0, according to Eq. (19). When the spectral frequency $1/L$ is equal to the fundamental frequency at crossing f/v, the transfer function of the maximum response is

$$T^2 = \frac{b}{c} v^{n-m} \frac{f^{1+m}}{f_0^{1+n}} \tag{42}$$

It is determined experimentally. For a vehicle moving on a ground surface of a known roughness spectrum, $\phi(1/L)$, the maximum response to be developed in the vehicle is

$$\overline{DI} = \left(\frac{b}{c}\right)^{1/2} v^{(n-m)/2} f^{(1+m)/2} \frac{\phi(1/L)^{1/2}}{f_0^{(1+n)/2}} \tag{43}$$

This means that the dynamic response of a moving vehicle increases with increasing velocity, fundamental frequency, and PSD of surface roughness.

In the Newark experiment, the observed fundamental frequency of the test vehicle did not change with its crossing velocity and was constantly in the range of 3.2 to 4.0 cps, having an average of 3.6 cps. The transfer function T^2, by Eq. (41), was equal to 20. The corresponding coefficient of damping computed by Eq. (19) was 0.14, and the logarithmic decrement, by Eq. (22), was $e^{-\delta} = 0.41$.

In the Kennedy test, the fundamental frequency of the Convair 440 was 1.1 cps, and its oleo-pneumatic shock absorber was known to be more efficient than that of the ground vehicle. The observed transfer function varied from 0.5 to 0.8. The corresponding coefficient of damping was greater than 1.0. The motion became aperiodic. The aircraft was more efficient in absorbing impact than the ground vehicle.

9.10. STRAIGHTEDGE METHOD

In the standard specifications of pavement contracts, there is usually a clause such as "the surface is smooth and free from irregularities greater than y fractions of an inch from its true plan when tested with an x-foot straightedge in any direction." This is purely an empirical approach, developed in accordance with local construction practice and the past performance of pavements. Due to the advancements in high-speed road vehicles and jet transport, the safety aspect of pavement surfaces has become a matter of deep concern to many transportation engineers.

From the viewpoint of the pavement user, such as the driver of a vehicle or the pilot of an airplane, the quality of a pavement is judged by the level of vibration developed in his vehicle while it is rolling over the pavement. The dynamic response of the passengers or pilots largely depends on the mass-spring-dashpot system, where the observers are located. The vertical accelerations developed at the center of gravity (c.g.) and at the fore and after positions of the vehicle are significantly different. The c.g. vibration, however, is usually a reliable indication of the degree of comfort of the passengers. In his study of NASA's roughness test, Houbolt suggests that the normal tolerance of the mean dynamic increment at the c.g. be $0.12g$ and that the upper-limit pavement roughness is likely to be the beginning of the development of a c.g. peak response of $0.30g$. A similar range of dynamic response has been reported by other researchers. Leonard, of the British Road Research Laboratory, reported that the acceptable level of vehicle vibration at the c.g. should be $0.10g$ for normal traffic conditions. In the Newark test, the human tolerance of vehicle response was not included in the study. There is a need for administrative agencies such as the FAA and the BPR to coordinate the requirements of airport and road users and to conduct a more comprehensive research program to define the tolerance of c.g. vibration and the corresponding dynamic response at the wheel-pavement interface.

For vehicles moving on a pavement surface of a known roughness spectrum, the $\overline{DI}$ value at the wheel-pavement interface can be determined by Eq. (43). In new pavement design, the roughness spectra are governed primarily by the initial construction tolerance and the subsequent surface deterioration due to vehicle loadings. Studies of many aircraft operations by Houbolt have suggested that the average acceleration response at the c.g. of the aircraft should be $0.12g$ at a normal taxiing speed and the peak acceleration should not exceed $0.3g$ when the runway surface is considered to be rather rough. This implies that $0.3g$ is the upper limit of the roughness spectrum, by which the pavement design and construction tolerance should be governed. The relation between the c.g. response and the response at the wheel-pavement interface varies with various aircraft and vehicles. For pavement design, the interface response should be used. Because of the lack of experimental data, the dynamic response at the interface is assumed to be equal to the c.g. response of the aircraft. By introducing the straightedge concept, as defined by Eq. (38), the roughness spectrum can be expressed as

$$\phi\left(\frac{1}{L}\right) = \frac{\Delta^2}{L} \frac{(v/f)^2}{8} \tag{44}$$

By substituting the above relation in Eq. (43), the geometric configuration of the straightedge criteria becomes

$$\Delta = \left(\frac{8c}{b}\right)^{1/2} f_0^{(1+n)/2} v^{(m-n)/2} f^{(2-m)/2} \frac{1}{v} \overline{DI} \left(\frac{1}{f}\right)^{1/2} L^{1/2} \tag{45}$$

In his study of runway roughness, Houbolt simplified the above relation to

$$\Delta = KL^{1/2} \tag{46}$$

9.11. EFFECTIVE GROUND LOAD OF AIRCRAFT

For airfield construction, the wing lift of the airplane should be considered in the evaluation of aircraft-pavement interaction. The wing lift F_L of an airplane is commonly expressed as

$$F_L = cA\rho \frac{v^2}{2} \tag{47}$$

where c = coefficient of uplift
A = area of wings
ρ = density of air
v = relative velocity of air and airplane

For normal runway takeoffs, full wing lift (completely airborne) will occur at a speed ranging from 130 to 150 knots. By assuming that the full wing lift occurs at 150 knots, the effective weight F of the aircraft can be expressed as

$$F = F_0(1.0 - 4.44 \times 10^{-5}V^2) \tag{48}$$

where F_0 is the static weight of the airplane and V is the velocity, in knots. As the dynamic response of an airplane is measured by the magnitude of the total force variation above the static-load level, the straightedge criteria, as given in Eq. (46), should be modified to include the parameter of the wing lift:

$$K = \frac{(8c/b)^{1/2}f_0^{(1+n)/2}v^{(m-n)/2}f^{(2-m)/2}}{(1 - 4.44 \times 10^{-5}V^2)} \frac{1}{v}\overline{DI}\left(\frac{1}{f}\right)^{1/2} \tag{49}$$

TABLE 9.2 List of Significant Wavelengths

Type of movement	Ground movement	Taxiing	Landing and takeoff
Velocity of airplane	20–30 fps	30–70 knots	120–150 knots
Assumed effect of wing lift	0.00	0.05	0.50
Significant wavelength v/f, ft	4–10	40–100	140–200
Smooth pavement, $DI = 0.12g$:			
Length of straightedge L, ft	10	25	50
Surface deviation, in.	0.34	0.21	0.24
Rough pavement, max. $DI = 0.3g$:			
Length of straightedge L, ft	10	25	50
Surface deviation, in.	0.60	0.37	0.42

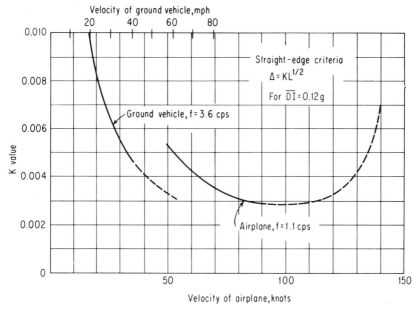

Fig. 9.14   Straightedge criterion for pavement construction.

The significant range of aircraft velocity is shown in Table 9.2, and the variations of the K value determined from the Newark and Kennedy tests are plotted in Fig. 9.14. The nature of the curve plottings suggests the use of a simplified expression:

$$K = C_0 \frac{1}{v} \overline{DI} \left(\frac{1}{f}\right)^{1/2} \tag{50}$$

where $C_0 = 4.32$ for a ground vehicle with velocity v, in feet per second, and $C_0 = 2.28$ for a taxiing aircraft, with velocity in knots.

SUMMARY

With the straightedge method, thus developed, it becomes feasible for engineers to get a much closer relationship between the physical geometry of the pavement surface as expressed by Eq. (46), where L is a function of aircraft velocity and Δ is a function of dynamic response. The value $\overline{DI}^2$ represents the total power in the aircraft vibration. The transfer function is reflected by the coefficient, K, as expressed by Eq. (50), in which v is the velocity of the aircraft, $\overline{DI}$ is the dynamic response at the interface of wheels and pavement, and f is the natural frequency of the aircraft at the interface with the pavement. The constant C_0 represents the damping and all other related factors in the vibrating system. In considering the wing lift at high speeds, the effective weight of the airplane is reduced. The Newark-Kennedy study indicates that the critical speed of an aircraft responding to the roughness of the pavement is in the range of 80 to 100 knots. The corresponding K value is 0.0029 for a smooth ride $(\overline{DI} = 0.1g)$ and a natural frequency of the aircraft of 1.1 cps. With this background, the significant wavelengths of pavement roughness can be related to the speed of the vehicle. For a country road (the farmer-to-market type of road), with a vehicle traveling at a speed of between 10 and 30 mph, the significant wavelengths are in the range of 4 to 12 ft. For a modern express highway, the vehicle speed ranges from 50 to 80 mph and the significant wavelengths of the pavement surface roughness are in the range of 25 to 40 ft. For airport pavements, the normal taxiway speed ranges from 30 to 70 knots, with consequent wavelengths of 40 to 100 ft. For runway pavement, the touchdown and takeoff speeds range from 120 to 150 knots. Because of its wing lift, the most critical velocity contributing to the vibration of the aircraft is likely to be in the range of 80 to 100 knots. The corresponding wavelengths of surface roughness are in the range of 140 to 200 ft. It has been the trend of aircraft construction to develop low-frequency landing gear. Therefore, the significant wavelengths of the pavement might be increased. For instance, for an aircraft having a natural frequency at interface of 1.5 cycles, the significant wavelengths are about 150 ft. For soft-sprung landing gear (natural frequency at interface of 0.8 cps), the significant wavelengths are increased to about 300 ft. Thus it can be seen that designing a good pavement to fulfill its functional requirements requires study of the interaction of aircraft and pavement.

REFERENCES

1. "Reports on Pavement Design and Tests—Redevelopment Program, Newark Airport," The Port of New York Authority, June, 1967.
2. H. C. Houbolt, Runway Roughness Studies in the Aeronautic Field, *Trans. ASCE,* vol. 127, pp. 427–447, 1962.
3. G. J. Morris, "Response of a Turbojet and a Piston-engine Transport Airplane to Runway Roughness," NASA TN-D3161, National Aeronautics and Space Administration, December, 1965.

4. J. C. Houbolt, R. Steiner, and K. G. Pratt, "Dynamic Response of Airplanes to Atmosphere Turbulence Including Flight Data on Input and Response," NASA TR R-199, National Aeronautics and Space Administration, pp. 61–115, June, 1964.
5. R. B. Blackman and J. W. Tukey, "The Measurement of Power Spectral from the Point of View of the Communications Engineer," Dover Publications, Inc., New York.
6. J. S. Bendat, "Principles and Applications of Random Noise Theory," John Wiley & Sons, Inc., New York, 1958.
7. B. G. Hutchinson, "Analysis of Road Roughness Records by Power Spectral Density Techniques," Department of Highways, Report 101, Ontario, January, 1965.
8. W. T. Thompson, "Vibration Theory and Application," pp. 314–344, Prentice-Hall, Inc., Englewood Cliffs, N. J., 1965.

APPENDIX TWO
Mathematical Model of Random Process

1. PROBABILITY OF RANDOM PROCESS

For a random time function $x(t)$, the probability of the value of x lying in the interval between x_1 and $x_1 + \Delta x$ can be expressed by the sum of the time intervals Δt during which $x(t)$ occupies the range x_1 to $x_1 + \Delta x$. This sum, divided by the total time of the record, represents the fraction of the time that $x(t)$ will satisfy the condition $x_1 \leq x \leq x_1 + \Delta x$ for the particular record length being considered.

If x_1 is chosen large enough so that it lies beyond the measured amplitudes of all x values, $(1/t)\Sigma\Delta t$ will be zero. Now, if the value of $x_1 + \Delta x$ is held constant while x_1 is reduced, we are in effect increasing Δx, which is expanding the range of acceptable x values. Therefore, $(1/t)\Sigma\Delta t$ tend to increase with decreasing values of x_1. Ultimately, we reach a point where x_1 is of such a low value that $(1/t)\Sigma\Delta t = 1$. The curve of $(1/t)\Sigma\Delta t$ is called the *cumulative probability curve,* and stated in mathematical terms, we have

$$0 \leq P(x) \leq 1 \tag{A1}$$

where $P(x)$ is the probability of x lying between x_1 and $x_1 + \Delta x$.

If the probability curve is smooth, it is possible to define another function $P(x)$ representing the slope of the cumulative probability curve by

$$\frac{d}{dx} P(x) = p(x) \tag{A2}$$

The curve of $p(x)$ is called the *probability density curve*. The probability of $x(t)$ having an amplitude lying in the range between x_1 and $x_1 + \Delta x$ may therefore be expressed by

$$P(x_1 + \Delta x) - P(x_1) = p(x_1)\, dx \tag{A3}$$

2. GAUSSIAN DISTRIBUTIONS

If a set of records is statistically uniform (i.e., its statistical properties remain constant) with a sufficient length of sampling, the record is said to be *stationary*. Such a record is said to have a probability distribution which we call a *gaussian* (or normal) distribution. When any random phenomenon is the result of the cumulative effects of a large number of independent contributions, the probability distribution tends to become gaussian.

The probability density of a random function can be related to the mean, and to the mean-square, values of that function. The mean value is given by the first moment of the function,

$$\bar{x} = \int_{-\infty}^{\infty} x p(x)\, dx \tag{A4}$$

For a stationary random function, the mean value becomes

$$\bar{x} = \lim_{T \to \infty} \frac{1}{2T} \int_{-T}^{T} x(t)\, dt \tag{A5}$$

The mean-square value of a stationary function is given by the second moment of the function

$$\bar{x}^2 = \lim_{T \to \infty} \frac{1}{2T} \int_{-\infty}^{\infty} x^2(t) \, dt \qquad (A6)$$

If, in turn, we consider the mean-square value of this function about its mean, we obtain the so-called *mean-square deviation,* which is known also as the *variance* and is expressed by

$$\sigma^2 = \int (x - \bar{x})^2 p(x) \, dx = \bar{x}^2 - (\bar{x})^2 \qquad (A7)$$

The positive square root of the variance σ is known as the *standard* deviation and, for a zero mean value, is called the *root=mean-square* (rms) value. A gaussian distribution in terms of its standard deviation is given by

$$p(x) = \frac{1}{\sigma\sqrt{2\pi}} e^{-x^2/2\sigma^2} \qquad (A8)$$

where e is the base of natural logarithms and is equal to 2.71828+.

As we can see, the standard deviation is a measure of the spread of the random distribution about its mean value. Therefore, as the σ value decreases, the gaussian curve will become narrower and taller. However, the area under the curve is always unity.

3. RANDOM-FREQUENCY TIME FUNCTIONS

If a record set consists of a number of random subsets, each of which is a time-frequency function, the final configuration will depend upon the average properties of these subsets.

For a simple harmonic function of the form $x = x_0 \sin 2\pi ft$, the variance with reference to the zero mean value is

$$\bar{x}^2 = \frac{f}{n} \int_0^{n/f} x_0 \sin^2 2\pi ft \, dt = \frac{x_0^2}{2} \qquad (A9)$$

where f is the cyclic frequency and $T = n/f$ is the total sampling time within which n cycles of the harmonic function take place.

In applying this principle to a multifrequency function $x(t)$, we use a Fourier series which, with respect to the zero mean value, is given by

$$x(t) = \frac{1}{2}\sum x_n e^{in2\pi ft} + \frac{1}{2}\sum x_n^* e^{-in2\pi ft} \tag{A10}$$

where x_n^* is the conjugate of the complex number x_n and $i = \sqrt{-1}$.

The integration of this Fourier transformation, when substituted into Eq. (A6), yields

$$\overline{x}^2 = \lim_{T\to\infty}\frac{1}{T}\int_0^T \frac{1}{4}\left[\sum x_n e^{in2\pi ft} + \sum x_n^* e^{-in2\pi ft}\right]^2 dt \tag{A11}$$

$$= \sum \frac{x_n^2}{2} = \sum \overline{x}_n^2$$

Thus we see that the mean-square value of a multifrequency periodic function, with respect to the zero mean value, is simply the sum of the mean-square values of all the individual harmonic components encountered in the record set. We may, therefore, express the contribution to the total mean-square value of each frequency interval Δf (representing the mean-square value in the interval Δf about some f) by

$$\phi(f) = \frac{\Delta(\overline{x}^2)}{\Delta f} \tag{A12}$$

where $\phi(f)$ is known as the *discrete spectral-density function*.

The discrete spectral-density function becomes a continuous spectral-density function for small values of Δf, so that

$$\phi(f) = \frac{d(\overline{x}^2)}{df} \tag{A13}$$

or

$$\bar{x}^2 = \int_0^\infty \phi(f) \, df \qquad\qquad \text{(A14)}$$

In other words, the area under the spectral-density curve $\phi(f)$ versus f is equal to the variance of a random periodic function.

While this simple procedure has proved to be very valuable in statistically describing a random periodic function, it is evident at once that Eq. (A14) is not able to account for the contribution to the variance of individual portions of a single harmonic component. In other words, we have so far, merely a method for summing up variances of individual harmonic components. These individual variances, however, are obtained by breaking up a record length into discrete segments, computing the variance of each segment, and averaging the differences. Therefore, for a nonperiodic record, in particular, we still are dealing with individual approximations of variances instead of a continuous function of them. Consequently, the integrity and reliability of the random-process treatment is seriously affected.

4. CORRELATION FUNCTIONS

When we have a random-function record set which is nonperiodic, the Fourier series in Eq. (A10) is not applicable. Instead, we must use the Fourier integral, which may be viewed as a limiting form of the Fourier series as the period $2T$ is extended to infinity. The Fourier integral is generated by a real function $x(t)$ whose absolute value $x(t)$ is integral over the interval $-\infty < t < \infty$. Thus, $\int_{-\infty}^{\infty} x(t) \, dt$ exists. Therefore, the coefficient of the Fourier series in Eq. (A10) is equal to

$$x_n = \frac{1}{2T} \int_{-T}^{T} x(t) e^{-in2\pi ft} \, dt \qquad\qquad \text{(A15)}$$

so that $x(t)$ may be written as

$$x(t) = \sum \frac{1}{2T} \int_{-T}^{T} x(t) e^{-in2\pi ft} e^{in2\pi ft} \, dt \qquad\qquad \text{(A16)}$$

Since the increment of the frequency Δf is given by $\Delta f = 1/2T$, Eq. (A16) becomes

$$x(t) = \int_{-\infty}^{\infty}\left[\int_{-\infty}^{\infty} x(t)e^{-in2\pi ft}\,dt\right]e^{in2\pi ft}\,df \qquad (A17)$$

Equation (A17) is the Fourier integral. Furthermore, since the quantity within the brackets is a function of only $i2\pi f$, Eq. (A17) can be rewritten in two parts as

$$X(i2\pi f) = \int_{-\infty}^{\infty} x(t)e^{-i2\pi ft}\,dt \qquad (A18)$$

and

$$x(t) = \int_{-\infty}^{\infty} X(i2\pi f)e^{i2\pi ft}\,dt \qquad (A19)$$

The quantity $X(i2\pi f)$ is the Fourier transformation of $x(t)$, and the two equations are referred to as the *Fourier transform pair*.

Substituting Eq. (A19) into Eq. (A6), we get

$$\overline{x^2} = \lim_{T\to\infty}\frac{1}{2T}\int_{-T}^{T} x^2(t)\,dt = \lim\frac{1}{2T}\int_{-T}^{T} x(t)\int_{-\infty}^{\infty} X(i2\pi f)e^{i2\pi ft}\,df\,dt$$

$$= \int_{-\infty}^{\infty}\lim_{T\to\infty}\frac{1}{2T}\left[\int_{-T}^{T} x(t)e^{i2\pi ft}\,dt\right]X(i2\pi f)\,df$$

$$\qquad\qquad (A20)$$

$$= \int_{-\infty}^{\infty}\lim_{T\to\infty}\frac{1}{2T}X(-i2\pi f)X(i2\pi f)\,df$$

$$= \int_{0}^{\infty}\lim_{T\to\infty}\frac{1}{T}X(-i2\pi f)X(i2\pi f)\,df$$

By comparing the integrands of Eqs. (A14) and (A20), it is established that

$$\phi(f) = \lim_{T \to \infty} \frac{1}{T} X(-i2\pi f)X(i2\pi f) \qquad \text{(A21)}$$

If we take the values of $x(t)$ to be multiplied together at a lag time τ apart, we have what is known as the *covariance* at the time lag, expressed by

$$R(\tau) = \lim_{T \to \infty} \frac{1}{2T} \int_{-T}^{T} x(t)x(t + \tau) \, dt = \lim_{T \to \infty} \frac{1}{T} \int_{0}^{T} x(t)x(t + \tau) \, dt \quad \text{(A22)}$$

This concept of an approach to a statistical expression of a random gaussian function is known as *autocorrelation,* and $R(\tau)$ is often called the *autocorrelation function.*

For a single sinusoid, $R(\tau)$ retains the same frequency as that of the sinusoid and does not approach zero as a limit. For a record with a wide frequency band, $R(\tau)$ is a peaked curve at $\tau = 0$ and approaches zero quickly. For a narrow frequency band, $R(\tau)$ approaches zero more slowly, so that it will achieve zero only if τ is relatively large. These properties make it possible to use the autocorrelation function to detect hidden periodicities in a random record of large T.

5. POWER-SPECTRAL-DENSITY FUNCTIONS

A most important property of autocorrelation has been discovered in recent years. If we use a Fourier transformation, Eq. (A22) becomes

$$R(\tau) = \lim_{T \to \infty} \frac{1}{2T} \int_{-T}^{T} x(t) \left[\int_{-\infty}^{\infty} X(i2\pi f)e^{i2\pi f(t+\tau)} \, df \right] dt$$

$$\qquad \text{(A23)}$$

$$= 2 \int_{0}^{\infty} \lim_{T \to \infty} \frac{1}{T} X(-i2\pi f)X(i2\pi f)e^{i2\pi f\tau} \, df$$

If we substitute Eq. (A21) into the above, we see that

$$R(\tau) = \int_0^\infty \phi(f)e^{i2\pi f\tau} \, df \tag{A24}$$

The Fourier transform of $R(\tau)$ can be obtained as follows:

$$\int_{-\infty}^\infty R(\tau)e^{-i2\pi f\tau} \, d\tau = \int_{-\infty}^\infty e^{-i2\pi f\tau}\left[\lim_{T\to\infty}\frac{1}{2T}\int_{-T}^T x(t)x(t+\tau)\,dt\right]d\tau$$

$$= \lim_{T\to\infty}\frac{1}{2T}\int_{-\infty}^\infty x(t+\tau)e^{-i2\pi f(t+\tau)}\,d(t+\tau)$$

$$\times \int_{-\infty}^\infty x(t)e^{i2\pi ft}\,dt \tag{A25}$$

$$= \lim_{T\to\infty}\frac{1}{2T}X(i2\pi f)X(-i2\pi f)$$

$$= \phi(f)$$

The above equation can be rewritten as

$$\phi(f) = 2\int_0^\infty R(\tau)e^{-i2\pi f\tau} \, d\tau \tag{A26}$$

Furthermore, since $x(t)$ is confined to a real time domain, the imaginary portion of the Euler identity ($e^{-ix} = \cos x - i\sin x$) can be omitted, with the result that Eq. (A24) becomes

$$R(\tau) = 2\int_0^\infty \phi(f)\cos 2\pi f\tau \, df \tag{A27}$$

and Eq. (A26) becomes

$$\phi(f) = 2\int_0^\infty R(\tau)\cos 2\pi f\tau \, d\tau \tag{A28}$$

If we wish to use the angular frequency ω, instead of f, we use the expression $\omega = 2\pi f$, noting that $\phi(f) = 2\pi\phi(\omega)$, to obtain

$$\phi(\omega) = \frac{1}{\pi} \int_0^\infty R(\tau) \cos\omega\tau \, d\tau \qquad (A28a)$$

We now have a continuous expression for $\phi(\omega)$, so that our determination of the variance of a random gaussian function will be more precise than was previously possible. In addition, $R(\tau)$ is convergent to zero in a real situation of sufficient lag value, so that the integral of $R(\tau)$ is capable of being evaluated exactly. Equations (A27) and (A28) are known as the *Wiener-Khintchine equations,* named after the American and Russian scientists who independently first noted their relationship.

6. COMPUTER COMPUTATIONS

Since, in actual practice, computers cannot perform integration, we must evolve a summation procedure for the digital computation of the autocorrelation and power-spectral-density functions. For a finite record length T, composed of N counting points at a unit distance t apart, equation (A22) becomes

$$R(J) = \frac{1}{(N-J)} \sum_{I=1}^{N-J} XS(I)XS(I+J) \qquad J = 1, 2, \ldots, M \qquad (A29)$$

where $XS(I)$ and $XS(I+J)$ are deviations from the mean at points I and $I + J$, respectively, and M is the maximum number of shifts of the autocorrelation function.

Therefore, the autocorrelation function corresponding to a J shift is the sum of the products of the deviations at a space J apart, divided by $N - J$, which is the range. For example, if $N = 5$ and $J = 2$, we have

$$R(2) = \frac{[XS(1)XS(3)] + [XS(2)XS(4)] + [XS(3)XS(5)]}{3}$$

After computing all the values of autocorrelation for a given record, we perform the summation corresponding to Eq. (A28):

$$\frac{1}{\epsilon} SL(K) = R(1)$$

$$+ \sum_{J=2}^{M} 2R(J) \cos\left(\frac{J-1}{N-1}\frac{K-1}{N-1}\pi\right) \quad K = 1, 2, \ldots, M \quad \text{(A30)}$$

which gives us a slightly weighted version of a raw estimate of the power spectrum $SL(K)/\epsilon$ at the frequency corresponding to K. The ϵ value represents the sampling interval in digital computation. Since it is known that in a stationary record, the central value appears only once and all other values appear twice, we shall use only once the value of $R(1)$ for our half record. Moreover, we shall not use the last value of R at all, since theory indicates that its value should be zero. We, therefore, shall be forcing the function to approach zero in the limit.

As is obvious from the discussion, J is being used as a counter of T and K counts w. In reality, since counters in digital computation always start from 1 instead of zero, we must adjust some of our parameters accordingly, as has been done with the cosine term in Eq. (A30). At this point, though, remember that when we speak of J, we actually mean $J \Delta t$ or $J(T/N)$. K counts the number of $\Delta\omega$, intervals of length ω/N or $(\pi/N)(1/T)$.

The use of π to define our record interval deserves some explanation, since it serves to eliminate the divisor factor in Eq. (A28). As is obvious from the above discussion of the angular function, ωt is actually defined by $\left(K \frac{\pi}{N}\frac{1}{T}\right)\left(J \frac{T}{N}\right)$, or $\left(\frac{J}{N} K \frac{\pi}{N}\right)$. Thus we see that the T terms cancel each other, and the value of the spectral density is achieved directly.

Finally, indications from actual study of power-spectral-density computations seem to indicate that these values should be smoothed by the use of averaging techniques known as *windows*. These windows have the effect of filtering the spectral-density estimates so that the spectral density for a given frequency interval is composed of a weighted average of a range of spectral densities about the particular one of interest. This is basically the same procedure as is involved in the common averaging techniques of engineering, such as the trapezoidal rule. While much theoretical work has been done in this area, it appears that trial-and-error procedures have evolved the best window to date. For nonsmooth samples, such as those which normally come from random problems of a civil engineering nature, the hamming window has been most recommended. This gives results as follows:

$$\phi(1) = 0.54SL(1) + 0.46SL(2) \tag{A31}$$

$$\phi(r) = 0.23SL(r - 1) + 0.54SL(r) + 0.23SL(r + 1) \tag{A32}$$

$$\phi(N) = 0.46SL(N - 1) + 0.54SL(N) \tag{A33}$$

A discussion of this technique, as well as a detailed mathematical treatment of the nature of spectral windows, has been reported by Blackman and Tukey.

It should be noted that in Eq. (A30), the cosine term involves the expression $\pi/(N - 1)$, which expresses the radians per interval of record, rather than π/N. This allows us to more meaningfully speak of frequency ranges and avoid the unclear situations at the beginning and the end of the record interval.

7. COMPUTER PROGRAM

The following discussion is not intended to present the precise computer program for the digital computation of the power spectral density since each computer system contains certain individualities which a program must account for. What is presented here is merely a basic setup, which is general with respect to the type of problem being solved by spectral density and the particular type of system being used. The flow chart and operation statement are outlined on the following:

Statement of Operation
(Equivalent Operation)

1. Dimension space for arrays of sample function values, autocorrelation values, and raw spectral-density estimates.
2. Read in values of the sample function X5 and the value of N, & M (the limit of J & K counters).
3. SUM = 0.0
4. DO 5 I = 1, N
5. SUM = SUM + XS(I)

6. XBAR = SUM/N
7. DO 8 J = 1, N
8. XS(J) = XS(J) − XBAR
9. DO 17 J = 1, M
10. CJ = 0.0
11. J1 = J−1
12. N2 = N−J1
13. DO 15 I = 1, N2
14. IJ = I + J1
15. CJ = CJ + XS (I) XS (IJ)
16. D2 = N2
17. R(J) = CJ /D2
18. AM = N−1
19. PM = 3.1415927 / AM
20. M1 = M−1
21. MN − M−1
22. DO 27 K = 1, M
23. SUM = 0.0
24. DO 26 J = 2, M
25. P = (J−1) · (K−1) / M N
26. SUM = SUM + 2. x R(J) x COS (P x PM)
27. SL(K) = R (1) + SUM
28. UJ = .54 x SL (1) + .46SL (2)
29. FREQ = 0.0
30. OUTPUT STATEMENTS # 28 and 29
31. SAM = 0.0
32. DO 34 K = 2, M1
33. SAM = SAM + 1.0
34. UJ = .23 x SL (K−1) + .43 x SL (K) + .23 x SL (K+1)
35. FREQ = (3.1415927) x SAM/AM)
36. OUTPUT STATEMENTS # 34 and 35
37. UJ = .46 x SL (M1) + .54 + SL (M)
38. FREQ = 3.1415927
39. OUTPUT STATEMENTS # 37 and 38
40. END

APPENDIX THREE

Power Estimation by Folding-frequency Method

As the power spectral density obtained by the folding-frequency method represents the expansion of a power series, the acquired record does not need to be of a random gaussian nature. The local disturbance of the record will be reflected in the estimation of the power spectral density. For this reason, the smooth technique employed at the autocorrelation function is not required for the frequency analysis. The statistical reliability of the analysis depends on the number of samples entered in the lower range of frequency resolution. For the profile analysis, the sampling interval will range from 2 to 4 ft, which corresponds to a significant shortwave length of 4 to 8 ft. For a significant long-wave length of 60 to 80 ft, the minimum record length will be at least 8 to 16 times the significant long-wave length, that is, in the range of 600 to 1,000 ft long or more. For data processing, the number of samples will be in the form of 2^n, and n is desirable to range from 7 to 9. For the Newark pavement test, the sampling interval is 3.75 ft, n is 8, the total number of samples is 256, and the total record length is 960 ft.

Two longitudinal profiles along the wheel path were taken during the test. The profile readings have been condensed into spatial frequency in cycles per foot and power spectral density in square feet per cycle per foot. The actual computations are shown in the following.

```
                        LIST
101. =      CF      PROGRAM SPEDEN
102. =              DIMENSION X(2048)
103. =              PRINT 1
104. =              10FORMAT(68H     THIS PROGRAM IS DESIGNED TO COM
                    1PUTE VALUES OF THE POWER SPECTRAL/64H   DENSI
                    2TY, BY THE FOLDING FREQUENCY METHOD, OF A
                    3 TIME FUNCTION/31H   SAMPLED AT EQUAL INTERVA
                    4LS.//51H   YOU WILL NOW BE ASKED FOR CERTAIN BI
                    5TS OF DATA./57H   PLEASE ENTER THESE IN ORDER
                    6 AND IN MODE AS INDICATED.//)
105. =              PRINT 2
106. =              20FORMAT(62H     ENTER: 'N', THE NUMBER OF SAMPLE
                    1 READINGS TO BE PROCESSED./70H     N MUST BE EIT
                    4HER 2, 4, 8, 16, 32, 64, 128, 256, 512, 1024 or 2048./)
107. =      CF      READ 0,N
108. =              PRINT 50
109. =              500FORMAT(65HENTER THE CONVERSION FACTOR WHIC
                    1H MULTIPLIES SPECTRAL DENSITY AND/72H DIVIDES
                    4 THE FREQUENCY. THIS FACTOR SHOULD REFLECT
                    THE SAMPLING INTERVAL/22HAND ANY SCALE FAC
                    TORS.)
110. =      CF      READ 0,FACTOR
111. =              PRINT 3
112. =              30FORMAT(71H     YOU WILL NOW FEED IN, IN ORDER
                    4 OF SAMPLING SEQUENCE, VALUES OF 'X',/34H
                    WHERE X IS THE SAMPLE READING./)
113. =      CF      READ 0,(X(I),I=1,N
114. =              AN=N
115. =              BN=AN*1.5
116. =              AM=ALOG10(BN)/ALOG10(2.)
117. =              M=AM
118. =              53SUM=0.0
119. =              AVE=0.0
120. =              N1=2**M-1
121. =              52DO 4 I=1,N1,2
122. =              SUM=SUM+(-(X(I))+X(I+1))**2
```

```
123. =        4AVE=AVE+X(I)+X(I+1)
124. =         AVE=AVE/AN
125. =         PRINT 51,AVE
126. =        51FORMAT(//10X,32HMEAN VALUE OF SAMPLE READI
                1NGS =,FLO.5)
127. =         SPEDE=SUM/AN*FACTOR
128. =         FREQ=1./2./FACTOR
129. =         PRINT 5,FREQ,SPEDE
130. =        50FORMAT(//10X, 9HFREQUENCY,28X, 8HSPECTRAL/
               4 12X, 5H(CPS), 30X, 7HDENSITY/9X,11(1H*), 26X,
                10(1H*)/ 11X, F7.5, 8X, F30.8/)
131. =         IF(M-1)100,100,6
132. =        6DO 7 1=2,M,1
133. =         N2=2**(M+1-1)
134. =         SUM=0.0
135. =         DO 8 I=1,N2
136. =         I2=2*I
137. =        8X(I)=X(I2-1)+X(I2)
138. =         DO 9 I=1,N2,2
139. =        9SUM=SUM+(-(X(I))+X(I+1))**2
140. =         SPEDE=SUM/AN*FACTOR
141. =         FREQ=1./2.**1/FACTOR
142. =         PRINT 10,FREQ,SPEDE
143. =        10FORMAT(11X,F7.5, 8X, F30.8/)
144. =        7CONTINUE
145. =      100CONTINUE
146. =         END
```

LONGITUDINAL PROFILE OF N.A. TEST PAVEMENT
SURVEYED 9/14/67   1' RIGHT OF CENTER LINE
STATIONS 10+18.75 TO 19+75.00   SAMPLING INTERVAL =
3.75'   ELEVATIONS IN FEET

```
111. = O 03   YOU WILL NOW FEED IN, IN ORDER OF SAMPLING
111. =     2  SEQUENCE, VALUES OF 'X', WHERE X IS THE SAMPLE
111. =     3  READING.
111. =     4
113. = I 00   .557/.552/.548/.544/.549/.552/.557/.571/.582/.590/.592/
113. =     2  .597/.599/.615/.620/.635/.629/.619/.627/.640/.652/
113. =     3  .624/.585/.555/.535/.539/.546/.559/.566/.556/.536/.520/
113. =     4  .515/.534/.533/.517/.539/.562/.577/.578/.579/.574/
113. =     5  .573/.572/.589/.601/.611/.615/.627/.625/.625/.629/.626/
```

```
113. =      6  .619/.616/.618/.629/.649/.647/.638/.636/.635/.619/
113. =      7  .602/.587/.546/.538/.470/.502/.489/.515/.516/.506/.493/
113. =      8  .464/.460/.459/.458/.460/.453/.439/.428/.418/.418/
113. =      9  .424/.440/.468/.499/.544/.593/.656/.690/.658/.646/.644/
113. =     10  .618/.587/.549/.534/.516/.511/.493/.473/.463/.442/
113. =     11  .438/.434/.435/.445/.450/.467/.474/.440/.484/.499/.498/
113. =     12  .497/.485/.486/.488/.471/.448/.437/.412/.408/.411/
113. =     13  .410/.412/.418/.426/.423/.401/.372/.376/.373/.383/.392/
113. =     14  .385/.377/.372/.386/.385/.398/.402/.415/.418/.423/
113. =     15  .433/.436/.447/.441/.422/.430/.444/.451/.448/.450/.439/
113. =     16  .443/.437/.439/.440/.443/.438/.432/.411/.393/.385/
113. =     17  .354/.333/.313/.295/.296/.297/.310/.298/.296/.315/.343/.
113. =     18  .378/.362/.377/.386/.392/.383/.384/.373/.367/.375/
113. =     19  .369/.374/.372/.368/.355/.336/.333/.342/.376/.393/.395/
113. =     20  .397/.410/.420/.422/.420/.426/.425/.422/.428/.423/.
113. =     21  .429/.422/.419/.416/.417/.416/.422/.418/.426/.431/.440/
113. =     22  .441/.439/.446/.444/.448/.451/.462/.477/.486/.479/
113. =     23  .467/.460/.453/.443/.427/.414/.396/.395/.393/.379/.373/
113. =     24  .379/.364/.361/.360/.369/.354/.351/.335/.330/.314/
113. =     25  .300/.285/.274/.275/
125. =O 51
125. =      2
125. =      3          MEAN VALUE OF SAMPLE READINGS = 0.46755
129. =O 05
129. =      2
129. =      3          FREQUENCY                 SPECTRAL
129. =      4            (CPS)                   DENSITY
129. =      5         ***********               *********
129. =      6          0.13333                 0.00047565
129. =      7
142. =O 10          0.06667                 0.00267898
142. =      2
142. =O 10          0.03333                 0.00585760
142. =      2
142. =O 10          0.01667                 0.07269710
142. =      2
142. =O 10          0.00833                 0.07742983
142. =      2
142. =O 10          0.00417                 0.07647771
142. =      2
142. =O 10          0.00208                 0.56133284
142. =      2
142. =O 10          0.00104                 5.07485926
```

142. =   2
146. =HALT END STATEMENT ENCOUNTERED DURING EXECUTION
147. +READY

CHAPTER TEN
Full-scale Pavement Tests

Based on analyses given in the preceding chapters, it can be seen that variations of material, construction, and vehicle performance and the inadequacy of the theoretical analysis have a significant effect on the design inputs of a pavement. It is necessary to conduct full-scale pavement tests to bridge the gap between theory and reality.

In the last 30 years, several full-scale pavement tests have been conducted in the United States. The most important one is the early airfield pavement test conducted by the Ohio River Division of the U.S. Army Corps of Engineers. The primary purpose was to validate and to modify the theoretical analysis advanced by Westergaard. Although much valuable information has been derived from that test, the inherent limitation of the scope did not allow the development of a new design concept.

The next series of tests was conducted by the Water Experiment Station of the U.S. Army Corps of Engineers. The primary purpose of the WES tests was to develop the design method for asphalt pavement construction utilizing compacted aggregate and soil as pavement base courses. The introduction of modern soil mechanics in pavement design was one of the outstanding contributions of the WES program. However, the lack of mechanistic analysis together with the application of CBR method in classifying the subgrade will affect the reliability of the pavement design if the boundary conditions are beyond the original scope of the WES test. In the last decade, the rapid advance of civil air transport has, in effect, caused a serious crisis in pavement design because the weight and traffic intensity of modern air transport have far exceeded the original limit of the WES test. No extrapolation of CBR method is reliable.

In the late 1950s, the American Association of State Highway Officials (AASHO) sponsored a comprehensive national pavement test program in Illinois. The program cost more than $26 million. Its primary purpose was to develop a method for evaluating the performance of pavement construction and to advance a rational pavement design method. However, there was no theoretical and conceptual study at the beginning of the pavement test, nor was any mechanistic concept introduced in processing the monitored data. The statistical process of random events was used in data reduction. There is no correlation between theory and test results. Moreover, the testing program was politically divided into (1) rigid pavement dealing with concrete construction and (2) flexible pavement dealing with asphalt material. The advance of modern pavement engineering may have been temporarily hindered.

10.1. NEWARK TEST

The redevelopment of Newark Airport will require the placing of approximately 2 million sq yd of new pavement. Economic considerations in the selection of the pavement materials are as important as the research for an appropriate design method. In an attempt to achieve maximum economies, a series of development tests was conducted in the laboratory which was aimed toward the utilization of inexpensive local materials for the construction of new pavement. Meanwhile, with the advent of the larger stretch-out subsonic aircraft, the probability of having to accommodate much heavier loads than are experienced today became obvious. It was imperative to develop a new concept in the design of a functional pavement to meet this challenge. An elaborate test program was conducted to determine the physical properties of the pavement structure and its supporting subgrade, as well as their contribution to pavement performance.

The test pavement was installed in a remote area of Newark Airport. It consisted of a 1,200 by 30 ft strip segmented into 16 test sections, each 75 ft long. Each section contained a different composition of pavement material. Seven materials, ranging from conventional to new, were used. The as-built compositions of the test pavements are shown in Fig. 10.1.

The heavy load was a test vehicle contributed by the Corps of Engineers. This vehicle weighed 187,000 lb and was carried on a four-wheel truck in simulation of the main gear of a Boeing 747 aircraft. The wheel configuration is shown in Fig. 10.2. Both the test pavements and the load vehicle were heavily instrumented to monitor the stress-strain in the pavements as well as the vertical vibration of the test vehicle. An aerial view of the test pavement in operation is shown in Fig. 10.3.

For a better understanding of the test results, some background information is summarized herein; a more detailed report can be found in "Reports on Pavement Tests—Redevelopment Program—

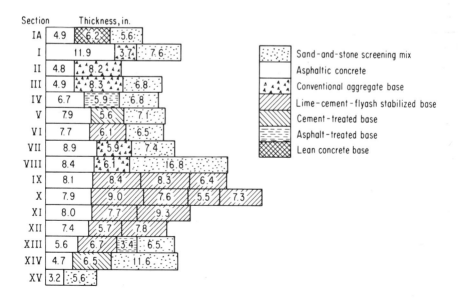

Fig. 10.1 Composition of test pavements.

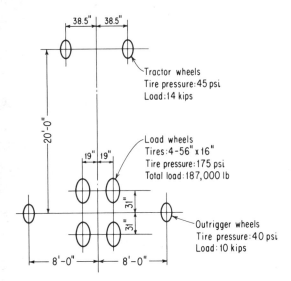

Fig. 10.2 Wheel configuration of landing-gear simulator.

Fig. 10.3 Aerial view of test pavement.

Newark Airport," issued by the Port of New York Authority, June, 1971. The test site was originally a marginal area behind the terminal moraine and formed a part of the glacial lake at Hackensack. The basement, consisting of red cemented sandstone and shale, is encountered 70 to 90 ft below the mean sea level. The overburden is primarily of red-brown boulder sand, which is a well-graded and compacted morainal deposit. After the retreat of the last glacier, the area became a tidal lowland. An organic mat varying from 4 to 10 ft in thickness has accumulated over the entire lowland of the airport. The organic deposit is highly compressible. Its moisture content ranges from 80 to 150 percent, and the rate of secondary consolidation is practically a linear function of time. A soil profile of the test site is shown in Fig. 10-4.

10.2. INSTRUMENTATION SYSTEM

For the instrumentation portion of the test program, measurements both within the pavement and on the vehicle were obtained. Within

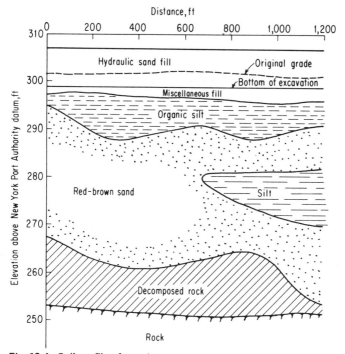

Fig. 10.4 Soil profile of test site.

the pavement, pressure, deflection, and temperature measurements were taken. On the vehicle, vertical and horizontal roll accelerations, vertical velocity, tire pressure, speed, vehicle path, tire deflection, and wheel axial strains were measured. The recording equipment and most of the transducers used were standard and consisted of balancing, calibrating, and amplifying components and photographic light-beam recorders. However, three types of transducers used are worth mentioning here. These are the bearing-plate, pavement-deflection, and pressure transducers.

Bearing Plate

For the bearing-plate tests, a transducer was made by attaching an SR4 strain-gage rosette to the counterbored center bottom of a steel plate 16 in. in diameter by 1 in. thick. Loads up to 10,000 lb were applied through a load cell, and the corresponding static bending strains were measured on two SR4 strain indicators.

Deflections

For the pavement deflections, linear variable differential trans-formers (LVDT) were used to measure total deformation, structural deflection, and vertical strain.

1. The total-deformation LVDT had an effective range of measurement of ±0.25 in. It was attached on the top of a nominal 1-in. pipe, driven about 18 ft below the bottom of the pavement. The total length of the LVDT was such as to allow attachment of a top plate (4 in. in diameter by 1/8 in. thick) at the selected depth just below the top lift of pavement, which permitted the measurement of surface deflection with respect to a fixed point in the subgrade.

2. The structural-deflection LVDT had an effective range of ±0.1 in. Its bottom was attached to a steel plate laid down on the top of subgrade. Its total length was such as to allow the measurement of surface deflection with respect to the bottom of the pavement structure.

3. The strain LVDT had a range of ±0.025 in. It was placed in a retention consisting of two brass end plates, which supported the LVDT. The retention was then placed in the paving components.

Pressure Gage

Preliminary tests were made using commercial pressure gages, but it was found that their output when embedded was too low for the expected loads. Accordingly, a new pressure gage was developed in an attempt to meet the various requirements for proper functioning within the material. In appearance, the designed gage resembled an oyster. This shape was chosen to provide more uniform distribution of stress as well as to reduced sensitivity to lateral stress. The electrical elements (semiconductor strain gages) were placed on a diaphragm supported on knife edges at the internal center of the oyster (see Fig. 10.5). The gage was coated with epoxy and sand to provide a good contact surface in the paving material.

In attempting to satisfy the criterion that the modulus of elasticity of the gage match that of the paving material, tests were carried out with three thicknesses of diaphragms. Outputs of the diaphragms were then measured for both the gage under pneumatic pressure and the gage under pressure embedded in the soil-cement material. By observation of the ratio of embedded output to pneumatic output, it was determined that a particular diaphragm thickness would provide the best compromise between the sensitivity and the level of output. The modulus of elasticity of the gage was designed to be in the range of 1.0 to 1.5×10^6 psi.

An in-place check of the pressure-gage calibration was subsequently carried out in the field. The pressure gage was located at a depth of 13 in. below a 24-in. bearing plate assembly. The gage was embedded in sand. The unit pressure at the gage level, according to the Boussinesq stress distribution, was equal to $0.588p^0$, where p^0 is the unit pressure on the test bearing plate. In Fig. 10.6, the observed gage output was compared with the pneumatic calibration for the theoretical gage pressure of $0.588p^0$. The results were in fairly good agreement. At a theoretical gage pressure of 60 psi, the discrepancy was about 4 percent.

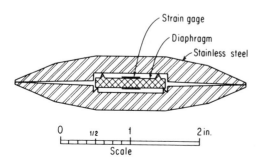

Strain gage
Diaphragm
Stainless steel

Fig. 10.5 Outline of pressure gage.

0   1/2   1   2 in.
Scale

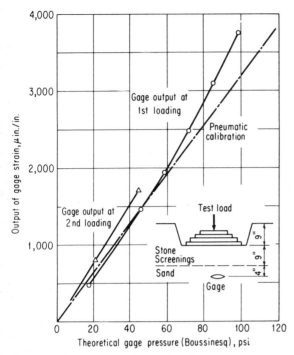

Fig. 10.6   Gage output under bearing-plate test.

10.3. PERMANENT LONGITUDINAL DEFORMATION OF PAVEMENT SURFACE

The performance of pavement under the influence of a moving aircraft is reflected by the interaction of the aircraft and longitudinal surface. As modern transportation grows rapidly in passenger capacity and vehicle speed, early pavement concepts become incapable of solving modern transportation problems. In the search for information, several tests were conducted in the Newark test program to establish transfer functions between (1) the dynamic response of a moving vehicle and the longitudinal profile of the pavement and (2) the permanent longitudinal and transverse deformation of the pavement surface. Thus the structural requirement of a pavement was related to a prescribed dynamic vehicle response and a corresponding deflection tolerance of the riding surface.

In Chap. 9, the theory governing the straightedge method was expressed by

$$\frac{\Delta}{L^{1/2}} = \frac{C_0}{vf^{1/2}} \overline{DI}$$

where Δ = surface deviation from straightedge
L = length of straightedge
$\overline{DI}$ = root mean square of vehicle response
v = crossing velocity of vehicle
f = fundamental frequency of moving vehicle
C_0 = transfer function depending on damping characteristics of traveling vehicle

In the Newark test, the fundamental frequency of the loading vehicle ranged from 1.6 to 2.0 cps. There was no shock absorber except the damping of the pneumatic tires. The operation speed of the test vehicle ranged from 4 to 5 fps. The longitudinal roughness was determined by rod-and-level measurement at intervals of 3.75 ft. Thus there were 20 readings per profile per test section, and three profiles were taken for each section.

The change of longitudinal roughness is governed by the number of load repetitions as well as the magnitude of settlement. The increase of roughness is indicated by the value of $\Delta/L^{1/2}$. For each test section, the processed $\Delta/L^{1/2}$ value was plotted against the number of load repetitions when the surface measurement was made (see Fig. 10.7). It was found that the roughness value $\Delta/L^{1/2}$ of LCF pavement increased at a linear rate of 0.20×10^{-3} per 1,000 load repetitions while the corresponding rate of increase was 0.54×10^{-3} for the aggregate-based pavement. The rigidity and bending strength of the LCF pavement make an important contribution in minimizing the effect of pavement roughness.

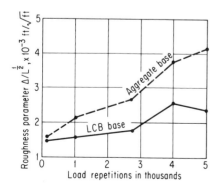

Fig. 10.7 Longitudinal roughness and number of load repetitions.

10.4. PERMANENT TRANSVERSE
DEFORMATION OF PAVEMENT

Permanent transverse deformation, commonly known as rutting, is a meaningful measurement in indicating the service condition of a pavement. A long deep rut may cause not only pitch of a moving vehicle but also a "birdbath" effect on the pavement surface, which can have a detrimental effect on vehicle operation during wet or icy conditions. In determining the progressive growth of surface deviation, a survey party constantly measured the elevation of the pavement surface, with readings at 1- to 2-ft intervals transversely and at cross sections spaced 5 ft apart. The true elevation of reference points in each cross section was measured by rod and level, with a sight distance of less than 100 ft. The best direct measurement was 0.001 ft deducted from multiple readings. In measuring the transverse cross sections, a straightedge with dial-gage attachment was employed. The dial gage, fixed on a steel rod, was used to measure the vertical distance between the pavement surface and the top of the straightedge at constant offset points. The dial gage had a total travel of 1 in. and gave a direct dial reading of 0.01 in.

According to the deflection theory given in Chap. 7, 85 percent of the total deformation takes place within a radius of $3.3a$ from the center of a load. The a value represents the radius of contact area of the tire, which was 9.25 in. for the Newark test. Therefore, a distance of $3.3 \times 9.25 + 19 = 49.5$ in. from both sides of the centerline of the test pavement was the significant width in measuring the depth of rutting. The entire survey records were compiled for an 8-ft straightedge, and these results are more indicative of the influence of traffic load.

When rutting or permanent deformation increases, the load-deformation curve in a semilog plot begins to deviate from a linear trend. The magnitude of deviation from the straight line represents the excessive deterioration of the pavement, which is usually associated with the presence of longitudinal crackings. In evaluating the performance of a pavement, it is necessary to define an upper limit of permanent rutting. Consequently, the initial tangent to the load-deformation curve would be a reliable starting point for predicting the service condition of a pavement (see Fig. 10.8).

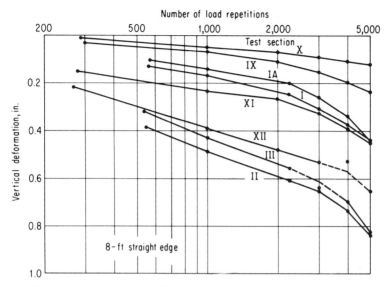

Fig. 10.8 Surface deformation of test pavements.

The development of rutting depends largely on the composition of the pavement base. In aggregate base and to some extent in asphalt-stabilized base, ruttings develop in the early stage of traffic loadings. The deflection configuration assumes a narrow, deep trough along the wheel paths, with a hump between the wheel paths. For pavements with LCF-stabilized sand as base courses, the entire pavement will maintain a definite flexibility but will be rigid enough to minimize the deep rutting along the wheel path. A wide and shallow deflection configuration was observed, and rutting along the wheel path was not pronounced.

There are two distinct patterns of pavement deformation. For pavements with LCF base thicker than 28 in., the magnitude of rutting decreases approximately in proportion to the thickness of pavement. The pavement structure of these sections seems to be in an elastic state of equilibrium condition, and the magnitude of permanent deformation increases proportionately with the increasing number of load applications. For LCF pavements thinner than 21 in., the surface deformation increases rapidly with the application of traffic load. The characteristics of surface deformation seems to represent the yield conditions in a plastic state of equilibrium.

As the service condition of a pavement under the influence of a moving load is directly reflected by the characteristics of the longitudinal profile, there is no way to integrate the design theories with the performance of a pavement unless a physical correlation between longitudinal and transverse deformations can be established. In the Newark test, such a correlation, or transfer function, was deduced from multiple-regression analysis (see Figs. 10.9 and 10.10). The result indicates that the stabilized-base pavements offer a much smoother riding surface, and consequently, a smaller dynamic response will result for moving aircraft.

10.5. STRESS CONDITION IN PAVEMENT UNDER A MOVING LOAD

Unlike surface deformation, which reflects the actual performance of a pavement, stress analysis is rather an academic study. In our present engineering practice, the method of measuring the stress in a body is based on the assumption that stress is a function of strain

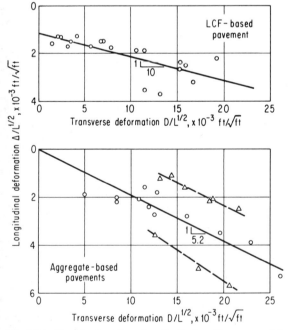

Fig. 10.9 Transfer function—longitudinal and transverse deformation.

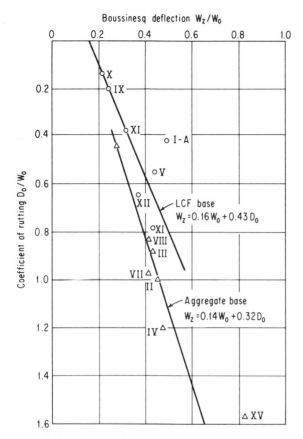

Fig. 10.10 Transfer function—rutting and Boussinesq deflection.

and that a change of strain output represents a variation of stress condition. Instruments are, therefore, developed for monitoring the strain. Even with a perfectly developed instrumentation system (which is not likely to be the case), the nonelastic stress-strain relationship, together with the gage-mass interaction under the influence of a load, is still an unknown factor in the reliability of stress calibration. For the Newark pavement test program, a new type of pressure transducer was developed. The gages had outstanding sensitivity, ruggedness, and reliability not found in commercially available gages. However, the interaction of the gage and mass was still a serious problem. The gages overregistered in some cases and underregistered in others. It was impossible to obtain an absolute value in the gage measurements. Therefore, the relative change of stress level was more reliable in indicating the strain-stress condition in the test pavement.

Thirty-one pressure gages were installed in the test pavements to determine (1) the stress pattern in various pavement materials, (2) the stress intensity at various depths below the loading surface, and (3) the stress distribution in the horizontal direction. While the test load was moving across the embedded gage, the electrical output of the gage was recorded on calibrated oscillographic paper. At the beginning of the recordings, the gages were electrically balanced by a known resistor. The electrical output was the equivalent strain of the bridge circuit. The equivalent-strain measurement was then converted into a stress reading based upon the pneumatic calibration. Thus any electrical outputs in the subsequent testings could be proportionally converted into pressure readings. Because of the random nature of the experimental data, at least four readings were taken in a single set of observations. The average of four to six sets of readings may give better information on the change of stress pattern in the pavement.

During the construction of the test pavements, layered components were compacted at various stages and the gage outputs were monitored when the pneumatic tire compactor moved directly over the gage. The relation between depth and stress coefficients is plotted in Fig. 10.11. It is interesting to note that the stress distribution in fill-sand subgrade followed very closely the Boussinesq pattern of stress distribution. The scatter of data below depths of $5a$ may have resulted from the small output of the gages.

During the traffic load test, the stress in different pavement materials was recorded. It was found that the asphalt-stabilized sand base exhibited a stress level identical to that encountered in an aggregate base and the maximum stress level in LCF-stabilized base was about 50 to 60 percent of that developed in the aggregate base. In terms of stress level, a 1-in.-thick layer of cement-stabilized base is as good as a 2-in.-thick layer of aggregate base. (In Sec. 10.6, we shall note that the result of the deflection measurements was entirely different.)

The stress level in a pavement is a function of (1) the elastic equilibrium in the mass and (2) the rigidity of pavement structure, such as the EI value in elastic analysis. For pavements having a conventional aggregate base, the modulus of deformation, E is not

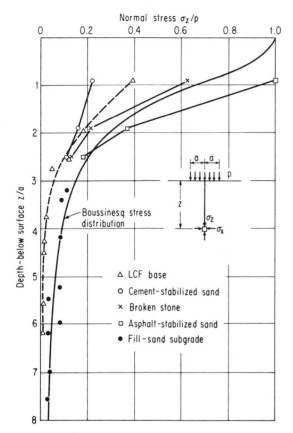

Fig. 10.11 Calibrated stress readings for gages at various depths.

very much better than that of the subgrade, and the measured stresses in the subgrade are in good agreement with the Boussinesq theory. For pavements having an LCF-stabilized base, the measured stresses are about one-half of those in the aggregate base. The lower stress level in the LCF-based pavements has resulted in less surface rutting and pavement deflection.

The original oscillographic record of the time-stress plot can be interpreted as the stress-distance function if the distance is assumed to be equal to the product of time and vehicle velocity. In general, the peak stress tapers off about 3 to 4 ft from the center of the wheel load. For pavements having an aggregate- or asphalt-stabilized sand base, a narrow and steep stress configuration is commonly encountered. The total width of the stress configuration is about 7 ft for a single wheel load. For the LCF-stabilized base, the stress configuration is wide and flat. The width of the stress curve is more than 8 ft for the same single wheel load.

In the Newark test program, three pressure transducers were installed vertically to monitor the stress in the horizontal direction. The horizontal compressive stress increased rapidly in the upper portion of the pavement structure. The peak horizontal compressive stress was always preceded by a small tensile stress when the vehicle was approximately 20 ft away from the gages. Then the compressive stress was registered within a distance of about 6 ft on both sides of the gage. Beyond this distance, the gage resumed its tensile-stress recording until the test load was about 23 ft away from the gage.

If the time-stress curve is a valid interpretation of the stress-distance function, the deflection basin of LCF pavement is almost 40 ft long and the maximum tensile stress is encountered about 10 ft in front of or behind the moving load. The alternating shift of stress resembles the cyclic fluctuation of wave propagation. For a takeoff or landing speed of 140 knots, the period of stress fluctuation from tension to compression to tension again is about 20/240 sec. If the natural frequency of the subgrade is about 10 to 15 cps, a resonant vibration may be developed in the pavement.

In actual pavement design, the normal stress has no meaningful application unless it is related to the performance of the pavement, which, at the Newark test, was reflected in the progressive development of surface ruttings. Their correlation, or transfer function, is shown in Fig. 10.12. Based on the surface-deformation criteria, the most desirable stress level at the bottom of the pavement (top of the subgrade) is in the range of 8 to 11 psi.

10.6. DEFLECTION OF PAVEMENT UNDER A MOVING LOAD

Under the influence of a moving load, the pavement deforms and rebounds like an elastic mat. The magnitude of deflection or strain is a direct indication of pavement performance. Under the influence of a static load, the pavement deforms elastically at the very beginning of load application and then undergoes a long period of stress-strain readjustment, which is commonly known as the *creeping* of

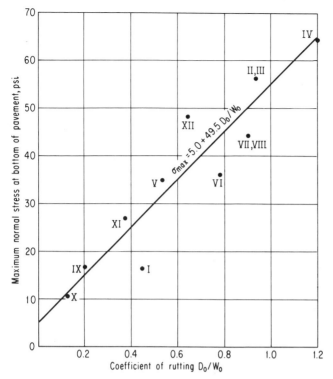

Fig. 10.12 Transfer function—maximum normal stress and rutting.

pavement materials. On removal of the applied load, a permanent deformation in the pavement components will remain. A moving vehicle's load history will affect the level of stress-strain in a pavement. The time required for a pavement to respond to an external load determines the magnitude of elastic strain and its associated plastic creeping.

The primary purpose of the Newark testing program was to determine the performance of test sections under a simulated traffic load. The physical limitations of the test site ruled out the use of any heavy aircraft as the test vehicle. The actual load vehicle consisted of a landing-gear simulator, a load box, and forward and backward power units of conventional earthmoving tractors. The rim pull governed the speed of the vehicle, which was in the range of 2 to 3 mph. The result of the test should therefore be interpreted as static deflection under a moving load. On the other hand, the loading period of the slow-moving test vehicle may have been short enough to prevent any significant plastic adjustment or creeping of the pavement materials. The instrument installed in the pavement was designed to measure the free-field strain in subgrade and stabilized bases. Because of the low rebound modulus of the embedding material, the measurement of compressive deformation was more reliable than that of tensile deformation.

Although the pressure transducer used in stress measurement was actually a strain indicator, the magnitude of gage deformation was regulated by its bending rigidity and the condition of installation. Its output did not reflect the true strain of the mass. In the early stage of the instrumentation design, two types of commercial strain gages were investigated. The linear vertical differential transformer (LVDT) showed more promising results. It consists of two electric coils in series with a floating magnetic core. The movement of the core is measured by the differential electrical output, which can be converted into the linear movement of the core. The LVDT itself is a very simple and reliable instrument. However, the LVDT unit has to be enclosed in a rugged housing, which affects the sensitivity of the core. When the housing unit is too flexible, damage of the LVDT gages is frequently experienced.

Eleven LVDTs were installed in the pavement with permanent reference to a steel rod driven to a nonyielding sand layer. The gage readings, in theory, reflected the deformation of pavement surface. The relationship between total deformation and the thickness of pavement is plotted in Fig. 10.13. It can be seen that the total deflection increases with decreasing thickness of the pavement. For pavements of equal thickness, the total surface deformation is almost identical regardless of the rigidity of the pavement. In the stress measurements, the stress in an asphalt-stabilized sand base was found to be twice as much as the stress in a soil cement base. This was not true in deflection measurement. The total deflection of these two pavements was practically the same.

Nine LVDTs were installed, with a reference plate, at the bottom of the pavement to measure the deflection taking place inside the pavement structure. The structural deflection decreased with increasing thickness of pavement. For pavements with identical thickness, the structural deflection in the aggregate-based pavement was twice that encountered with stabilized base. The remainder, subtracting the structural deflection from the total deflection, can be

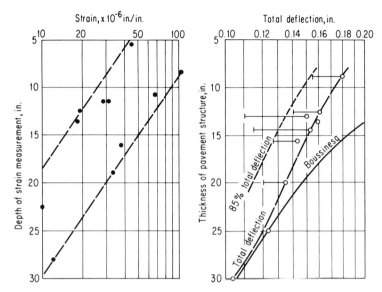

Fig. 10.13 Deflection of test pavements.

interpreted as the deflection of the subgrade, which is always a major portion of the total deflection. Therefore, pavement components do not have a significant effect on the total deflection of the pavement surface. The deflection in the subgrade is of prime importance in evaluating the performance of a pavement. Eleven LVDTs were installed in the pavement layers to measure the vertical strain of pavement components. As the electrical output of the strain LVDTs had been amplified to the capacity of the monitoring system, the strain readings were completely random. There were indications that (1) the strain of the asphalt-stabilized base was much higher than that of the soil cement or the LCF-stabilized bases, (2) the strain of the LCF-stabilized base was about the same as that of the soil cement base, and (3) the strain in the LCF base tended to decrease with increasing thickness of base structure.

In Fig. 10.14, four typical strain curves are reproduced from the original oscillographic records. The test results represent the strain of three different types of stabilized base with identical subgrade and top course. The deflection basin of the asphalt-stabilized base is deep and narrow, whereas the deflection basin of the cement or LCF-stabilized bases is relatively shallow. This indicates that the cement or LCF-stabilized base can distribute the wheel load over a

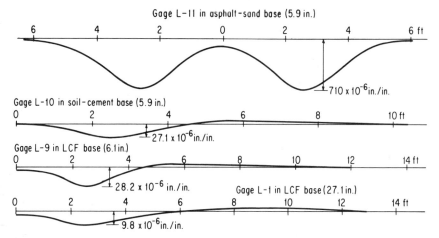

Fig. 10.14 Typical strain-distance curve for sections with various base materials.

wide subgrade area. The large rebound between the twin peaks indicates that the aggregate or asphalt-stabilized bases exhibit a high degree of resilience. However, the large fluctuation of pavement deflection will have a significant effect on the fatigue stability of pavement components. In the Newark test, longitudinal cracks were encountered only in sections where aggregate or asphalt-stabilized sand base was employed.

CHAPTER ELEVEN
Systems of Pavement Design and Analysis

In the present engineering practice of pavement design, there are two approaches: (1) the statistical-analysis, or empirical, approach and (2) the deterministic-analysis approach. Both approaches have their own merits and advantages, and many good methods have been developed for pavement design using either approach. However, when design problems are encountered for pavements at modern airports, it will be found that the present design practice is not sufficiently accurate. First of all, the empirical and theoretical methods were, in most cases, developed independently. There is no correlation between them. Second, no pavement design includes the parameters of vehicle response and pavement roughness and maintenance and safety criteria. Pavement theories developed in the early automobile age are not accurate for today's aircraft. During the pavement tests for the redevelopment program of Newark Airport, efforts were made by the staff of the Port Authority to obtain pertinent information which would contribute to the knowledge of pavement engineering. The conclusions drawn from those tests form the basis for the new pavement design system. However, the statistical relations and design analysis advanced are valid only for the construction practices and traffic pattern of Newark Airport. In applying this design system to other airports, the input parameters should be derived in accordance with local construction practices and airport functional requirements.

11.1. BASIC CONCEPTS

The basic concept of the design system involves (1) the introduction of vehicle response and pavement roughness in defining the functional requirements of a pavement surface, (2) the integration of design theories and methods governing so-called "rigid" and "flexible" pavements, and (3) the introduction of construction practices, material concepts, cost of construction, and maintenance requirements as governing parameters in deriving a pavement design.

The system flow chart, shown in Fig. 11.1 consists of four subsystems. The first subsystem deals with the interaction of aircraft and pavement and relates the response of the aircraft to the roughness of the pavement. Also, the traffic volume and the progressive deformation of the pavement are related. The functional criteria of the pavement surface are translated into limiting elastic deformation and subgrade stress during the service life of the pavement.

The second subsystem makes use of design theories to determine the pavement thickness which would distribute the aircraft load over the subgrade, causing an elastic deformation and a stress level in the subgrade to a tolerance defined in the first subsystem. This design procedure is very similar to the conventional flexible-pavement design method.

The third subsystem involves the analysis of the pavement elements or layers. The progressive deterioration of the material's stress-sustaining capacity under repetitive loading is known as the *fatigue* of elastic material. The allowable working stress of the pavement materials should be determined by the material's endurance limit, the number of load repetitions, and the stress due to the natural environment. The general method used for this analysis is an idealized condition similar to those advanced for concrete pavement design.

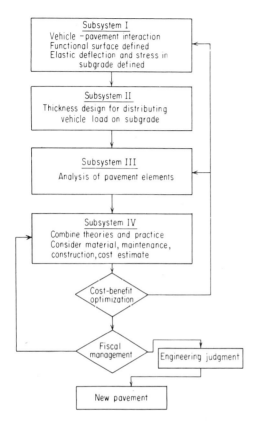

Subsystem I
Vehicle – pavement interaction
Functional surface defined
Elastic deflection and stress in
subgrade defined

Subsystem II
Thickness design for distributing
vehicle load on subgrade

Subsystem III
Analysis of pavement elements

Subsystem IV
Combine theories and practice
Consider material, maintenance,
construction, cost estimate

Cost-benefit optimization

Fiscal management

Engineering judgment

New pavement

Fig. 11.1 Flow chart of system of pavement design.

The fourth subsystem becomes the melting pot for theory, practice, and engineering judgment. Here the entire design comes together. The results derived from subsystems II and III are combined with the finding of the yield theory, which defines the lower limit of the plastic state of equilibrium. To achieve a smooth pavement, it is essential to maintain stress and deflection within the elastic state of equilibrium. There are other practical considerations, such as maintenance cost, construction practice, initial construction cost, and material concepts, which should be evaluated in developing a sound engineering judgment.

The system defies traditional design practice. All parameters used in the system analysis are either carefully evaluated by theoretical analysis or deduced from pavement tests. In the use of test results, more emphasis is placed on the actual observed data than on interpretation. The random nature of the test results suggests that the probabilistic approach is the best method for defining the test model. If the parameters in the actual model are different from those in the test, the correlation function deduced from the test should be modified either by another test or by a theoretical correction. Therefore, in the process of data reduction, the correlation function should always be related to a theoretical analysis which can be subsequently used to extend the scope of application of a test model. No correlation function developed for this study can be used for another pavement design unless the construction variations and traffic conditions of the new pavement are in full agreement with those of the study. Therefore, some tests may have to be performed to develop the appropriate parameters for the job under consideration.

11.2. FIRST SUBSYSTEM—FUNCTIONAL REQUIREMENTS OF PAVEMENT

The flow chart of the first subsystem is shown in Fig. 11.2. The first item involves defining the functional requirement of the pavement surface based on the consideration of aircraft and pavement interaction. The primary function of the pavement surface is to limit the response level of the aircraft riding on the pavement structure. The mathematical model can be expressed by

$$F(\Delta, L, N) \leq P(\overline{DI}, f, \beta, v) \tag{1}$$

The functional surface F can be characterized as a function of the surface deviation Δ, the wavelength L, and the operational service life represented by the number N of traffic coverages. The functional surface offers a service condition represented by the response of the riding vehicle P—characterized by the dynamic increment $\overline{DI}$ of the vehicle at the interface with the pavement—the natural frequency (mass-spring) f and damping β of the vehicle at the interface, and the velocity v of the vehicle traveling on the pavement surface. In the previous study of aircraft and pavement interaction, the theory of random vibrations was introduced to define the dynamic response of the vehicle:

$$\overline{DI}^2 = \phi\left(\frac{1}{L}\right)\frac{\pi f}{4\beta} \tag{2}$$

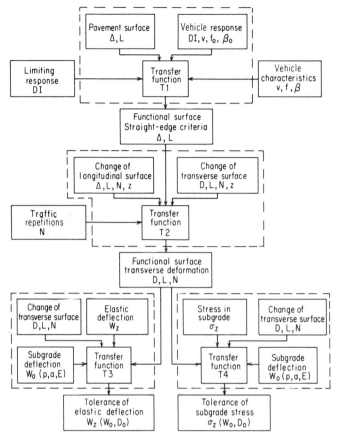

Fig. 11.2 Flow chart of first subsystem.

where $\overline{DI}$ = average dynamic response of aircraft

$\phi(1/L)$ = PSD of pavement surface for wavelength L

$\pi f/4\beta$ = transfer function of dynamic test

The peak response of the aircraft occurs when the wavelength of the pavement surface is equal to the crossing velocity of the aircraft. Thus,

$$\frac{1}{L} = \frac{f}{v} \tag{3}$$

For a discrete wavelength of pavement surface, the PSD is expressed by

$$\phi\left(\frac{1}{L}\right) = \frac{1}{8}\frac{\Delta^2}{1/L} \tag{4}$$

The functional surface of Eq. (1) can be defined by a straightedge criterion:

$$\Delta(N) \leq KL^{1/2} \tag{5}$$

where the K value is a function of the operational characteristics of the aircraft, expressed by

$$K = C_0 \frac{1}{v}\overline{DI}\left(\frac{1}{f}\right)^{1/2} \tag{6}$$

The transfer function $T1(f, \beta)$ is related to the result of the dynamic test as follows:

$$T1(f_0, \beta_0) = \frac{\overline{DI}_0}{\phi(1/L)^{1/2}} \tag{7}$$

for the test aircraft having fundamental frequency f_0, damping coefficient β, and dynamic increment $\overline{DI}_0$ at the test run. The validity of the test depends largely on instrumentation for monitoring the interface response of moving aircraft and a precise level survey of the pavement surface. Arbitrary disturbance of the pavement surface, such as from runway or taxiway crossings, will affect the meaning of the transfer function.

There is room for developing deterministic models and computer simulations to establish the parameters of aircraft-pavement interaction. However, formidable difficulties could be encountered in dealing with an overdamped system such as the oleo-pneumatic shock-absorber system of aircraft.

The above analysis represents the first attempt at introducing the dynamic response of aircraft in defining the functional requirement of a pavement surface. Consequently, there is little information available to define the tolerance of dynamic response. In his study of NASA's airfield test, Houbolt suggested that the normal tolerance of the c.g. response be $0.12g$ and that the upper limit of the pavement roughness be the development of a c.g. peak response of $0.3g$. Leonard of the British Road Research Laboratory reported that the acceptable level of vehicle vibration is $0.1g$ for normal traffic conditions. The research work conducted by Houbolt and Leonard does not differentiate the dynamic response encountered at the vehicle-pavement interface and the center of gravity of the aircraft. Since the structural characteristics of the landing gears of aircraft are entirely different from those of the fuselage and wing structure, the mass-spring and damping action above the pneumatic tires can be very different from that at the interface. Therefore, it should be anticipated that the vibration above the interface may not be identical to the vibration at the interface. There will be variations in dynamic response observed at different positions. Works conducted by Houbolt and Leonard represent the best judgment possible based on the present state of knowledge. There is a need for an administrative agency, such as the FAA or BPR, to determine the needs of airport and road users by conducting a more comprehensive research project in cooperation with aircraft and automobile manufacturers. For the Newark pavement design, the dynamic response at the interface is assumed to be $0.12g$ and $0.3g$ for pavements in the concentrated and infrequent traffic areas, respectively. The $0.12g$ response represents a smooth operation on pavement surfaces, and $0.3g$ represents the marginal smoothness for rarely used pavements. A dynamic response greater than $0.3g$ will necessitate repairs to maintain the smoothness of the pavement surface at an acceptable level of vehicle response.

In airport operations, the speed of aircraft varies within a very narrow range, depending on the area of operation. For ground movement at the holding pad and gate positions, the operational speed ranges from 10 to 30 knots. For aircraft operating on taxiways, the speed varies from 40 to 70 knots. Normal landing and takeoff speeds range from 120 to 150 knots. However, the critical landing or takeoff speed at which excessive vibration of the aircraft is registered, depends largely on the development of wing lift. The interaction tests conducted by the Port Authority indicated that this critical aircraft speed ranges from 90 to 110 knots. With such grouping of operational speeds, the pavement needs for the runways are different from those for taxiways and holding areas. The smoothness requirements for a high-speed pavement are more rigid than for low-speed operations. However, the bending stresses in the pavement for resisting the long-term static load on the pavement should be considered for both high- and low-speed pavements.

For most commercial aircraft, the fundamental frequency ranges from 1 to 1½ cps. Some planes may have a natural frequency as high as 2 cps, and new aircraft in the planning stage might have a frequency as low as 3/4 cps. The Convair 440 used for the Port Authority test had a fundamental frequency of 1.1 cps.

For modern aircraft, the efficiency of a shock-absorber system, which consists of a pneumatic tire and oleo-pneumatic strut, is also within a very narrow range. For pneumatic tires, the shock-absorbing efficiency ranges from 0.45 to 0.47. The shock-absorbing efficiency of oleo-pneumatic struts ranges from 0.75 to 0.80. The combined efficiency of the tires and struts is in the range of 0.85 to 0.92. It can be assumed that because of the narrow range of the fundamental frequency and shock-absorbing efficiency of the aircraft, the transfer function $T1$ may be reasonably valid for today's commercial aircraft. However, for future generations of large-bodied aircraft, the fundamental frequency tends to decrease and the efficiency of absorbers tends to increase. It is necessary to conduct more tests to determine the actual transfer function $T1$ for new aircraft.

Progressive Deterioration of Pavement Surface

The functional surface, as defined by the straightedge criterion, is the ultimate goal for pavement construction. At the pass of each traffic load, however, the pavement deforms and then rebounds, leaving a small permanent deformation on the pavement surface. Therefore, the progressive deterioration of the longitudinal surface should not exceed the limitation as defined by the straightedge criterion during the service life of the pavement. Under the influence of a traffic load, the pavement deforms not only in the longitudinal direction but also in the transverse direction. The load-deformation condition in the transverse direction is much closer to the theoretical stress-strain analysis of a pavement under a static load. In this study, the longitudinal pavement deformation will be translated into a functional surface represented by the transverse deformation of the pavement. The flow chart of this portion of the design work is shown in Fig. 11.2.

There are two major causes for the surface deterioration of the pavement. One is the environmental conditions, such as temperature, moisture, and differential settlement of the pavement support. These are random events, and local experience is the most reliable design parameter. The other major cause is the contribution of the stress-strain condition in the pavement system. Three factors contribute to the deterioration of the functional surface due to the moving load. First, the traffic load is of nonuniform distribution over the pavement surface. The rutting and excessive deformation in the heavily trafficked area will affect the functional service of the pavement surface. The second contributing factor is the heterogeneity of the subgrade and pavement components. The physical properties of the subgrade and pavement material are not evenly distributed. The relative density of the subgrade and the strength of the pavement components assume a random pattern of distribution. The third contributing factor is the inelastic behavior of the pavement material and subgrade. At different stress levels, the magnitude and extent of the permanent damage vary. Due to changes in operational aircraft, the stress level in the pavement is not uniform. Consequently, the degree of permanent deformation may vary widely.

For a channelized traffic pattern, the deterioration of the longitudinal profile is significantly affected by all three contributing factors mentioned above. However, for the deterioration of a transverse cross section, the inelastic behavior of material and subgrade has a greater influence than the other two contributing factors. For processing random events, the probabilistic analysis can provide a good correlation model. At the Newark test, the progressive deformation of a transverse cross section was observed to assume a relation such that

$$D_N = D_1 + D_0 \log N \tag{8}$$

where D_N is the transverse permanent deformation at N traffic coverages, D_1 is the initial deformation, and D_0 is the rate of progressive transverse permanent deformation, expressed in feet per log cycle of traffic coverages (see Fig. 11.3). This equation is very similar to the one used for evaluating the fatigue strength of material. For channelized traffic areas, the rate of progressive deformation D_0

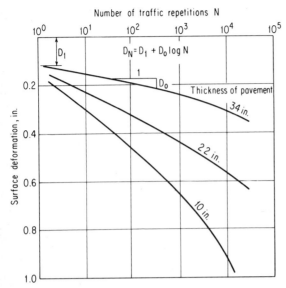

Fig. 11.3 Progressive deformation of pavement surface.

can be expressed as

$$D_0 = \frac{D_N - D_1}{\log N} \tag{9}$$

or approximately

$$D_0 \approx \frac{D_N}{\log N} \tag{9A}$$

In measuring the depth of the surface deflection, its magnitude is closely related to the length of the deflection dish selected in the study. Theoretically, the deflection of the pavement extends an infinite distance from the load. For practicality, it is necessary to define the significant transverse deflection basin. Since the subgrade contributes more than 85 percent of the total deflection of the pavement surface, it becomes logical to use 85 percent of the total deflection as a guideline in determining the significance of transverse deflections, which in theory correspond to a straightedge of $6.6a$, where a is the radius of the contact area of the load. As mentioned in Chap. 3, the E value for the design is equal to the arithmetic mean minus one standard deviation, which represents an 84 percent confidence that the actual E value will exceed the design value. Therefore, the selection of 85 percent of the theoretical deflection should be reasonable for evaluating the distribution of pavement load onto the subgrade.

The service life of a pavement is measured by its structural ability to withstand the traffic loads applied. Insofar as traffic coverages are concerned, it is the parameter used in evaluating the fatigue strength and progressive change of the surface configuration. At Kennedy International Airport, the average daily operation in 1967 was 1,400 aircraft movements, of which 750 were jet traffic. In considering the number of runways and taxiways, the full-load aircraft operations were likely to be less than 120 movements per day per runway. The probability is that the number of full-load repetitions at the same point of a pavement, simply called *load coverages,* would be less than 5,000 per year, and 100,000 such coverages should be anticipated during the 20-year service life of the pavement. For the new pavement at Newark Airport, the volume of air traffic as well as the weight of future aircraft were considered in the evaluation of traffic coverages. It was assumed that the traffic volume would increase about 7 percent a year and that the new aircraft would weigh 700,000 lb in 1970. The anticipated traffic volumes were estimated to be 1 million load coverages for the taxiways, 100,000 load coverages for normal operations on runways, and 10,000 load coverages for the pavement in infrequent traffic areas. With this discussion as background, it is now possible to evaluate the transfer function between the longitudinal permanent deformation Δ, the number of load coverages N, and the transverse permanent deformation D. The linear regression equation deduced from the test is in the form

$$D_N = a_1[\Delta(N) - a_2\sqrt{L}] \tag{10}$$

From the results of the Newark pavement test, the best-fitting equations are

$$D_N = 5.2\Delta(N) \qquad \text{for aggregate-based pavement} \quad (11a)$$

$$D_N = 10[\Delta(N) - 0.0012\sqrt{L}] \quad \text{for LCF-based pavement} \quad (11b)$$

The results of this transformation, known as transfer function $T2$, are shown in Fig. 11.4. It has been noted that the deformation at the early stages of traffic loading will be greater for pavements having an asphaltic-concrete top course on a stabilized base. However, the rate of progressive change of the longitudinal deformation of the stabilized base is about one-half that for the aggregate base. Therefore, the stabilized base will give much smoother performance over a prolonged service life.

The stabilized-base pavement assumes a long flat deflection basin; thus the bending stresses are apt to be low. However, where the original design stress is high, approaching the ultimate strength of the pavement, fatigue failure can become critical. A sudden break can be anticipated because of the brittle nature of the stabilized material. Consequently, an accelerated rate of longitudinal roughness will be experienced. It is good practice to have the stabilized base reasonably understressed.

Elastic State of Equilibrium of Pavement Structure

The translation of the straightedge criterion into transverse permanent deformation is an important step in the development of a rational pavement design method. The conventional engineering theories were developed based on the elastic state of equilibrium of the pavement structure. In order to utilize these theories, it is

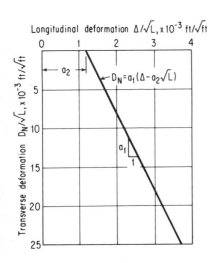

Fig. 11.4 Transfer function $T2$—Longitudinal - to - transverse deformation.

necessary to translate the permanent deformation developed in the preceding paragraph into the elastic deformation, or stress, in the pavement. In this study, the actual test results obtained from the Newark test will be used to review the validity of the elastic theory in pavement design.

In the discussion of transfer function $T2$, it can be seen that a good pavement can be achieved if the rate of the progressive deformation is small and a longer deflection configuration is achieved. Therefore, the pavement layers should have good rebound characteristics and the ability to distribute the wheel load to a wider area of the subgrade. In structural analysis, these requirements are closely related to the phenomenon of the elastic state of equilibrium. In actual construction, pavements must be built on natural subgrades, and the erratic formation as well as wide variation in supporting capacity of a natural subgrade result in a serious challenge to the validity of the elastic state of equilibrium. The question then becomes: Why should the elastic state of equilibrium be considered, and how should elastic theories be used? This question has divided pavement engineers. At the Second International Conference on Asphalt Pavement Design in 1967, the question came up for discussion. Engineers for and against using the elastic theory were about equally divided.

Since the rate of progressive change of deformation is so important to the performance of pavement, it is necessary to develop a theoretical method for rationalizing the load-deformation characteristics. The progressive change of the transverse deformation of a pavement surface consists of the accumulation of the nonrecoverable portion of the pavement deflection under the influence of a wheel load. This is not an event that reflects the elastic stress-strain condition of the pavement. Therefore, the elastic theory cannot be directly used in the pavement design unless it is modified.

In achieving a longer service life and smoother pavement surface, the recoverable deformation of the subgrade should be kept below a certain limit so that the rate of permanent deformation will not be excessive. There are many good elastic theories which can be used in their present form to predict the upper limit of the elastic state of equilibrium. Thus the nonrecoverable deformation can be kept at a tolerable rate.

The next question may be stated: Is there a real elastic stress-strain condition in the subgrade as well as in the pavement components?

The answer is emphatically no. However, within the lower range of the stress-strain curve, a large portion of the load deflection is recoverable. The small portion of nonrecoverable deformation will be accumulated during the service life, eventually to beyond the functional requirements of an operational surface. The relation between recoverable deformation and load intensity is considered to be the *elastic state of equilibrium.*

The validity of the assumption of the elastic state of equilibrium has been observed at the Newark pavement test (see Figs. 11.5 and 11.6). Test section VII consisted of 6-in. asphaltic concrete on a 6-in. crushed stone base laid on the subgrade. The subgrade modulus had an E value of 7,000 psi, and the E value of the aggregate base was in the range of 10,000 psi. Section VI had the same 6-in. asphaltic-concrete surface, but the base consisted of 6 in. of LCF-stabilized base on the subgrade. The E value of the subgrade was identical in both sections. The E modulus of the LCF-stabilized base was about 450,000 psi. Identical pressure and LVDT gages were installed in the pavements to measure stress and deflection. The record shows that the stress in the aggregate base was 32.5 psi, as compared with 17 psi in the stabilized base (see Fig. 11.5). The strain measurement of section VII was 71 μ in/in., whereas the strain of the LCF base was 28 μ in/in. (see Fig. 11.6). The difference in E modulus of the two bases had a significant effect on the stress and strain measurements in the base courses. As for the surface deformations, the surface of section VII deformed much more than that of section VI (see Fig.

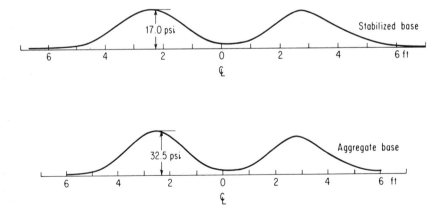

Fig. 11.5 Stress measurement in pavement base.

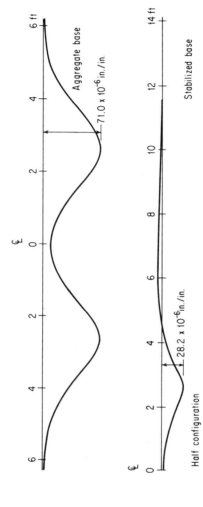

Fig. 11.6 Strain measurement in pavement base.

11.7). The surface of the aggregate-base pavement had a narrow, sharp deflection dish and had more surface cracks than that of section VI. As the subgrade and the top course were identical for both test sections, the only difference between them was the E modulus of the base.

There are many good elastic theories which can explain the difference between stress, strain, and surface deformation due to the change of the E values of the base. For instance, Burmister's layered theory can be effectively used in explaining the observed stress-strain relationship if all parameters are properly evaluated prior to the use of the theories. The elastic theory indicates that the pavement with a stabilized base will have a longer deflection dish with less elastic deflection and a large percentage of rebound. For the aggregate-base pavement, the stress in the base course is high and the total deflection is large. Consequently, a small rebound is experienced, and the sharp curvature of the pavement deflection dish results (see Fig. 11.8).

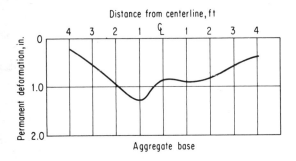

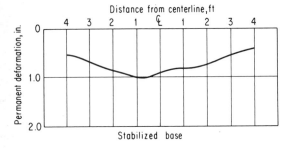

Fig. 11.7 Measurement of surface deformation.

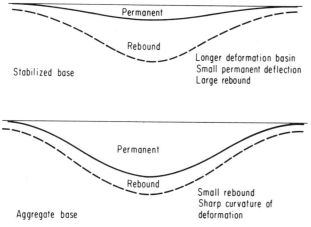

Permanent

Rebound

Stabilized base

Longer deformation basin
Small permanent deflection
Large rebound

Permanent

Rebound

Small rebound
Sharp curvature of
Aggregate base   deformation

Fig. 11.8 Deformation of pavement.

Application of elastic theories to pavement design requires the translation of experimental data from the rate of progressive deformation into the recoverable deflection. The rate of progressive deformation observed at the test is indicated by the coefficient D_0 and the recoverable deflection of a pavement is expressed by W_z. By using a common parameter W_0, two dimensionless coefficients, D_0/W_0 and W_z/W_0, are obtained. The transfer function $T3$ of the rate of progressive deformation and the Boussinesq elastic deflection is determined by multiple regression, in the form (see Fig. 11.9)

$$W_z = d_1 W_0 + d_2 D_0 \tag{12}$$

From the results of the Newark pavement tests, the best-fitting equations are

$$W_z = 0.14 W_0 + 0.32 D_0 \quad \text{for aggregate-based pavement} \tag{13a}$$

$$W_z = 0.16 W_0 + 0.43 D_0 \quad \text{for LCF-based pavement} \tag{13b}$$

Considering the parameters involved in $W_0(p, a, E)$ and $D_0(N, \overline{DI})$, it can be stated that the limiting value of elastic deflection of the pavement W_z is governed by the load parameters p and a, the dynamic response of the moving aircraft $\overline{DI}$, the anticipated service life of the pavement structure N, and the physical property of the subgrade E. It is a

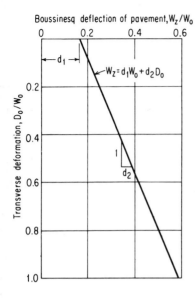

Fig. 11.9 Transfer function $T3$—rate of transverse deformation to Boussinesq elastic deflection.

comprehensive expression for pavement design and possibly represents a major step in the development of a pavement design theory.

Concept of Limiting Stress Level in Subgrade

Stress level, like surface deflection, reflects the actual performance of a pavement. Although stress analysis is purely an academic study, most of the elastic theory in engineering analysis is based on stress computation. Therefore, for the full utilization of present elastic theory, it is necessary to define a transfer function for the pavement stress. Such a transfer function has to be determined by actual test. In our present engineering practice, the method for measuring the stress in a body is based on the assumption that the stress fluctuation can be read out by the change of strain measurements. Instrumentation systems are therefore developed for monitoring the strain of a gage, not the material or the pavement itself. Even if the monitoring system is perfectly developed, the true interaction of the gage and mass, and the time history of the load, are still unknown factors in the calibration of the gage stress. For the Newark pavement tests, a great effort was made to refine the gage-mass interaction. A new type of electric transducer was developed for the job, which had a degree of accuracy, toughness, and reliability not found in commercially available gages. Although the new gage had practically no drift in its electrical output, the gages were balanced electrically at every measurement. The pressure calibration was done in a pneumatic chamber in the laboratory. The strain output of a transducer was converted into a quasi-static pressure. The stress-strain relationship in the field measurements was assumed to be reproducible in the laboratory.

There are inherent errors in the stress-strain conversion. For example, the creeping of material, which cannot be determined in the laboratory by a short-timed load calibration, is an important physical property of the pavement under service conditions. The actual stress-strain relation may significantly deviate from the laboratory proportions. Moreover, the transducer itself is foreign to the elastic mass; therefore, the strain of the gage does not necessarily represent the strain of the elastic mass in which the gage is placed. The interaction of gage and mass may result in overregistration of strain in some cases and underregistration in others. It is impossible to obtain a flawless gage measurement. However, the relative changes of the stress level may be more reasonably reflected by the relative variations of the gage readings. The inherent error in stress measurement can be greatly minimized.

With respect to the performance of the pavement surface, the limitation of the stress level should be applied to the subgrade below the pavement structure, where a large part of the deformation occurs. If the subgrade is subject to excessive normal stress, the direct effect on the pavement's performance is excessive deformation along the wheel path. In the Newark test, the measured maximum normal stress in the subgrade σ_z was correlated with the rate of progressive change of the transverse deformation D_0. The theoretical subgrade deformation W_0, by the Boussinesq equation, was used as a common parameter. The statistical relation, as plotted in Fig. 11.10, is known as transfer function $T4$ and is given by

$$\sigma_z = b_2 + b_3 \frac{D_0}{W_0} \tag{14}$$

With the development of transfer functions $T3$ and $T4$, it becomes possible to translate the transverse deformation into a elastic deflection of the pavement surface and an elastic stress in the subgrade. This completes the operation of the first subsystem of pavement design.

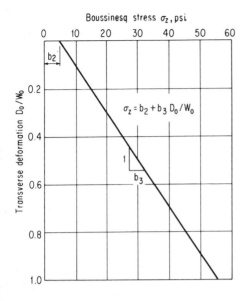

Fig. 11.10 Transfer function T4—rate of transverse deformation to Boussinesq stress in subgrade.

11.3. SECOND SUBSYSTEM—THICKNESS DESIGN

In the first subsystem, tolerance for limiting subgrade stress and pavement surface deflection was developed. This second subsystem involves predicting the thickness of the pavement to fulfill that tolerance. The flow chart of this subsystem is shown in Fig. 11.11. As all transfer functions are developed based on the statistical relations of one event to another, the limitation of the probabilistic approach is in its evaluation of only what happened in the past. It cannot be used to extrapolate for present and future events, nor can it provide a means of predicting the equilibrium of force and resistance. The deterministic approach dealing with the theory of equilibrium is naturally a favorable method for overcoming the deficiency of the probabilistic approach. However, the drawback of the deterministic approach is its determination of the parameters to be used in the theory. If the parameters are in error, the result of the deterministic model is in question. In the Newark pavement tests, large volumes of test data were condensed into several transfer functions. The data processing used a basically probabilistic approach, transferring one event to another. The events were selected from the theoretical studies in anticipation that the transfer functions could be profitably used as reliable parameters in the deterministic model. For this reason, every deterministic model, prior to its use in the final design, had to be tested for its compatibility with the transfer function developed from the test.

For distributing the aircraft load on the subgrade, there are three deterministic theories: the Boussinesq, the Burmister, and the finite-element method. The Boussinesq theory considers the stress distribution in an elastic mass of infinite width and depth and develops a simple algebraic equation predicting the stress distribution in three dimensions. The solution of the algebraic equation is simple and is readily accomplished for practical engineering application. The Burmister layered theory is a refined Boussinesq theory involving the introduction of multiple layers of elastic mass. Each layer is identified by its modulus of elasticity E, Poisson's ratio μ, and thickness h. Since more input parameters are required, the error in evaluating the input parameters is the major drawback to applying this theory. Too often the theory itself is valid but the parameters introduced into the equation are not. As the supporting capacity of the subgrade is widely scattered, the use of a refined theory is questionable.

A third elastic theory is the finite-element method for analyzing stress distribution and displacement of a half-space elastic mass. The

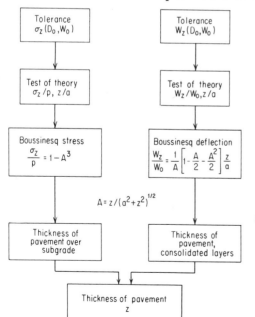

Fig. 11.11 Flow chart of second subsystem.

mass may consist of different layers of material of different E values. The method is in the development stage, and the concepts and principles employed in this method show great promise for future use. However, the reliability of the analysis depends largely on the size of the mesh; the finer the mesh, the more reliable the results. On the other hand, the finer the mesh, the more computation will be required. Therefore, the limitation of the computer and the computation cost become the major drawbacks to utilizing the finite-element method. The utilization of such a system in pavement design still depends on future research and development. However, the variation in pavement construction and subgrade may have a serious effect on the application of the refined theories.

Validity of Boussinesq Stress Theory

Pavement construction always involves the use of a massive volume of road materials, and economy dictates a reasonable variation in quality control. Because of this variation in materials, the simple Boussinesq theory should be accurate enough for practical pavement design computations.

During the construction of the Newark test pavement, layer components were compacted at various stages of construction and the gage outputs were monitored when the pneumatic tire compactor moved directly over the gages. Four readings were obtained for every gage at each stage of construction. The calibrations of the stress readings were similar to those obtained in the laboratory and were plotted against the depth of pavement between the surface and the gages. In order to make the plot more useful, dimensionless parameters were formed; the stress σ_z was divided by the tire pressure p, and the depth z to the gage was divided by a, the radius of the test load. The results of actual load tests, plotted in this manner, are shown in Fig. 11.12. Note that the stress distribution in the subgrade under the aggregate base closely follows the Boussinesq pattern of stress distribution, as shown by the solid line. The scatter of the data for z/a near 5 or 6 may be caused by variations in the gage performance. For the LCF-stabilized base, the magnitude of the stress readings in the LCF base range from 25 to 50 percent of the Boussinesq pattern of stress. The flexural strength of the LCF material should make a significant contribution to spreading the wheel load and, consequently, reducing the intensity of stress in the

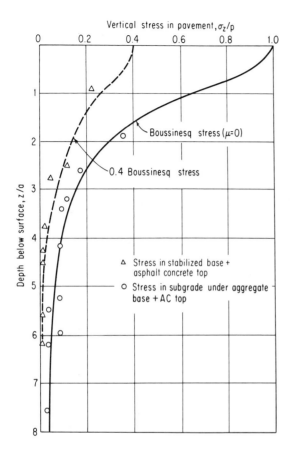

Fig. 11.12 Boussinesq stress as compared with field measurement.

pavement. The layered theory would give similar results, however. Because of the difficulty in determining the design parameters, it may be more appropriate to determine the transfer function between the Boussinesq stress and actual stress monitored in the field. In this case, the stress in the LCF pavement is about 40 percent of the Boussinesq stress and the stress in the subgrade under the aggregate base is about 100 percent of the Boussinesq stress. With this correction factor, the Boussinesq stress theory can still be used to determine the subgrade stress under the LCF- or aggregate-base pavement.

Validity of Boussinesq Deflection Theory

The original Boussinesq stress formula has been modified by integrating the strain in the elastic mass to reflect the displacement under external load. In the Newark pavement test, attempts were made to verify the validity of this theory. LVDT displacement gages were installed in 11 test pavements, with the permanent reference of a steel rod driven to a nonyielding layer. During the test, nine gages operated normally. The gage readings constituted the direct measurement of the surface deformation of the test sections. The measured surface deflection was divided by the parameter W_0, as expressed by the dimensionless parameter W_z/W_0, which is plotted against another dimensionless parameter z/a, in Fig. 11.13. On the same chart, the solid line represents the theoretical Boussinesq

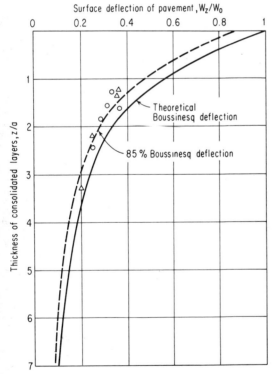

Fig. 11.13 Measured pavement deflection as compared with Boussinesq theory.

deflection distribution. The test results are in good agreement with 85 percent of the Boussinesq deflection.

As discussed in the characterization of the subgrade, the confidence level of the E value of the subgrade was designed to be 84 percent. Therefore, the theoretical Boussinesq deflection fits very well into the whole system of pavement analysis.

In the first subsystem, the tolerances of elastic deflection and stress in the subgrade were developed based on aircraft-pavement interaction and the functional requirements of a pavement surface. It is possible in the second subsystem to use Boussinesq stress and deflection equations to determine the thickness of the pavement. For the stress tolerance in the subgrade, the equation is

$$\sigma_z = p(1 - A^3) \quad \text{and} \quad A = \frac{z}{(a^2 + z^2)^{1/2}} \tag{15}$$

where a is the radius of the load area, p is the tire contact pressure, and z is the thickness of pavement layers above the subgrade. The deformation W_z at a depth z below the surface and along the vertical axis of the center of the surface load is obtained by integrating Eq. (15) and is expressed by

$$W_z = \frac{W_0}{A}\left(1 - \frac{A}{2} - \frac{A^2}{2}\right)\frac{z}{a} \tag{16}$$

W_z and σ_z represent the tolerances of Boussinesq deflection and stress, respectively, in the subgrade. The depth z in Eq. (16) becomes the thickness of the consolidated layers of pavement which satisfies the level of deflection tolerance. The design computation involves only the solution of the cubic equation.

In actual design, the solution of the real root of the cubic equation can be determined by the use of design charts. For the solution of Eq. (15), the design chart is similar to the curve for the Boussinesq stress distribution shown in Fig. 11.12 and displays two dimensionless parameters, z/a and σ_z/p. For the solution of Eq. (16), a similar chart (Fig. 11.13) can be plotted for the parameters z/a and W_z/W_0. However, it is more convenient in practical design to use three parameters: a, z, and $W_z E/p$. The graphic solution of Eq. (16) is shown in Fig. 11.14. The abscissa represents the radius of the contact

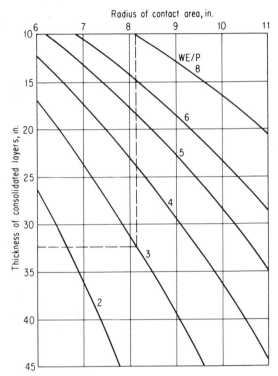

Radius of contact area, in.

Fig. 11.14 Curves for Boussinesq deflection.

area of aircraft load, and the ordinate represents the thickness of the consolidated layers of pavement. Across the chart, $W_z E/p$ is the parameter. For given $W_z E/p$ and a values, the thickness z of the pavement can be readily determined. For instance, if $a = 8.13$ in. and $W_z E/p = 3.0$ in., the corresponding thickness of pavement is 32.2 in. This 32.2 in. represents the total thickness of the consolidated layers to be placed on the subgrade for control of the surface deflection of the pavement within the tolerances required by the first subsystem. Here the thickness of the consolidated layers does not reflect the materials of the pavement composition. The total thickness of consolidated layers applies to asphaltic-concrete pavement having either LCF-stabilized base, asphalt-stabilized base, soil-cement base, or lean-concrete base. As all stabilized bases in the Newark test had a modulus of elasticity less than 1,000 ksi, any base material having a modulus of elasticity higher than that number was beyond the scope of the test.

At the conclusion of the second subsystem, the outputs are two pavement thicknesses, namely:

1. Based on limiting the normal stress in the subgrade. The output represents the total thickness of pavement above the subgrade, disregarding its composition.

2. Based on limiting the elastic deflection of the pavement surface. The output is the total thickness of the consolidated (or bonded) layers of the pavement over the subgrade. The physical composition of the consolidated layers is not considered in this subsystem as long as the consolidated layers do not contribute a significant permanent deformation to the pavement surface.

11.4. THIRD SUBSYSTEM—ANALYSIS OF PAVEMENT ELEMENTS

In the second subsystem, the thickness of pavement was determined for proper distribution of the aircraft load on the subgrade. The third subsystem involves the analysis of pavement elements or layers. The general theory used for this analysis deals with the idealized condition that a single wheel load is carried by an elastic plate which is supported by an elastic mass. The flow chart of the third subsystem is shown in Fig. 11.15. It is very similar to the conventional method of designing concrete pavement.

The purpose of the third subsystem is twofold: (1) to integrate the conventional and rigid pavement design methods and (2) to provide an optional choice for the design engineer to use pavement materials which agree with the subgrade conditions and for control of the pavement surface deflection. To some degree, this gives the engineer the motivation for developing pavement materials to suit the construction needs. This is a very ambitious program. The first input to this subsystem is the thickness of the consolidated layers over the subgrade, which is the output of the second subsystem. Another set of inputs to the third subsystem consists of the static load of the aircraft, the impact factors, and the horizontal forces due to braking and rolling if applicable.

There are two elastic theories which can be utilized in the analysis: (1) the stress-strain analysis of an elastic plate on an idealized elastic foundation, such as a heavy liquid, and (2) the analysis of the behavior of an elastic plate on a viscoelastic foundation. The first

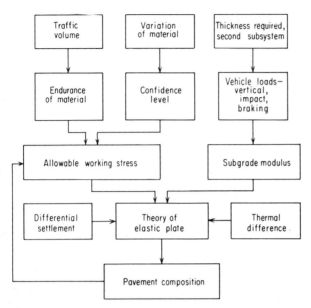

Fig. 11.15 Flow chart of third subsystem analysis of pavement elements.

theory is commonly known as the *Westergaard method*. Its validity has been tested by the Corps of Engineers and the Portland Cement Association (PCA). The theory itself is valid for predicting the bending stress in a concrete slab over compacted subgrade. The relation between the theoretical surface deflection of a concrete slab and the tested values is shown in Fig. 11.16. Osawa indicated that the test results of the PCA were in general agreement with the theoretical computations. The test conducted by the Ohio River Division of the Corps of Engineers indicated that the observed deflection was about 2 percent larger than the theoretical deflection. It can be assumed that the Westergaard theory is valid within the limits of the ORD and PCA tests.

The second theory, which involves an elastic plate on a viscoelastic foundation, is much more refined than the method which considers the elastic plate on an idealized elastic foundation. However, the reliability of the theory depends largely on the accuracy of the input parameters. Although there are several laboratory tests to define the viscoelastic characteristics of the subgrade, no full-scale test has been conducted to positively determine the reliability of the input parameters.

Bending Stress under Vehicle Load

The classic Westergaard theory represents one of the mathematical solutions of the general differential equation of the elastic plate, in the form

$$\nabla^2 \nabla^2 w + \lambda^4 w = 0 \tag{17}$$

in which ∇^2 is the differential operator, in the form

$$\nabla^2 = \frac{d^2}{dy^2} + \frac{d^2}{dx^2} \tag{18}$$

and

$$\lambda^4 = \frac{k(1 - \mu^2)}{EI} \tag{19}$$

The approximate solution of the above differential equation by Westergaard can be expressed by the bending moment in a

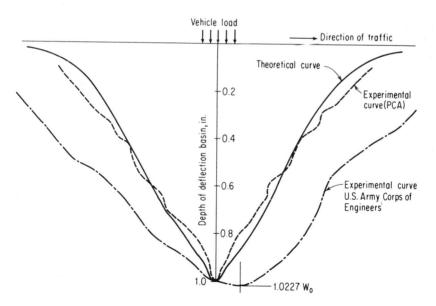

Fig. 11.16 Comparison of deflection profiles of concrete pavement.

homogeneous plate:

$$M = \frac{P}{4\pi}(1 + \mu)\left(\ln\frac{1}{\lambda a} + 0.616\right) \tag{20}$$

or by the bending stress in the plate:

$$\sigma = \frac{P}{z^2}\frac{3(1 + \mu)}{2\pi}\left(\ln\frac{1}{\lambda a} + 0.616\right) \tag{21}$$

For a thick plate, such as when z is thicker than $0.58a$, the a value used in the stress computation should be replaced by a b value, which is expressed by

$$b = (1.6a^2 + z^2)^{1/2} - 0.675z \tag{22}$$

Equation (20) is actually an exercise in mathematical manipulation. For practical application of this equation, the accuracy of the result is largely dependent on the k value substituted in Eq. (19). As Professor Westergaard was a structural engineer, it was convenient for him to solve the differential equation by introducing the k value. After his introduction of the solution to the differential equation, an empirical method, subsequently standardized by the ASTM and AASHO, was introduced in determining the k value. The testing method involves the use of a 30-in.-diameter plate in transmitting the test load. The deflection and load intensity are used to determine the k value. This method was developed when the thickness of the concrete pavement ranged from 6 to 10 in. and the radius of the wheel load was 4 to 5 in. Under this kind of loading, experience indicated that the k value determination was in agreement with the Westergaard methods for computing the bending stress of the concrete pavement. However, for today's airfield pavement construction, the thickness of concrete pavement ranges from 12 to 18 in. and the radius of the footprint of the aircraft tires ranges from 8 to 10 in. It is logical to question the validity of the standard test procedure and its application for determining the k value for airfield pavement construction. From his study of concrete pavement, Vesic has recommended a modification of the k value in the form

$$k = \frac{E_s}{1 - \mu^2} \frac{1}{z} \sqrt[3]{\frac{E}{E_c}} \tag{23}$$

At the Newark tests, the k value was observed to be of the form

$$k = \frac{E_s}{1 - \mu^2} \frac{1}{2(a + z)} \tag{24}$$

In order to comply with the above modifications, the size of the load plate should be in the range of 50 to 60 in. in diameter. According to the present procedure of the ASTM standards, the computed k value will be 50 to 100 percent over the actual performance of airport construction. This is a very dangerous assumption, and many airports have experienced distress of concrete pavement due to the inadequacy of this assumption.

The classic Westergaard solution is also limited to concrete pavement of a uniform strength, from top to bottom of the pavement layers. For modern airport pavement construction, however, it may be desirable to use a lower-strength material at the base and higher strength for the top. This involves composite construction of multiple layers. The resisting moment of the composite construction depends on the ratio of stress-strain between the bottom and the top layers of the pavement. The resisting moment can be expressed by (see Fig. 11.17)

$$M = \sigma \frac{n}{3(n + 1)} z^2 \tag{25}$$

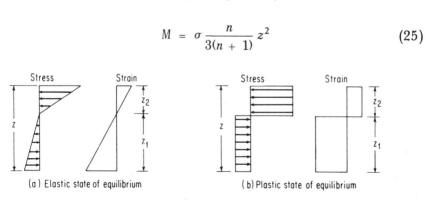

(a) Elastic state of equilibrium   (b) Plastic state of equilibrium

Fig. 11.17 Equilibrium of composite construction.

where n is the ratio of stress-strain as expressed by the ratio of the E value of the top layer to that of the bottom layer. The bending stress according to the classic Westergaard formula can be rewritten as

$$\sigma_w = \frac{3(n+1)}{4n}(1+\mu)(1+\overline{DI})p\frac{a^2}{z^2}\left[\ln\frac{z}{b}+\frac{1}{4}\ln\frac{E_c}{E_s}+\frac{1}{4}\ln\left(\frac{a+z}{z}\right)\right.$$
$$\left.+\,0.168\right] \tag{26}$$

where the single equivalent wheel load is represented by πpa^2 and the dynamic impact is represented by $\overline{DI}$.

Differential Settlement

Although the Westergaard equation can be used to establish the bending stress and thickness of a pavement due to the static load of an aircraft, there are several environmental factors which also influence the bending stress in the pavement. At Newark Airport most pavements are constructed on swampland. The subsidence of the ground is not even, and the resulting differential settlement of the pavement support creates a bending stress in the pavement. In the past, portland-cement concrete pavement was not used extensively for pavement construction on a settling field, on the assumption that asphaltic concrete would better follow the contour of the ground settlement. For deep-seated settlement, the pavement construction inherits the settlement problem during the service life of the pavement. If the pavement is continuous and strong enough to resist the bending moment caused by the sagging ground, the pavement may be in a better position to maintain its smoothness while the subgrade is settling. The first step in the determination of bending stress due to differential settlement is that the settlement configuration assumes a harmonic curve, such as that shown in Fig. 11.18. L is the wavelength, and Δ is the maximum differential settlement. For the coordinates shown in Fig. 11.18, the settlement y at a point x can be defined by the equation

$$y = \frac{1}{2}\Delta\cos\frac{2\pi x}{L} \tag{27}$$

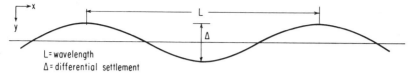

L = wavelength
Δ = differential settlement

Fig. 11.18 Surface configuration due to differential settlement.

The bending moment due to the deflection of the pavement layers can be represented by the equation

$$M = EI \frac{d^2 y}{dx^2} \qquad (28)$$

The maximum bending moment at the bottom of the settlement dish is

$$M_0 = 2\pi^2 EI \frac{\Delta}{L^2} \qquad (29)$$

It can be seen that the bending moment is reduced significantly when the wavelength L of the differential settlement is increased. Since the differential settlement takes place over a very long period, the progressive adjustment in the pavement material itself causes the bending stress in the pavement to occur in the plastic state of equilibrium. Therefore, the E value used in the above equation should reflect the plastic state of the E modulus, which is about one-third to one-quarter of the elastic modulus. The bending stress of the pavement therefore becomes

$$\sigma_0 = \frac{n + 1}{3n} \pi^2 E' \frac{\Delta}{L^2} z \qquad (30)$$

This equation was oversimplified for the settlement coordinates. In considering the magnitude of the stress developed in the pavement due to the differential settlement, the equation provides a simple but reasonable estimation of the bending stress.

Temperature Variation

Another environmental factor (other than the aircraft load) affecting the bending stress in a pavement is the fluctuation of temperature in the pavement. The surface of a pavement is exposed to changes of ambient temperature, while below the pavement surface the temperature is more stable. Therefore, a thermal gradient is encountered in the pavement from the surface down to the subgrade, and beyond. In the Newark tests, thermocouples were installed in the pavement to measure the thermal variation in the pavement at different depths. The experiment was carried out from summer through to the following spring. The maximum daily variation of the temperature in the pavement was about $0.5°F/in.$, and the seasonal variation of the temperature in the pavement was $1.5°F/in.$ (see Fig. 11.19). For studying the daily variation, the elastic state of the E value should be used, and for studying seasonal variation, the adjustment of pavement deformation dictates the use of the E' value. The change of temperature with depth will cause warping of the pavement, with resulting bending stress in the pavement components. The bending moment is given by

$$M = EI\epsilon \frac{dt}{dz} \tag{31}$$

where ϵ represents the coefficient of expansion (per degree per inch per inch). In the Newark tests, the observed value of ϵ was $5.8 \times 10^{-6}/(°F)(in.)(in.)$. The bending stress in the pavement caused

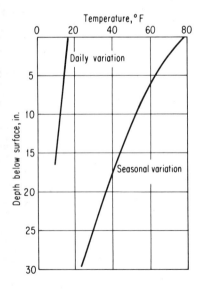

Fig. 11.19 Temperature variation in pavement.

by the seasonal temperature variation is

$$\sigma_t = \frac{n+1}{6n} zE'\epsilon \frac{dt}{dz} \tag{32}$$

It should be pointed out that if dt/dz represents the daily variation of the temperature, the E value should be used in the above equation.

The engineer must now design the pavement material to conform to the requirements of the second and third subsystems, that is, total pavement thickness and total bending stress. It is a challenge to the engineering profession to design pavement materials to conform to the requirements for limiting subgrade deflection, which will result in a smoother and more durable pavement. In attempting to design the pavement material, the first consideration is the endurance of the material, or its ability to withstand the repetition of aircraft loads. According to the laboratory test on lime-fly-aggregate material, the progressive deterioration of the material can be represented by

$$\sigma_n = \sigma_b(1 - 0.092 \log N) \tag{33}$$

where N is the number of load repetitions (traffic volume), σ_b is the initial bending stress, and σ_n is the fatigue strength of the material at the nth repetition of the traffic load.

Of next consideration in the design of material is the confidence level, or quality control, of the material. As discussed in the evaluation of subgrade reaction, a reasonable design figure for the subgrade is the mean value minus one standard deviation, which represents an 84 percent reliability in the subgrade support. For ordinary pavement construction, the standard deviation of base materials varies from 10 to 20 percent of the mean value. Therefore, for an identical reliability in the pavement components and the subgrade, the design value for the base materials should be 80 to 90 percent of the mean strength of the material. If the standard deviation is assumed to be 0.20, the designed stress should be 80 percent of the mean strength. This is equivalent to a factor of safety (F.S.) of 1.25. Thus the combined bending stress in the pavement with respect to the average bending stress of the material is indicated by the relation

$$f_b \geq \frac{\sigma_w}{1 - 0.092 \, logN} \, (\text{F.S.}) + \sigma_d + \sigma_t \qquad (34)$$

The relationship between pavement thickness and average bending stress, for example, is shown in Fig. 11.20. If the output in the second subsystem requires a total thickness of the consolidated layers of 28.5 in., the corresponding average bending stress of the material is 250/psi (according to the example in Fig. 11.20). This is the average bending strength required for the pavement. If conventional portland-cement concrete having a compressive strength of 4,000/psi is used, there is a big reserve strength in the material in order to fulfill the requirement of the second subsystem. If the concrete strength is to be utilized, such as with a bending stress of

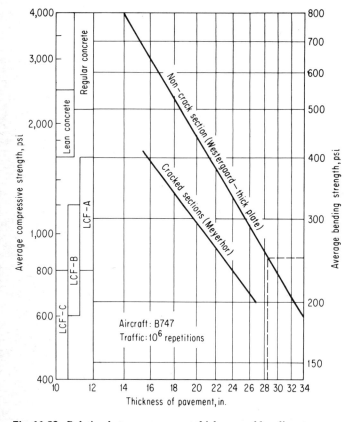

Fig. 11.20   Relation between pavement thickness and bending stress.

800/psi, the thickness required would be only 14 in. However, the subgrade would deform excessively, and the functional requirements outlined in the first and second subsystems could not be fulfilled. The pavement itself might not break, but the subgrade would give. To balance the requirements for strength and thickness is an engineering challenge which will lead to a new concept in pavement design.

11.5. FOURTH SUBSYSTEM—THE FINAL STAGE OF DESIGN

In the second and third subsystem analyses, the thickness and stress requirements of the pavement were determined. The fourth subsystem requires the use of engineering judgment in deciding upon the appropriate thicknesses and corresponding strengths of materials in the pavement construction. Here the entire design comes together. The results derived from elastic theories in the second and third subsystems are combined with the yield theory, which defined the lower limit of the plastic state. To achieve a smooth pavement, it is essential to maintain stress and deflection within the elastic state of equilibrium. The flow chart of the fourth subsystem is shown in Fig. 11.21.

Yield Theory

The first process in this fourth subsystem is to define the upper limit of the elastic state. As discussed previously, it is difficult to make a precise delineation between the elastic state of equilibrium and the plastic state of equilibrium. Engineers often understress the structural components by considering only the elastic state of equilibrium. To avoid this waste of engineering material, it may be desirable to use a different theoretical approach to define the boundary of the elastic and plastic states of equilibrium. Along these lines, Lobsberg and Meyerhoff have conducted several tests to define the plastic state of equilibrium of an elastic plate on a heavy liquid foundation. Based on their expcrimentation, the yield condition truly represents the lower limit of the plastic state of equilibrium, defined as the upper boundary of the elastic state of equilibrium. The empirical formula developed by Meyerhoff, based on test results, is in the form

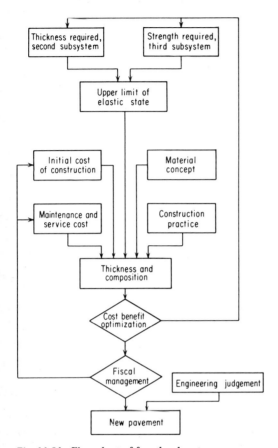

Fig. 11.21   Flow chart of fourth subsystem.

$$M = (1 - 2\lambda a) \cdot \frac{P}{4.6} \quad \text{for } 0.15 \le \lambda a \le 0.25 \tag{35}$$

With modifications to include the dynamic impact of the wheel load and the effect of composite construction, the bending stress σ_y can be expressed by

$$\sigma_y = \frac{2(n + 1)}{n} (1 + DI)p \frac{a^2}{z^2} (1 - 2\lambda a) \tag{36}$$

where the λ value is given by Eqs. (19) and (24). The σ_y value represents the boundary condition of equilibrium. If the σ_y value exceeds the material strength, the elastic plate is in the plastic state of equilibrium. As the stress encountered in the plastic state of equilibrium is always higher than the elastic stress-strain condition, the upper limit of the elastic state of the pavement exists if, for the same thickness of pavement,

$$\sigma_y \leq \sigma_w \tag{37}$$

where σ_w is the bending stress computed by the Westergaard formula. If σ_y is substantially smaller than σ_w, it is possible to reduce the thickness of pavement for a more economical design. However, in deciding the thickness and stress requirements of the pavement, it should be kept in mind that the requirements in the first subsystem should be met and variation in the pavement construction should be considered. As mentioned in the previous discussion, the reliability of the pavement design is 84 percent, which is the mean value minus one standard deviation. Thus it should be realized that in five samples out of six there is a variation from the mean value of more than 10 percent. Therefore, in deciding the thickness of pavement and strength of the materials, a variation of 10 percent from the computed value may not seriously affect the performance of the pavement.

Material Concept

Theory alone is not sufficient, however. There are other practical considerations. For example, the materials used in conventional pavement construction at Newark Airport represent more than 60 percent of the total construction cost. A substantial saving can therefore be achieved if a local material is fully utilized. Each paving job should be considered as a unique entity in developing an appropriate material concept. Two-thirds of material costs involve the handling of the material; therefore, it is necessary to eliminate the rehandling of materials as much as possible. There are several important considerations in selecting an economical pavement material:

1. Utilizing local materials to reduce the cost of transportation.
2. Minimizing rehandling of materials to reduce the cost of labor.
3. Optimizing the cost-strength ratio of the material in considering a tolerance in quality variations. Usually the material costs less when the variation is large, and there should be a compromise between the

cost, strength, and variation of the material. For instance, the coefficient of variation of concrete ranges from 0.08 to 0.12, and the cost of the concrete is about $15 to $20/cu yd. On the other hand, for stabilized material such as LCF, the cost of stabilization varies from $3 to $5/cu yd, with a coefficient of variation ranging from 0.15 to 0.20. It may not be economical to use a certain kind of pavement material, either high- or low-grade, without considering its role in the pavement performance.

4. As dictated by the second and third subsystems, it is necessary to design material strength to conform to the requirements for limiting the deflection of the subgrade. This radical change in pavement design concept obliges the engineer to design material strength to suit the needs of the pavement.

5. Along these lines, the elastic property of a material is a great asset for achieving a good pavement. A little strength stabilization of the subgrade, subbase, and base pays a big dividend in pavement performance.

Construction Practice

Besides economics, an important consideration is the construction practice in the area the pavement is to be constructed. There are several important parameters involved in the construction:

1. Labor, and its relation to the cost, production, and performance of construction. In large metropolitan areas, the labor cost is high; therefore, the cost-production ratio is low. Because of shorter working hours and the large dependence on mechanical equipment, manual labor should be reduced in the pavement construction.

2. For construction of a pavement which involves the handling of a large volume of material, the uniformity of the construction will more or less depend on the mechanical output of the construction equipment. However, each piece of construction equipment has its outstanding features and its limitations. The engineer should consider the mechanical equipment available in the area for construction.

3. The cost of pavement construction, in many cases, is largely reflected by the size of the job. Regardless of the size of the job, the contractor has to pay for the mobilization and demobilization of equipment. For mechanized pavement construction, the contract should be reasonably large to absorb the depreciation of the equipment investment.

4. Another item affecting the construction cost is the standard of quality control. The degree of quality control and construction tolerance should be developed in the design stage by balancing construction cost against the structural requirement. For the construction of pavement bases in the New York area, material could have a coefficient of variation ranging from 0.15 to 0.25. High-quality material is limited to the pavement surfacing, where high strength is needed. The use of relatively low-strength base material ensures not only resilient pavement but economical construction as well. In quality control, the conventional specifications dictate the performance of the contractor and the end result of the job. On many occasions, because of ambiguity in the specification, the construction cost of the pavement job is somewhat higher than necessary. In recent years, the use of a quality-assurance program has been introduced in highway construction. It is a sound and logical program. However, there are still many things to be done. The most important is that the engineer and contractor must both be well educated concerning use of quality control for pavement construction. Without such education, there can be misunderstandings of quality-assurance programs.

5. Most pavement bids are solicited in an open competitive manner. Under this condition the construction cost largely depends on the availability of material and the contractor's experience in executing the construction. If new materials and construction procedures are introduced, the bid price may be unreasonably high or certain premiums may be required for this innovation. However, this should not discourage the introduction of new materials or construction procedures if the program is carefully managed to guide the contractor and eliminate uncertainties in the contract which can arise.

6. Another important item affecting the construction cost is restriction on construction operation. At a busy airport, there are many restrictions to the contractor's operation. If the operator of the airport can remove some of the restrictions, the construction cost can be slightly reduced.

7. Another factor involved in the construction cost is the overtime requirement of the construction. For instance, in concrete construction there is always a need for longer working hours to finish the concrete pavement. Therefore, a large percentage of the cost is incurred by overtime payments to labor.

8. The construction cost is also affected by the cost escalation during the time span of the contract. For longer contract periods, the contract usually provides allowance for cost escalation to compensate for inflation. On the other hand, medium-size contracts will invite more competition than large-size contracts.

Economic Study and Fiscal Policy

In the final analysis of cost, the flow of the economic study roughly follows this outline:

1. In the feasibility study of the project, a preliminary estimate is established and the engineering study is performed. Here, engineers have an obligation to prepare a ball-park estimate for a pavement which will be smoother and last longer.

2. The next step is to prepare a budget estimate for the project.

3. Project authorization is received.

4. After the budget estimate is reviewed and authorized, the engineer starts design and preparation of the pavement contract. It is the obligation of the engineer to keep the initial construction cost either less than or equal to the budget estimate.

At this point, the study involves the question of what is the true cost of pavement construction. As mentioned earlier in the discussion, the difference between the farmer-to-market road and a superhighway is reflected in the comfort, safety, and speed of vehicle. For the farmer-to-market road, there is little initial construction cost but the performance is poor and maintenance costs may be large. For superhighways, the initial construction cost is high but maintenance and repair costs are low. The true cost of a pavement consists of (1) the initial construction cost of the pavement, (2) the direct cost of repair and maintenance, (3) the value of smooth and safe pavement, and (4) the indirect cost due to interruption of service, which is very important for today's busy airports. In considering the service requirement and maintenance cost, the true cost of the pavement tends to shift toward a heavier initial construction cost, which in the long run can be a more profitable solution. Sound fiscal policy is to reduce the maintenance cost by increasing the strength of the pavement, although this may raise the initial capital investment. If a pavement can be designed in harmony with the best local practice, pavement cost can be substantially lower.

Engineering Judgment

The most important item in the whole design system, however, is the engineering judgment of the designer. There is no substitute for the sound judgment of a well-informed engineer. All human beings are liable to make mistakes, however, and an appropriate factor of safety should be used to cover human ignorance. To improve the reliability of our engineering judgment, the system of pavement design and analysis is developed to review various important aspects relating to the performance and construction of a pavement.

The End Product

At Newark Airport, the new pavement was designed to accommodate B747 aircraft. Progressive pavement deterioration is limited to a dynamic response of the aircraft not exceeding $0.12g$ in heavy traffic areas and $0.3g$ in infrequent traffic areas. The service life of the pavement is assumed to be 20 years without major maintenance. An example of manual design computations is shown in Appendix 4, and the computerized program is shown in Appendix 5. In general, the heaviest pavement is placed at the gate positions and holding pads. The next heaviest pavement is in the center strip of the taxiways and runway ends. A light pavement is designed for the side of the runways and taxiways and the center portion of the runways. In all cases, the pavement consists predominantly of LCF-stabilized base having an average compressive strength varying from 800 to 2,000 psi and covered with a single 4-in. lift of asphaltic concrete. The layout is rather unconventional; however, it is believed that a balanced and economical design has been achieved. The economic advantage of the Port Authority's research and engineering effort can be illustrated by the following statistics:

1. The average cost of new airport pavement for 1968's jet airplane in the mid-Atlantic states compiled by the FAA is $12.55/sq yd.

2. The LCF pavement used at Newark Airport, designed to accommodate 700,000-lb aircraft, has an initial construction cost of $7.60 for an anticipated service life of 20 years.

By converting local materials into useful pavement components and by using the systems approach in pavement design and analysis, a cheaper and better pavement system has been achieved for New York's Airports.

APPENDIX FOUR

Example of Design Computations

Part 1: Load-distribution Concept

Ground Load and Landing Impact

Type of Airplane—Boeing 747

Main gear wheel configuration—twin-tandem 44 by 58 in.

Main gear coordinates—four trucks (6.25, 10.83 ft) (18.08, 0.75 ft) symmetrical

Size of tires—46 by 16 in. at normal inflation pressure (200 psi)

Ground load: Taxiing speed—30 to 70 knots

Takeoff speed—130 to 150 knots

Takeoff length—5,000 to 7,000 ft at sea level, $90°$F

1. Maximum ramp weight—683,000 lb
2. Main gear load—166,000 lb
3. Ground turning (approx.)—262,000 lb/gear
4. Rolling friction (approx.)—$0.07\,g$ (horiz.)
5. Brake stop (approx.)—$0.3\,g$ (horiz.)
6. Static landing load—124,000 lb/gear
7. Hard landing (approx.)—300,000 lb/gear
8. Ultimate gear load—500,000 lb/gear
9. Maximum brake force (approx.)—$0.8\,g$ (horiz.)

Equivalent Single Wheel

Tire contact radius, $a = \left(\dfrac{166,000}{4 \times 200 \times \pi} \right)^{1/2} = 8.13$ in.

Distance from reference point	Influence factor with respect to surface deflection of a single tire
$x/a = 0.00$	1.000
5.41	0.094
7.13	0.070
8.95	0.056
18.4	4 x 0.027
23.1	4 x 0.021
39.0	4 x 0.013
	1.464

Equivalent single-wheel tire pressure = $200 \times 1.464 = 293$ psi

Modulus of Deformation of Subgrade, E

Tangent modulus at a stress range of 40 to 60 psi
Mean value of 33 plate-bearing tests   = 7,000 psi
Standard deviation   = 1,300 psi
Optimum design value   = 5,700 psi
Reliability of design value:
    84% of random tests will have an E value $> 5,700$ psi

Limit of Dynamic Response of Airplane

Maximum limit during the service life of the pavement over and above the roughness tolerance for pavement construction:
Mean response = $0.12g$ in normal operational area
Max. response = $0.30g$ in infrequent traffic area
Traffic volume during 20-year service life to be designed:
    7,000,000 movements/airport
    3,000,000 movements/terminal
    4,000,000 movements/runway
    Fully loaded aircraft anticipated to be 45% of the total movements

Deformation Parameter

$$w_0 = \frac{2pa}{E} = \frac{2 \times 293 \times 8.13}{5,700} = 0.835 \text{ in.}$$

Significant Width of Transverse Deflection Dish

$$L = 6.6a = 53.7 \text{ in.} = 4.48 \text{ ft}$$

Symbols

z = thickness of pavement, in.

a = radius of tire contact area = 8.13 in.

E = tangent modulus of subgrade = 5,700 psi

$f'c$ = compressive strength of pavement components

σ_b = flexural strength of bottom layer of pavement components

E_c = modulus of elasticity of LCF subbase = 450,000 psi

E'_c = creep modulus of elasticity of subbase = 150,000 psi

n = square root of the ratio of modulus of elasticity between top and bottom fibers of pavement components

R/W = runway

T/W = taxiway

TA = terminal aprons

AP = aircraft parking

G = gate position

C = concentrated traffic area

N = normal traffic area

I = infrequent traffic area

	Symbol	R/W – C	R/W – N	T/W – C	T/W – N	TA – G	TA – N	AP – I
Traffic coverages	N	10^5	10^4	10^6	10^5	10^6	10^4	10^3
Limit of dynamic response, g	DI	0.12	0.30	0.12	0.30	0.12	0.12	0.12
Operational speed, knots	v	150	150	70	70	30	30	30
Fundamental frequency, cps	f	1.1	1.1	1.1	1.1	1.1	1.1	1.1
Straightedge constant	K	0.0028	0.0070	0.0036	0.0090	0.0090	0.0090	0.0090
Significant width	L	4.48	4.48	4.48	4.48	4.48	4.48	4.48
Permanent long. deformation, ft	Δ	0.0059	0.0148	0.00762	0.0191	0.0191	0.0191	0.0191
LCF-base Pavement								
Mode I, limiting surface deflection:								
Permanent trans. deformation								
$D_N = 10(\Delta - 0.0012\sqrt{L}) \times 12$, in.	D_N	0.415	1.470	0.610	1.99	1.99	1.99	1.99
Rate of progressive deformation, in.	D_0	0.081	0.365	0.100	0.396	0.330	0.495	0.660
$D_0 = (D_N - 0.01)\log N$	$0.43 D_0$	0.0348	0.157	0.043	0.170	0.142	0.213	0.284
Subgrade parameter, $w_0 = 0.835$, in.	$0.16 w_0$	0.1136	0.134	0.134	0.134	0.134	0.134	0.134
Deflection tolerance, in.	w_z	0.168	0.291	0.177	0.304	0.276	0.347	0.418
Parameter $w_z/p\,(1+DI)$		2.92	4.35	3.08	4.55	4.80	6.03	7.26
Thickness of consolidated layers, in.	z	33.5	22.0	31.5	20.5	19.2	15.0	11.5
Mode II, limiting subgrade stress:								
$(\sigma_z = 5.1 + 49.4\, D_0/w_0)$	D_0/w_0	0.097	0.437	0.120	0.474	0.395	0.593	0.790
Tolerance of subgrade stress, psi	σ_z	9.9	26.7	11.0	28.5	24.6	34.4	44.1
Stress parameter $2.5\, \sigma_z/p\,(1+DI)$	z/a	0.076	0.175	0.084	0.187	0.187	0.262	0.336
Thickness parameter		4.5	2.75	4.2	2.65	2.65	2.12	1.82
Total thickness of pavement including compacted subbase, in.	z	36.6	22.4	34.2	21.6	21.6	17.2	14.8

	Symbol	R/W – C	R/W – N	T/W – C	T/W – N	TA – G	TA – N	AP – I
Traffic coverages	N	10^5	10^4	10^6	10^5	10^6	10^4	10^3
Limit of dynamic response, g	DI	0.12	0.30	0.12	0.30	0.12	0.12	0.12
Operational speed, knots	v	150	150	70	70	30	30	30
Fundamental frequency, cps	f	1.1	1.1	1.1	1.1	1.1	1.1	1.1
Straightedge constant	K	0.0028	0.0070	0.0036	0.0090	0.0090	0.0090	0.0090
Significant width	L	4.48	4.48	4.48	4.48	4.48	4.48	4.48
Permanent long. deformation, ft	Δ	0.0059	0.0148	0.00762	0.0191	0.0191	0.0191	0.0191
Aggregate-base pavement								
Mode I, limiting surface deflection:								
Permanent trans. deformation								
$D_N = 5.2\Delta \times 12$, in.	D_N	0.368	0.924	0.475	1.20	1.20	1.20	1.20
Rate of progressive deformation, in.	D_0	0.071	0.228	0.0775	0.238	0.199	0.298	0.397
$D_0 = (D_N - 0.01)\log N$	$0.32 D_0$	0.0227	0.0730	0.0248	0.0762	0.0637	0.0954	0.1270
Subgrade parameter, $W_0 = 0.835$, in.	$0.14 W_0$	0.1168	0.1168	0.1168	0.1168	0.1168	0.1168	0.1168
Deflection tolerance, in.	W_z	0.1395	0.1898	0.1416	0.1930	0.1805	0.2122	0.2438
Thickness parameter $W_z E/p(1+DI)$		2.42	2.84	2.46	2.89	3.14	3.69	4.23
Thickness of consolidated layers, in.	z	40.0	34.5	39.6	33.5	31.0	26.0	22.2
Mode II, limiting subgrade stress:								
$(\sigma_z = 5.1 + 49.4\,D_0/W_0)$	D_0/W_0	0.085	0.273	0.093	0.285	0.238	0.357	0.474
Tolerance of subgrade stress, psi	σ_z	9.3	18.6	9.7	19.2	16.9	22.7	28.6
Stress parameter $\sigma_z/p(1+DI)$	z/a	0.028	0.049	0.030	0.050	0.052	0.069	0.087
Thickness parameter	z/a	>8.0	6.0	>8.0	5.9	5.7	4.8	4.2
Total pavement thickness including aggregate subbase, in.	z	>65	49	>65	48	46	39	34

Part 2: Working-stress Concept

Bending Stress Due to Differential Settlement (Plastic State)

Observed maximum differential settlement: 4 in. in 200 ft
Bending stress of pavement:

$$\sigma_d = \frac{n+1}{3n} \pi^2 E_c' \frac{\Delta}{L^2} z$$

or

$$\sigma_d = 0.52z \quad \text{for } n = 2$$

Bending Stress Due to Thermal Differential

Maximum bending stress:

$$\sigma_t = \frac{n+1}{6n} z E_c' \epsilon \frac{dt}{dz}$$

Thermal gradient—Seasonal variation: $\dfrac{dt}{dz} = 1.5° \, \text{F/in.}$ $\qquad$ plastic state

$\qquad\qquad$ Maximum daily variation: $\dfrac{dt}{dz} = 0.5° \, \text{F/in.}$ $\quad$ elastic state

Coefficient of volumetric change: $\epsilon = 5.8 \times 10^{-6} \, \text{in./(in.)(°F)}$

or

$$\sigma_t = 0.33z \quad \text{for } n = 2$$

Bending Stress Due to Wheel Load (Elastic State)

Bending stress:

$$\sigma_w = \frac{3(n+1)}{4n} (1 + \mu)(1 + \overline{DI})p\frac{a^2}{z^2}\left(\ln\frac{z}{b} + \frac{1}{4}\ln\frac{a+z}{z} + \frac{1}{4}\ln\frac{E_c}{E} \right.$$

$$\left. + 0.168\right)$$

$$= 28{,}100 \; \frac{\ln(z/b) + \frac{1}{4}\ln[(8.13 + z)/z] + 1.261}{z^2} \quad \text{for } n = 2$$

$$b = (106 + z^2)^{1/2} - 0.675z$$

Poisson's Ratio: $\mu = 0.15$
Dynamic increment: $\overline{DI} = 0.12g$

Bending Stress Due to Wheel Load— Yield-line Method

Bending stress at yield state:

$$\sigma_y = \frac{2(n+1)}{n}(1 + DI)p\frac{a^2}{z^2}(1 - 2\lambda a)$$

where

$$\lambda a = \left(\frac{6E}{E_c}\right)^{1/4}\frac{a}{z}\left(\frac{z}{a+z}\right)^{1/4} = \left(\frac{4.27}{z}\right)\left(\frac{z}{8.13+z}\right)^{1/4}$$

Valid range: $0.15 < \lambda a < 0.25$

$$28.5 \text{ in.} > z\left(\frac{a+z}{z}\right)^{1/4} > 17 \text{ in.}$$

$$\sigma_y = 65,000\frac{1 - 8.54/z[(a+z)/z]^{1/4}}{z^2}$$

Fatigue Stress Due to Traffic Coverages

Normal working stress:

$$\sigma_N = \sigma_b(1 - 0.092 \log N)$$

Equivalent bending stress:

$\sigma_b = 2.25\sigma_N$ for 10^6 traffic coverages
$\sigma_b = 1.85\sigma_N$ for 10^5 traffic coverages
$\sigma_b = 1.58\sigma_N$ for 10^4 traffic coverages
$\sigma_b = 1.38\sigma_N$ for 10^3 traffic coverages

Total Bending Stress

$$f_b = \frac{\sigma_w \text{ or } \sigma_y}{1 - 0.092 \log N}1.25 + \sigma_d + \sigma_t$$

Table of Working-stress Computation

Aircraft: Boeing 747
Volume of traffic: 1,000,000 coverages
Factor of safety: 1.25

Thickness of pavement, in.	Computed stress in pavement, psi				Bending strength required, psi	
	σ_d	σ_t	σ_w	σ_y	f_b (elastic)	f_b (yield)
12	6	4	358	---	1,018	---
14	7	5	288	---	795	---
16	8	5	223	133	640	388
18	9	6	181	115	525	338
20	10	7	151	100	441	297
22	11	7	127	87	376	263
24	12	8	108	76	324	234
26	14	9	93	67	285	211
28	15	9	81	---	253	---
30	16	10	72	---	227	---
32	17	11	63	---	204	---
34	18	11	56	---	187	---
36	19	12	50	---	173	---

APPENDIX FIVE

Computer Program— Thickness Design

```
      SUBROUTINE PAVDEF
      DIMENSION DIBAR(10),F(10),V(10),AN(10),SIGMAZ(10)
      COMMON FACTOR,RADIUS,PSI,N,E,DIBAR,C1,     D1,D2,A1,A2,
            CONST,XX,AN,F,V,FUDGE,B1,B2,B3,SIGMAZ
      READ (12,4) N,E,(DIBAR(I),AN(I),F(I),V(I),I=1,N)
4     FORMAT(I2,F8.1/(F5.3,F10.1,F6.2,F7.2))
      WRITE(13,5)E
5     FORMAT(///20X,43HSUBROUTINE PAVDEF — E (SUBGRADE
            MODULUS) = ,F9.1,
   X4H PSI//)
      READ (12,6)C1,D1,D2,A1,A2,CONST,XX
6     FORMAT(3F4.3,F6.2,F6.2,F6.3,F5.3)
      PRESS = PSI * FACTOR
      WO = 2.*PRESS*RADIUS/E
      D3  = D1*WO
      DO 1 I=1,N
      AK = CONST/V(I)*DIBAR(I)/SQRT(F(I))
      DL=12.*AF*SQRT(XX*RADIUS/12.)
```

```
          DN=12.*A1*(AK-A2)*SQRT(XX*RADIUS/12.)
          AAN=AN(I)
          DODEF= (DN - C1) / ALOG10(AAN)
          WZ = D3 + D2 * DODEF
          PESACH = WZ * E /(PRESS * (1.+DIBAR(1)))
          DO 2 J=10,1000
          AJ=J
          Z=AJ/10.
          TERM= RADIUS**2 +Z**2
          YAKOV=SQRT(TERM)*(2.-Z**2/TERM-Z/SQRT(TERM))
          IF(YAKOV-PESACH)3,3,2
2         CONTINUE
3         CONTINUE
          WRITE (13,9)I,AN(I),DIBAR(I),F(I),V(I),DL,DN
9         FORMAT(/20X,21HPAVEMENT TYPE NUMBER ,I2/10X,F10.1,2X,13H
         XDESIGN PASSES/10X,28HLIMIT OF DYNAMIC RESPONSE = ,F5.3,2H
         XG/10X,29HVEHICLE RESPONSE FREQUENCY = ,F6.2,4H CPS/10X,
         X19HVEHICLE VELOCITY = ,1X,F7.2//10X,37HLONGITUDINAL PERM
         XANENT DEFORMATION = ,F7.5,7H INCHES/10X,35HTRANSVERSE P
         XERMANENT DEFORMATION = ,F7.5,7H INCHES/)
          WRITE (13,10)DODEF,WZ,Z
10        FORMAT(10X,40HRATE OF CHANGE OF TRANS. PERM. DEFOR. =
         X,F5.3,6HINCHES/10X,31HLIMIT OF ELASTIC DEFORMATION = ,F5.3,
         X7H INCHES//20X,29HDEPTH OF PAVEMENT REQUIRED = ,F5.1,1X,6
         XHINCHES///)
1         CONTINUE
          RETURN
          END


          SUBROUTINE PAVBEN
          DIMENSION DIBAR(10),F(10),V(10),AN(10),SIGMAZ(10),XN(10)
          COMMON FACTOR,RADIUS,PSI,N,E,DIBAR,C1,D1,D2,A1,A2,CONST,
         X        XX,AN,F,V,FUDGE,B1,B2,B3,SIGMAZ
          READ (12,9)N
9         FORMAT (I2)
          ENTRY FSTBEN
          READ(12,16)(AN(I),I=1,N)
16        FORMAT(10F10.1)
          ENTRY NXTBEN
          READ (12,10)E,(DIBAR(I),I=1,N)
10        FORMAT(F8.1,10F5.3)
          GO TO 15
          ENTRY THRBEN
```

```
         READ(12,17)(AN(I),I=1,N)
17       FORMAT(10F10.1)
15       ENTRY LSTBEN
         WRITE(13,11)E
11       FORMAT(///20X,43HSUBROUTINE PAVBEN — E (SUBGRADE MODU
         XLUS) = ,F9.1,4H PSI//)
         READ (12,12)EC,ECX,XMU,EPSIL
12       FORMAT(2F10.2,F6.3,F10.8)
         READ (12,13)TEMDEL,CF,SF,DELTA,XLEN
13       FORMAT(F5.2, F6.2,2F5.2,F6.2)
         READ (12,14)(XN(I),I=1,N)
14       FORMAT(10F6.2)
         PRESS = PSI * FACTOR
         BF     = CF**.75*1.6
         WRITE(13,20)CF,BF,SF
20       FORMAT(10X,33HCOMPRESSIVE STRENGTH AT BOTTOM = ,F9.2,
         X4H PSI/10X,37HALLOWABLE BENDING STRESS AT BOTTOM = ,F7.2,
         X4H PSI/10X,29HSAFETY FACTOR FOR THE LOAD = ,F6.2//)
         C      = DELTA/XLEN**2./144.*ECX *3.14159**2
         D      = ECX * EPSIL * TEMDEL
         FE     = EC/E
         DO 1 I =1,N
         BLIV   = 3.*(XN(I)+1.)/(4.*XN(I))
         APRIME = BLIV*(1.+XMU)*PRESS*(1.+DIBAR(I))
         BPRIME = .25*ALOG(FE)+.168
         CPRIME = (XN(I)+1.)/(3.*XN(I))
         DPRIME = (XN(I)+1.)/(6.*XN(I))
         AAN=AN(I)
         AA = 1.-.092*ALOG10(AAN)
         DO 2 J=100,1000
         RADIAS=RADIUS
         AJ=J
         H=AJ/10.
         RPRIME=RADIAS/1.724
         IF(H-RPRIME)3,3,4
4        DD = 1.6*RADIAS**2.+H**2
         RADIAS = SQRT(DD)-.675*H
3        CONTINUE
         EE = H/RADIAS
         SIGMAW = APRIME/AA*(RADIUS/H)**2*(ALOG(EE)+.25*ALOG((H+
         XRADIUS)/H)+ BPRIME)
         SIGMAD = CPRIME * C * H
         SIGMAT = DPRIME * D * H
```

```
      BEBOP = SIGMAW * SF + SIGMAD+SIGMAT
      IF(BF-BEBOP)2,6,6
6     CONTINUE
      WRITE (13,7)I,AN(I),DIBAR(I),XN(I)
7     FORMAT (20X,21HPAVEMENT TYPE NUMBER ,I2/10X,F10.1,2X,13H
XDESIGN PASSES/10X,28HLIMIT OF DYNAMIC RESPONSE = ,F5.3,2H
XG/10X,27HRATIO OF ELASTIC MODULI = ,F5.2/)
      WRITE(13,8)     SIGMAW,SIGMAD,SIGMAT,H
8     FORMAT(10X,
      X          29HWESTERGAARD FATIGUE STRESS = ,F9.2,4H PSI/10X,
X40HSTRESS DUE TO DIFFERENTIAL SETTLEMENT = ,F9.2,4H PSI/
X10X,37HSTRESS DUE TO TEMPERATURE GRADIENT = ,F9.2,4H PSI/
X/20X,26HDEPTH OF PAVEMENT REQUIRED/25X,F5.1,7H INCHES///)
      GO TO 1
2     CONTINUE
1     CONTINUE
      RETURN
      END
```

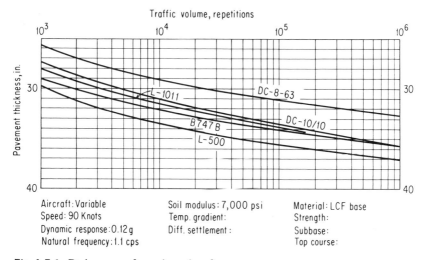

Fig. A.5.1 Design curves for various aircraft.

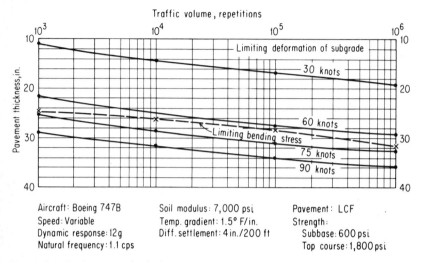

Fig. A.5.2 Design curves for 747B, variable speed.

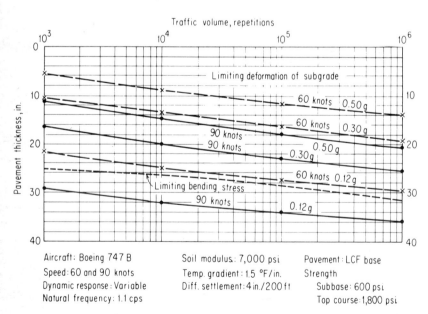

Fig. A.5.3 Design curves for 747B, variable dynamic response.

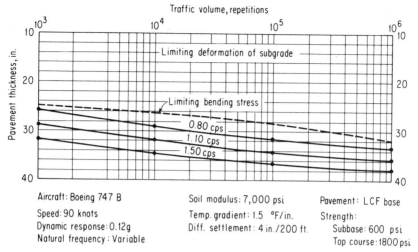

Fig. A.5.4 Design curves for 747B, variable natural frequency.

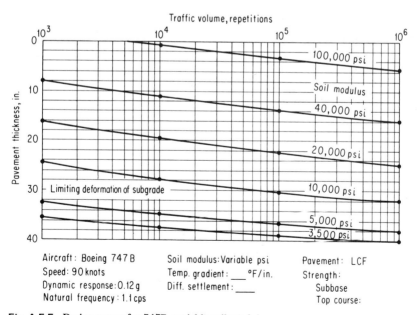

Fig. A.5.5 Design curves for 747B, variable soil modulus.

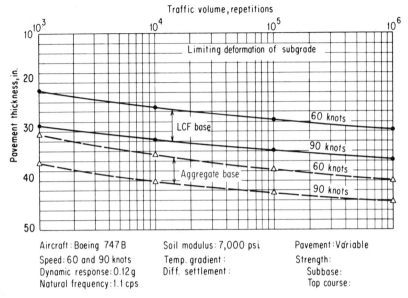

Fig. A.5.6 Design curves for 747B, variable pavement base material.

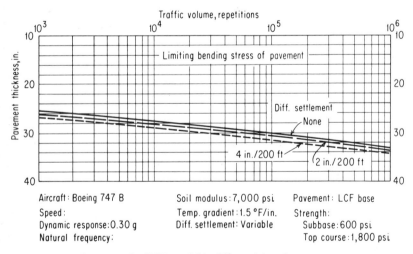

Fig. A.5.7 Design curves for 747B, variable differential settlement.

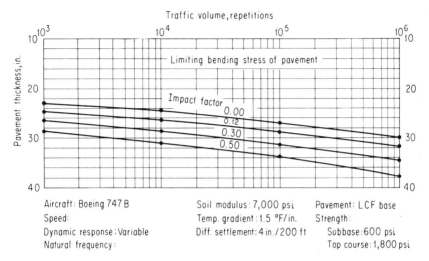

Fig. A.5.8   Design curves for 747B, variable dynamic impact.

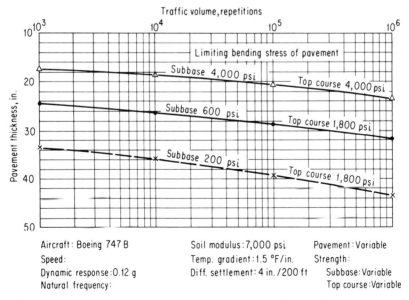

Fig. A.5.9   Design curves for 747B, variable composition of pavement base.

CHAPTER TWELVE
Redesign of Existing Pavement

Airports must keep pace with the rapid growth of air transportation. The need to expand existing airports and to develop new airport facilities has become urgent. Public necessity compels an industry-wide program of vast improvement. For instance, in 1948, the year the Port Authority assumed responsibility for the development and operation of Newark Airport, the airport served 800,000 passengers and handled some 40,000 tons of cargo. Twenty years later, in 1968, Newark accommodated an annual volume of 6,700,000 passengers and 160,000 tons of cargo. By 1980, it is anticipated that 12,000,000 passengers a year will stream through Newark Airport.

This is not the case for Newark Airport alone. Most commercial airports were developed and constructed as early as the 1930s, when the weight of airplanes was less than 100,000 lb. In the last 10 years the rapid growth of air transportation has increased airplane weight to 350,000 lb, and the passenger and cargo volumes have multiplied many, many times over. As projected for 1980, the most conservative estimate of aircraft weights will be in the neighborhood of

850,000 lb and the passenger volume may be double or triple that of today. With the heavy traffic volumes and increased aircraft weights, there have been frequent reports that pavements are not smooth and that the surfaces are cracking. These are natural results of a growing air transport industry. Not only are maintenance and expansion in themselves costly to the airport operator, but the interruption of service and inconvenience to the traveling public also has a negative effect on industry revenues.

Although pavement problems are a natural outgrowth of the expanding air transport industry, some of them have been inherited from the original pavement design methods, which lacked a defined concept of the functional requirement of a pavement surface. Conventional structural concepts for static loads were common in pavement design analysis, and therefore, the stress-strain condition of the pavement could be overemphasized. The airplane was not considered to be moving, and the geometry of the pavement surface was not studied for its effect on aircraft vibration and the safety of the traveling public. In the design computations, the physical properties of the subgrade were not characterized according to actual performance under load. An arbitrary k value was used for concrete pavements, an empirical CBR value was used for asphalt pavements, and soil classification was used by government agencies. Because of this lack of a unified concept of pavement design, the performance of existing pavement varies widely in quality.

12.1. CONDITION OF EXISTING PAVEMENT AND SUBGRADE

Before doing anything with existing pavement, the condition of the pavement and subgrade should be evaluated. The first step is to characterize their physical properties, which can be described by two parameters: Young's modulus and Poisson's ratio of the pavement materials. For practical design and tests, the recoverable deflection of pavement and subgrade is the most useful parameter. For the subgrade, the relationship between the deflection, Young's modulus, and Poisson's ratio can be expressed by

$$W_0 = \frac{2pa}{E} (1 - \mu^2) \tag{1}$$

The surface deflection W_z of existing pavement can be expressed by

$$W_z = \frac{2pa}{E}(1 - \mu^2)F \tag{2}$$

The factor F can be determined by Burmister's layered theory.

According to the test results at Newark Airport and research conducted by others, W_z can be expressed in the form

$$W_z = W_0\left[1 + \left(\frac{z}{a}\right)^2\right]^{1/2}\left[1 - \frac{z^2}{2(a^2 + z^2)} - \frac{z}{2(a^2 + z^2)^{1/2}}\right] \tag{3}$$

where z is the total thickness of consolidated layers of the pavement structure. When the ratio of W_z to W_0 is known, the thickness ratio z/a can be determined. As was discussed in Chap. 3, Part B (on nondestructive tests), the use of the output ratio will increase the reliability of the test results.

The reliability of one good test, however, does not eliminate the variation due to pavement construction, subgrade condition, traffic density, and maintenance operations. These are natural events in random pattern. The best way to improve the test reliability of the entire system, therefore, would be to increase the number of test samples, as represented by

$$\sigma_n = \sigma\frac{t}{\sqrt{n}} \tag{4}$$

The acquisition of quantitative data will improve knowledge of the performance of the existing system. The mean value of the pavement system or its subsystem is used to measure the level of pavement quality, and the degree of variation, as represented by the standard deviation, is used to indicate the scatter of pavement performance. The combined values of mean and standard deviation represent the degree of confidence of the pavement performance.

As the deflection by nondestructive testing and the E value from plate-load tests represent two different aspects of the tests, the application of these two tests on the pavement design analysis should be established independently, with the correlation function deduced from the test for each job.

In conducting the nondestructive tests, there are several approaches to determining surface deflection of the subgrade. For instance, the testing cannot be conducted on the initial subgrade where the surface has not been compacted. Movement of the machine sinking into the unconsolidated surface prevents the measurement of absolute surface deflection of the subgrade. Another way of measuring the surface deflection of the subgrade is by vibrating a pavement surface under the initial construction compaction. This represents the area in which the pavement does not receive a significant volume of traffic. A test in this area would indicate the deflection of the subgrade under the original construction condition. The third method represents the existing pavement under heavy use. The deflection of the subgrade reflects the influence of the traffic load. This is the area in which pavement strengthening is really required.

On the original uncompacted subgrade, the test will determine the surface deflection W_i and the velocity of wave propagation V_1 in the subgrade. Tests on existing pavement at the edge of the runway, outside the traffic area, consist of the determination of the surface deflection W_z of the existing pavement and the velocity of wave propagation V_2 in the subgrade. A test sample is shown in Figs. 12.1

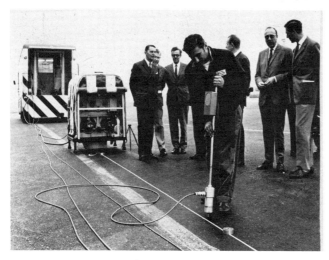

Fig. 12.1 Determination of wavelength.

and 12.2. The surface deflection of the subgrade under initial construction compaction is computed to be

$$W_0 = \left(\frac{V_1}{V_2}\right)^2 W_i \tag{5}$$

This equation is based on the assumption that the dynamic modulus of elasticity E is

$$E = c\rho v^2 \tag{6}$$

where ρ is the density of the subgrade, c is a constant, and v is the velocity of wave propagation in the subgrade. For the same subgrade, that is, c and ρ values not changed, the deflection times the velocity squared should be equal to a constant. The ratio W_z/W_0 from these two tests can be used to determine the thickness of the consolidated layers of existing pavement. On existing pavement at the central

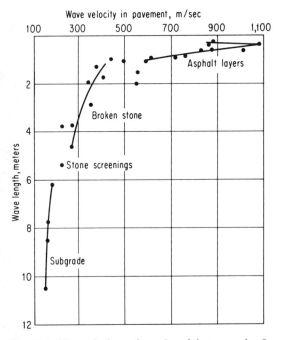

Fig. 12.2   Wave velocity and wavelength in test section I.

portion of the runway or taxiways, where traffic has a significant influence on the deflection of the pavement and the subgrade, tests should consist of determination of the surface deflection W'_z of existing pavement and the velocity of wave propagation V_3 in the subgrade. The possible increase in wave velocity and the possible decrease in surface deflection of the pavement may contribute to the effect of the traffic loads. The equivalent deflection of the subgrade in this area is

$$W'_0 = \left(\frac{V_1}{V_3}\right)^2 W_i \tag{7}$$

The ratio of W'_z to W'_0 can be used to determine the equivalent thickness of a consolidated layer of existing pavement under the influence of traffic load.

Thus these two different test groups, representing the initial construction condition and the condition after the traffic loads, can be used in the analysis for new pavements. W'_0 should be used as the reference for pavement overlay in heavy-traffic areas, and W_0 should be used for the overlay analysis for infrequent-traffic areas.

12.2. DATA ACQUISITION

The actual test program for the New York airports will be used for this discussion. In a period of 3 months, 9,000 measurements were made at 650 locations, which covered practically all aircraft pavements at Newark, Kennedy, and LaGuardia Airports. The effective operation as well as reliable instrument monitoring should be credited largely to the high caliber of the Shell scientists and the close cooperation of the airport facilities and engineers.

The direct output of the Shell machine, through an internal process of double integration, was the deflection-frequency function. For the majority of the tests, the output resembled the characteristics shown in Fig. 12.3. Therefore, an average deflection in the range of 15 to 25 cps was used in the evaluation. When the monitored output was significantly different from that in Fig. 12.3, a detailed review of the patterns of variation, wave propagation, and temperature fluctuation was made to determine the appropriate deflection of the pavement.

Dynamic Deflection Test							Airport: Newark	

Date: 10-1-68	Temp						Area: RW 22/4	
Weather: clear	Air 17							
	Top 15						Location: 22	
night	-3" 22.5							
	-6" 22.5						Meas No: 120	

Print code	1	2		3	4	5	6		
Symbol	Freq	Code	Fe	Fw	Zw	S	φ	Sc	1/Sc
Lo	10	10	1070	800	31	260	20	258	39
	12	10	1050	800	29	205	14	276	36
	14	10	1150	795	31	265	15	257	39
	16	10	1230	820	34	250	19	242	41
	18	10	1300	800	33	255	24	243	41
Hi	20	11.25	800	810	32	252	26	253	39
	24	11.25	720	800	32	255	34	250	40
	28	11.25	630	810	33	253	38	245	41
	32	11.25	560	790	29	280	46	272	37
	36	11.25	570	815	27	300	46	302	33
	40	11.25	460	800	26	305	50	308	32
	44	11.25	370	820	24	350	48	342	29

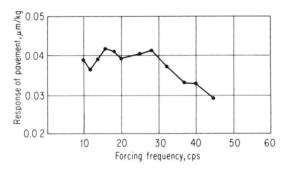

Fig. 12.3 A typical record of vibratory test.

The machine output of pavement deflection was then adjusted for nonlinear performance of the monitoring system and the calibration of static load. This adjustment was made to reduce the error of the monitoring system. The final data deduced from the test represent the dynamic response (deflection) of the pavement surface in micrometers per kilogram. The calibrated test results of taxiway A and runway 4-22 at Newark are shown in Fig. 12.4. It should be pointed out that the cross taxiways demonstrated a much larger deflection than taxiway A and runway 4-22. The original construction of these facilities was identical. However, due to the concentrated airport traffic as well as more frequent overlay maintenance, the load-carrying capacity of taxiway A and runway

4-22 have improved to far exceed their original design condition. The discovery of such an improvement, quantitatively for the first time, is one of the outstanding contributions of the Shell test in developing an economic pavement-strengthening program.

For the nondestructive test on existing asphaltic pavement, the machine output of pavement deflection is sensitive to the pavement temperature at the time the measurement is made. In the Newark and Kennedy tests, the machine output of pavement deflection at 80°F was 50 percent higher than that at 50°F. There is no question that the variation was largely due to the change of E modulus of the asphalt layer, which is one of the major disadvantages of the asphalt pavement. In order to improve the reliability of using deflection as the basic parameter in overlay pavement design, it is necessary (1) to conduct pavement tests in a normal temperature range and to avoid

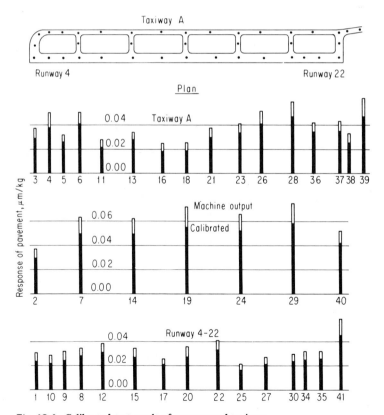

Fig. 12.4 Calibrated test result of runway and taxiway.

tests in extreme cold and warm weather and (2) to fully utilize the ratio of deflection output in the overlay design analysis. The fluctuation of deflection output due to temperature variation can be considered as an error in monitoring, and therefore, the use of relative values of deflection output will increase the reliability of the data processing.

12.3. DESIGN SEQUENCE

In our present engineering practice, there is no rational method which can be utilized to determine the requirements of strengthening an existing pavement. The design method given in the following represents an extension of an effort in research and development to upgrade pavement design philosphy.

1. The first step in the design involved improvement of the statistical reliability of the test results. The existing pavements were grouped together according to the distribution of airport traffic. For instance, runway 4-22 was divided into three segments—two end sections, each 1,500 ft long, and the remaining center portion (see Fig. 12.10). The deflection data in each group were then processed for the arithmetic mean and standard deviation. The representative deflection of the group was considered to be the mean value plus one standard deviation. This reflected a statistical reliability such that in one out of six observations, the measured deflection data would be larger than the representative deflection. Thus the performance of new pavement would be more reliable than that obtained by the averaging method. (Reference should be made to the optimization of the E value in Chap. 3).

2. The next step involved determination of the dynamic deflection of the subgrade under the pavement. There were three areas in which to test the subgrade response: (a) the field area, where neither aircraft traffic nor construction compaction had been received; (b) under the sides of taxiways and runways, where construction compaction had been applied but which received no appreciable traffic load; (c) under the center strip of runways and taxiways, where channelized traffic was encountered. The tests were conducted in this way: The Shell machine was used in the field area to determine the actual deflection reading W_i, as well as the wave propagation, V_1, in the subgrade. The machine was then moved to

the sides of the taxiway or runway to monitor the actual deflection W_z and the wave propagation V_2 in the subgrade under the pavement. The deflection reading in the field area was then adjusted by the velocity of the two measurements according to Eq. (5). The adjusted subgrade deflection value W_0 was used in establishing the deflection ratio W_z/W_0 between the existing pavement and the original compacted subgrade.

The deflection ratio becomes an important parameter for indicating the thickness of consolidated layers according to Eq. (3), which reflects the performance and service condition of the existing pavement.

3. The Shell machine was then used to determine the dynamic deflection of all 16 sections of the test pavement. The relation between the measured deflection and the thickness of the test pavement is shown in Fig. 12.5. This test served as a basis for

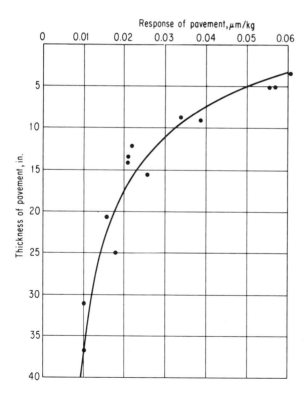

Fig. 12.5 Measured deflection of test pavements.

establishing the dynamic-deflection measurement and the static plate-load test. In this case, the average E value was 7,000 psi and the dynamic-deflection reading was 85 μm/kg. According to Eq. (1) or (2), the product (WE) is constant; a family of E curves can be established as shown in Fig. 12.6.

4. A pavement design computation, as outlined in the flow chart in Fig. 12.7 and discussed in Chap. 11, was made for the Boeing B747 aircraft. The results are shown in Figs. 12.8 and 12.9.

5. Based on the functional requirements, such as traffic coverages, and the acceptable level of aircraft vibration, the thickness of a new pavement was determined from these charts.

6. By using the ratio W'_z/W'_0 or W_z/W_0, an equivalent thickness z was assigned to the existing pavement for the area studied. The rigidity of existing pavement material is reflected by the deflection measurement W_z or W'_z.

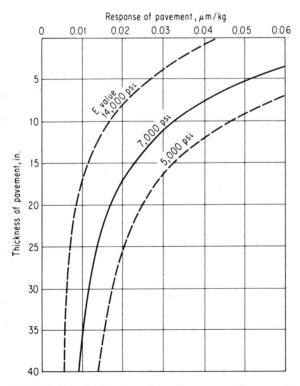

Fig. 12.6 Transfer function–dynamic response of pavement and E value of subgrade.

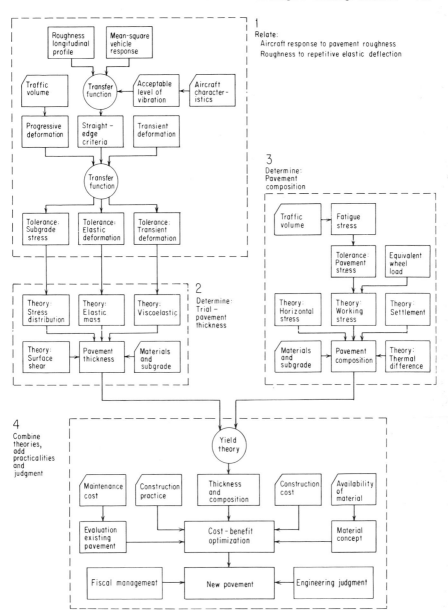

Fig. 12.7 System of pavement analysis.

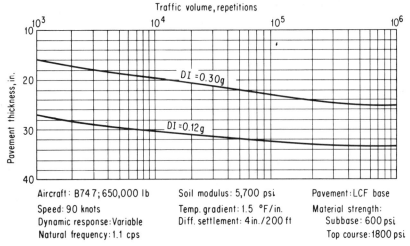

Fig. 12.8 Pavement design chart—runways.

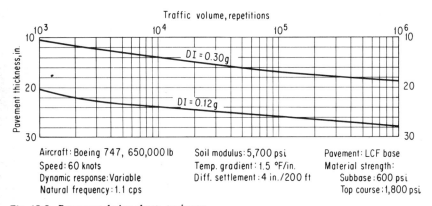

Fig. 12.9 Pavement design chart—taxiways.

7. The difference in thickness between the computed new pavement and the equivalent existing pavement represented the requirement for overlay thickness to strengthen the existing pavement to meet the same functional requirements as those for a new pavement (see Figs. 12.10 and 12.11).

12.4. PRACTICAL APPLICATION OF DESIGN ANALYSIS

The design analysis given above is actually the first step in establishing a good engineering design. In Figs. 12.10 and 12.11, the thickness of the overlay required along the centerline of runway 4-22 has been computed. These two design charts form the basic information for determining the thickness of the overlay. For practical application, the design analysis must be modified to conform to the variation of surface geometry of the existing pavement. In many cases, the existing pavement has deteriorated and, because of surface roughness, does not meet the functional requirements for operational surfaces. It is then necessary to rebuild

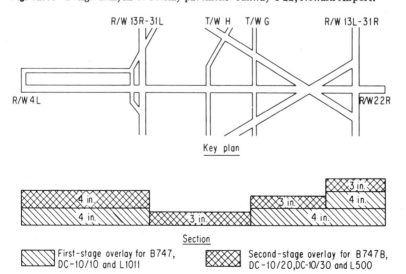

Test No.	1	10 9 8	12 15	17 20 22	25 27	30 34 35	41
Mean value of pavement response	0.0236	0.0241	0.0298	0.0270	0.0186	0.0246	0.0462
Standard deviation, σ		0.0015	0.0012	0.0048	0.0019	0.0007	
Mean plus 1.0 σ	0.0236	0.0256	0.0310	0.0318	0.0205	0.0253	0.0462
Equivalent pavement thickness for E = 5,700 psi	20 in.	18.6 in.	15.0 in.	14.7 in.	23.5 in.	18.6 in.	9.3 in.
Thickness required DI = 0.13 g, N = 10^5	31.6 in.	31.6 in.	- -	- -	- -	31.6 in.	31.6 in.
Thickness required 75% static load	- -	- -	23.7 in.	23.7 in.	23.7 in.	- -	- -
Thickness of overlay	13 in.	13 in.	9 in.	9 in.	0 in.	13 in.	22 in.

Fig. 12.10 Design analysis of overlay pavement—runway 4-22, Newark Airport.

Key plan

Section

First-stage overlay for B747, DC-10/10 and L1011

Second-stage overlay for B747B, DC-10/20, DC-10/30 and L500

Fig. 12.11 Thickness of overlay pavement—runway 4L-22R, Kennedy Airport.

the pavement surface to meet these newly defined functional requirements.

For the longitudinal profile, the surface should be smooth enough to meet the requirements for limiting the dynamic response of aircraft and the anticipated service life of the pavement. The transverse profile, or cross section of the pavement, should meet the requirements for proper drainage and maintenance of a clean pavement surface. For airports having runway and taxiways crossings, a priority for the smooth pavement surface must be selected among the runway, the taxiway, and the crossing runways. At Kennedy Airport, the bay runway 31L carries 52 percent of the takeoff traffic and it seemed reasonable to give first priority to the runway pavement. Therefore, smooth grading was made for that runway, and grading for drainage was pushed to the crossing runway and taxiways. After establishing the theoretical elevation of the runway and side slopes, it is very often found that the deepest point on the runway is where the maximum overlay is required. By carefully studying the surface and thickness requirements, a final runway elevation and slope can be established.

In establishing the new profile and side slope, it is necessary to consider (1) the minimum thickness for the overlay, (2) practical considerations for construction, and (3) the durability of the materials. For all practical purposes, the thickness of asphaltic-concrete overlays should not be less than 3 in. and portland-cement concrete overlays should not be less than 4 in. thick. Also, for maximum utilization of equipment, the maximum lift of an asphaltic-concrete overlay should be limited to 4 in. The performance of the compactor should be considered in deciding the effective thickness of each lift.

Since the thickness design for the overlay computation denotes the requirement for the total thickness of the consolidated layer, no unbound materials should be introduced in the overlay construction. The environmental condition also dictates that no sandwich construction be allowed in order to avoid the destruction of the pavement structure due to freezing and thawing cycles. As overlay construction is normally conducted at existing airports with heavy air traffic, many overlays must be constructed under limited operating conditions, such as at night or under certain unfavorable working conditions. This limitation may affect the quality of the overlay construction. Therefore, if at all possible, the overlay operation should be conducted in the normal daytime hours and without other restrictions on the construction.

Another consideration is the mixing formula for the overlay material. It should follow the general principle that the mix should be stable with respect to temperature and have a minimum volumetric change under load and environment. Since the overlay is always exposed to temperature variations and moisture changes, as well as to high load intensity, the material requirement of stability under load and temperature is one of the prime considerations for a successful overlay design.

Index

D

Damping:
  coefficient of, 10, 314
    critical, 294
  of vibration, 244, 325
Damping characteristics, 37, 275
Darcy's law, 151
Decision criteria, 85
Deep-strength black base, 26
Deflection:
  dynamic, 245
  of pavement: structure, 381
    surface, 381, 398, 444
  recoverable, 13
  static load, 74
Deflection profile, 70
Deflection tolerance, 28, 46, 399
Deformation:
  cumulative, 13, 63
  elastic, 15, 39
  under load, 111
  longitudinal, 12, 370, 390
  modulus of, 29, 45, 46, 50, 54, 83
  permanent, 13
  of subgrade, 21
  transverse, 13, 372, 390
Design analysis:
  new system of, 11, 383
  optimization of, 33
Design manual, Navy's 28
Design methods, pavement, 25
Design tolerance, 101
Deterministic model, 7
Dewatering, 175
Differential settlement, 40, 176, 414
Digital computation, 23
Distribution of wheel load, 26
Drainage, side, 152
Drainage condition (FAA), 49
Drainage system for pavement, 158
Driver's response, 2
Drop test, 279, 284, 291
Ductility, 4, 22, 113
Durability, 21, 118
Dyna-Flex machine, 70
Dynamic deflection, 245
Dynamic impact, 10, 303
Dynamic increment, 10, 305, 335
Dynamic interaction, 3
Dynamic load test, 74
Dynamic magnification, 294
Dynamic response:
  maximum, 338
  of pavement, 313
  of vehicle, 37, 313

E

Early English method, 29
Economic study, 424
Elastic constants, 46
  linear, 111
Elastic solid foundation, 222
Elastic state of equilibrium, 396, 413
Elastic theories:
  Boussinesq, 16, 188
  finite-element method, 8, 212
  layered method, 16, 201
  visco-elastic analysis, 16, 207
Empirical design methods, 6, 30
Endurance of material, 21, 40
Energy absorbing efficiency, 280, 315, 390
Engineering judgment, 35, 41, 425
Engineering models:
  deterministic, 7
  experiment, 8
  observation, 9
  probabilistic, 5
  simulation, 8
Environment design method, 156
Environment factors, 12, 22, 30, 138
Equilibrium of pavement on subgrade, 22,
    26, 39, 56
Equivalency concept, 30, 31
Equivalent number of repetitions, 298
Existing pavement:
  evaluation of, 443
  redesign of, 443
  thickness of, 83

F

Federal Aviation Administration (FAA),
    U.S., 34, 331
    method, 35, 49
Field control, 132
Finite-element method, 8
Fiscal policy, 40, 106
Flyash, 38, 127
Folding frequency method, 329
Force:
  body, 183
  surface, 183
Forcing function, 6, 37, 74, 99
  amplitude, 71, 74
  frequency, 71
  range, 74
Fourier series of time function, 72, 351
Freezing and thawing, cycles of, 23, 155–
    157
Frequency, fundamental, 10, 37, 83, 314,
    315

M